Drug Disposition and Pharmacokinetics
From Principles to Applications

Drug Disposition and Pharmacokinetics
From Principles to Applications

Stephen H. Curry

Professor of Pharmacology and Physiology, University of Rochester, USA

Robin Whelpton

Senior Lecturer, Queen Mary University of London, UK

WILEY-BLACKWELL

A JOHN WILEY & SONS, LTD., PUBLICATION

This edition first published 2011
© 2011 John Wiley & Sons, Ltd.

Wiley-Blackwell is an imprint of John Wiley & Sons, formed by the merger of Wiley's global Scientific,
Technical and Medical business with Blackwell Publishing.

Registered office: John Wiley & Sons Ltd, The Atrium, Southern Gate, Chichester, West Sussex, PO19 8SQ, UK

Other Editorial offices: 9600 Garsington Road, Oxford, OX4 2DQ, UK
111 River Street, Hoboken, NJ 07030-5774, USA

For details of our global editorial offices, for customer services and for information about how to apply for permission to reuse
the copyright material in this book please see our website at www.wiley.com.

Library of Congress Cataloging-in-Publication Data

Curry, Stephen H.
  Drug disposition and pharmacokinetics : from principles to applications /
Stephen H. Curry and Robin Whelpton.
     p. ; cm.
  Includes bibliographical references and index.
  ISBN 978-0-470-68446-7 (cloth)
  1. Pharmacokinetics. 2. Biopharmaceutics. 3. Drugs–Metabolism. I.
Whelpton, Robin. II. Title.
  [DNLM: 1. Pharmacokinetics. 2. Biopharmaceutics. 3. Pharmaceutical
Preparations–metabolism.
QV 38 C976da 2010]
  RM301.5.C863 2010
  615'.7–dc22
                                                    2010003126

A catalogue record for this book is available from the British Library.

This book is published in the following electronic formats: ePDF 9780470665220; Wiley Online Library 9780470665190

Set in 10/12pt, Times-Roman by Thomson Digital, Noida, India
Printed and Bound in Singapore by Markono Print Media Pte Ltd.

First Impression 2011

Contents

Preface

The origins of this book can be traced to a previous book with the same title written by one of us (S.H.C.) in 1974, second and third editions being published in 1977 and 1980. At the time, we were both in the early stages of what has become a very enjoyable career-long collaboration. Since that time newer approaches to the subject have been developed and many other books have been published. Mostly these have tended to be either very basic texts on pharmacokinetics, or weightier tomes, although some have been brief introductory texts, while yet others have concentrated on clinical applications. We have aimed to produce a book that takes the middle road, providing sufficient information and background to make it informative, clear, readable and enjoyable, without unnecessary complexity, maintaining the philosophy of the original *Drug Disposition and Pharmacokinetics*. Consequently, this book should be of benefit, in particular, to undergraduate and postgraduate students of science, including pharmacology, toxicology, medicinal chemistry and basic medical science, and students preparing for, and in, pre-professional programmes such as those in pharmacy, medicine and related disciplines, including dentistry and veterinary science, and environmental and public health. However, we are all life-long students, and thus this book is for anyone at any stage in his or her career wishing to learn about drug disposition and pharmacokinetics, that is, what happens to drug molecules in the body, but with strong emphasis on the pharmacological and clinical consequences of drug consumption, and so we expect our book to find readers among researchers, teachers and students in universities, in research institutes, in the professions, in industry, and in public laboratories employing toxicologists and environmental scientists in particular.

Clearly pharmacokinetics cannot be taught without recourse to mathematics. However, understanding the equations in this book requires little more than a basic knowledge of algebra, laws of indices and logarithms and very simple calculus. Anyone wishing to refresh his or her knowledge in these areas is recommended to read the Appendix. In a practical sense, it is important to be able to match standard equations to common graphical displays. There is little need in this context for the ability to derive complex equations. We believe in the old maxim, that a picture is worth a thousand words, and have noted that many of the principal pharmacokinetic relationships can be demonstrated empirically by the movement of dye into and out of volumes of water. We have used this approach to illustrate the validity of several models and further examples and colour plates can be found on the companion web site. This site also contains more mathematical examples, further equations and worked examples for readers who require them.

With few exceptions, we have adopted the system of pharmacokinetic symbols recommended by Aronson and colleagues (*Eur J Clin Pharmacol* 1988; 35: 1) where C represents concentration, A, is used for quantities or amounts, V for volumes of distribution, Q for flow rates, etc. First-order rate constants are either k or, if they are *elimination* rate constants, λ. Numbers have been used to designate compartments rather than letters (letters can lead to confusion, for example between plasma and peripheral) and for the exponents in multiple-compartment models. Thus, the variables for the biphasic decline of a two-compartment model are C_1, C_2, λ_1 and λ_2 rather than A, B, α and β, which may be familiar to some readers. It has not been possible to select a single convention for drug nomenclature. Rather, the names most likely to be familiar to readers around the world are used, and so we have leaned towards recommended International Non-proprietary Names (rINN). In most instances this should not cause problems for the majority of readers: cyclosporin, cyclosporine and ciclosporin clearly refer to the same drug. Where names are significantly different, alternatives are given in parentheses, e.g. pethidine (meperidine), and in the Index.

We have designed this book to be read from beginning to end in the order that we have presented the material. However, there is extensive cross-referral between sections and between chapters – this should aid those readers who prefer to 'dip in,' rather than start reading from Page 1. Thus, Chapter 1 is a brief presentation of the general chemical principles underlying the key mechanisms and processes described in the later chapters, effectively a mini-primer in medicinal chemistry. Drug disposition and pharmacokinetics is a discipline within the life sciences that depends entirely on these and other chemical principles. Chapters 2 and 3, which detail distribution and fate of drugs, are largely descriptive. Pharmacokinetic modelling of drug and metabolites, including more advanced concepts of clearance can be found in Chapters 4– 7. Chapter 8 is devoted to bioavailability, particularly the influence of tablet formulation on concentrations of drugs in plasma and therefore on clinical outcome. The next four chapters (9–12) deal with what can be referred to as 'special populations' or 'special considerations': sex, disease, age and genetics in particular. The relationships between pharmacokinetics and pharmacological and clinical effects (PK-PD) are the topic of Chapters 13 and 14, whilst extrapolation from animals to human beings is considered in Chapter 15. The kinetics of macromolecules, including monoclonal antibodies, are considered in Chapter 16. The final chapters exemplify the importance of pharmacokinetics in three clinical areas, considering aspects of drug interactions, toxicity and therapeutic drug monitoring. Thus our sequence is from scientific preparation, through relevant science, to an introduction to clinical applications. The logical extension of the learning process in this area would be obtainable through one or more of the excellent texts available that focus on patient-care orientated pharmacokinetic research and practice.

Our examples come from our own experience, from literature of pivotal significance in the development of the subject, and from drugs that will be especially familiar to readers. Certain drugs stand out as demonstrating basic principles of widespread significance throughout the subject. They include propranolol, warfarin, digoxin, aspirin, theophylline and isoprenaline. It should be noted that these, and several other examples, can be considered as 'old' drugs. Some, indeed, such as isoprenaline and guanethidine, are obsolete as therapeutic agents, but still of paramount importance historically and as models. It was with these and other long-established drugs that principles of lasting significance were discovered. Of relevance to this is the fact that interest in this area of science undoubtedly existed among the ancients. More recently, Shakespeare referred to the risk–benefit ratio associated with alcohol consumption in *Macbeth*, and to the duration of action of the fantasy drug consumed by Juliet in *Romeo and Juliet*. Henry Bence Jones was probably the first to describe the rates of transfer of drugs between tissues, in work conducted in the 1860s, after he had developed assays for lithium and quinine. Awareness of the first-order removal of digoxin from the body originates from Gold *et al.* in the 1920s. However, it was in the 1930s that Widmark and Teorell first examined concentrations in blood. In the 1950s and 1960s, Brodie and Williams focused our interest on metabolism and metabolites, and then on quantitative pharmacology related to concentrations in blood. The big explosion of interest resulted from stirrings of pharmacokinetic thought in the colleges of pharmacy, and from the development of Clinical Pharmacology primarily in medical schools, in the 1960s, 1970s, and 1980s. It has been exciting to be associated with this dramatic development in medical science. Mathematical pharmacokinetics first gained prominence in relation to dosage form design, then to profiling of drugs in humans and control of clinical response, and, remarkably, only recently, in the process of new drug discovery.

We have not discussed bioanalysis, apart from a brief consideration of assay specificity in Chapter 19. However, it is important that the reader be aware that pharmacokinetic information can be no better than the quality of the concentration–time data provided. Thus, the pharmacokineticist should ensure that concentration data are from specific, precise and accurate assays, including their application to error-free timing of sample collections. Notes on other methods of drug investigation are to be found throughout the book, and readers interested in particular methods should be able to access relevant information through the index and references.

We hope that you enjoy this book. We thank our various students in London, Gainesville, Rochester, and too many other locations to mention, for their help in formulating our understanding of our readers' needs in this subject area, and dedicate this book to them. We are immensely grateful to our publishers for their sage advice, and to our families for their support, tolerance and encouragement during the writing of this book.

Stephen H. Curry, Rochester, New York,
Robin Whelpton, London,
January 2010

About the Authors

Stephen Curry

Stephen has been Professor of Pharmacology at The London Hospital Medical College, Professor of Pharmaceutical Science at the University of Florida, and Adjunct Professor of Pharmacology and Physiology at the University of Rochester. He has also spent ten years with AstraZeneca and predecessor companies. He was honoured by the Faculty of Medicine of the University of London with the Doctor of Science Degree and is a Fellow of the Royal Pharmaceutical Society. He currently works in the field of technology transfer and translational science with early stage companies based on discoveries at the University of Rochester (PharmaNova) and Cornell University (ADispell). He can be contacted at www.stephenhcurry.com.

Robin Whelpton

After obtaining his first degree in Applied Chemistry, Robin joined the Department of Pharmacology and Therapeutics, London Hospital Medical College, University of London as research assistant to Professor Curry. Having obtained his PhD in pharmacology, he became lecturer and then senior lecturer before transferring to Queen Mary University of London, teaching pharmacology to preclinical medical students. He is currently a member of the School of Biological & Chemical Sciences and has a wealth of experience teaching drug distribution and pharmacokinetics to undergraduate and postgraduate students of medicine, pharmacology, biomedical sciences, pharmaceutical chemistry and forensic science.

1

Chemical Introduction: Sources, Classification and Chemical Properties of Drugs

1.1 Introduction

Pharmacology can be divided into two major areas, pharmacodynamics (PD) – the study of what a drug does to the body and pharmacokinetics (PK) – the study of what the body does to the drug. Drug disposition is a collective term used to describe drug absorption, distribution, metabolism and excretion whilst pharmaco-kinetics is the study of the rates of these processes. By subjecting the observed changes, for example, in plasma concentrations as a function of time, to mathematical equations (models), pharmacokinetic parameters such as elimination half-life ($t_{1/2}$), volume of distribution (V) and plasma clearance (CL) can be derived. Pharmacokinetic modelling is important for the:

- Selection of the right drug for pharmaceutical development
- Evaluation of drug delivery systems
- Design of drug dosage regimens
- Appropriate choice and use of drugs in the clinic.

These points will be expanded in subsequent chapters.

A drug is a substance that is taken, or administered, to produce an effect, usually a desirable one. These effects are assessed as physiological, biochemical or behavioural changes. There are two major groups of chemicals studied and used as drugs. First, there is a group of pharmacologically interesting endogenous substances, for example acetylcholine, histamine and noradrenaline. Second, there are the non-endogenous, or 'foreign' chemicals (xenobiotics), which are mostly products of the laboratories of the pharmaceutical industry.

There are numerous ways in which drugs interact with physiological and biochemical process to elicit their responses. Many of these interactions are with macromolecules, frequently proteins and nucleic acids. *Receptors* are transmembrane proteins, with endogenous ligands typified by acetylcholine and noradrenaline (norepinephrine). Although substances may be present naturally in the body, they are considered drugs when they are administered, such as when adrenaline is injected to alleviate anaphylactic shock. Drugs can either mimic (agonists) or inhibit (antagonists) endogenous neurotransmitters. Salbutamol is a selective β_2-agonist whereas propranolol is a non-selective β-blocker. Some receptors are *ligand-gated ion channels*, for example the cholinergic nicotinic receptor, which is competitively antagonized by (+)-tubocurarine. *Enzymes*, either membrane bound or soluble, can be inhibited – for example neostigmine inhibits acetylcholinesterase

Drug Disposition and Pharmacokinetics, By Stephen H. Curry and Robin Whelpton
© 2011 John Wiley & Sons, Ltd

and aspirin inhibits cyclooxygenase. Other proteins that may be affected are *voltage-gated (regulated) ion-channels* – a typical one being voltage-gated sodium channels which are blocked by local anaesthetics such as lidocaine (lignocaine). Antimalarials, chloroquine, for example, intercalate in DNA. Some drugs work because of their physical presence – often affecting pH or osmolarity – for example antacids to reduce gastric acidity or sodium bicarbonate to increase urinary pH and thereby increase salicylate excretion (Section 3.3.1.5).

1.1.1   *Source of drugs*

Primitive therapeutics relied heavily on a variety of mixtures prepared from botanical and inorganic materials. The botanical materials included some extremely potent plant extracts, with actions for example on the brain, heart and gastrointestinal tract, and also some innocuous potions, which probably had little effect. The inorganic materials were generally alkalis, which did little more than partially neutralize gastric acidity. Potassium carbonate (potash, from wood fires) was chewed with coca leaves to hasten the release of cocaine. Inevitably, the relative importance of these materials has declined, but it should be recognized that about a dozen important drugs are still obtained, as purified chemical constituents, from botanical sources and that alkalis still have a very definite value in certain conditions. Amongst the botanical drugs, are the alkaloids: morphine is still obtained from opium, cocaine is still obtained from coca leaves, and atropine is still obtained from the deadly nightshade (belladonna). Although the pure compounds have been prepared synthetically in the laboratory, the most economical source is still the botanical material. Similarly, glycosides such as digoxin and digitoxin are still obtained from plants. These naturally occurring molecules often form the basis of semisynthetic derivatives – it being more cost-effective than synthesis *de novo*.

Similar considerations apply with some of the drugs of zoological origin. For instance, while the consumption of raw liver (an obviously zoological material) was once of great importance in the treatment of anaemia, modern treatment relies on cyanocobalamin, which occurs in raw liver, and on hydroxycobalamin, a semisynthetic analogue. Another zoological example is insulin, which was obtained from the pancreatic glands of pigs (porcine insulin) but can now be genetically engineered using a laboratory strain of *Escherichia coli* to give human insulin.

Most other naturally occurring drugs, including antibiotics (antimicrobial drugs of biological origin) and vitamins, are generally nowadays of known chemical structure, and although their synthesis in the laboratory is in most cases a chemical possibility, it is often more convenient and economical to extract them from natural sources. For the simpler molecules the converse may be true, for example chloramphenicol, first extracted from the bacterium, *Streptomyces venezuelae*, is totally synthesized in the laboratory. For some antibiotics, penicillins and cephalosporins for example, the basic nucleus is of natural origin, but the modern drugs are semisynthetic modifications of the natural product.

Amongst the naturally occurring drugs are the large relative molecular mass (M_r) molecules, such as peptides, proteins (including enzymes), polysaccharides and antibodies or antibody fragments. Some of these macromolecules, snake venoms and toxins such as botulinum toxin ($M_r \sim 150,000$) have long been known. The anticoagulant, heparin, is a heavily sulfated polysaccharide ($M_r \sim 3,000–50,000$). Peptide hormones used as drugs include insulin and human growth hormone. Streptokinase, urokinase and tissue plasminogen activator (tPA) are enzymes used as thrombolytic agents. Other therapeutic enzymes are the pancreatic enzymes given to sufferers of cystic fibrosis. Antibodies are a recent addition to macromolecular drugs. Digoxin-specific antibodies, or light-chain fragments (F_{ab}) containing the specific binding site, are used to treat poisoning by cardiac glycosides. Advances in molecular biology have led to the introduction of a number of monoclonal antibodies with a range of targets: various cancers, viruses, bacteria, muscular dystrophy, the cardiovascular and immune systems, to name but some. Furthermore, the antibodies can be modified to carry toxins, cytokines, enzymes and radioisotopes to their specific targets. Monoclonal antibodies have the suffix *-mab* and the infix indicates the source of the antibody and the intended target; *-u-* indicates human and *-tu(m)-* that the target is a tumour. Thus, trastuzumab is a monoclonal antibody directed at (breast) cancer that has been 'humanized', *-zu-*; that is over 95% of the amino acid sequence is human. The prefix is unique to the drug.

With only minor exceptions, drugs are chemicals with known structures. Some of them are simple, some complex. Some of them are purely synthetic; some are obtained from crude natural products and purified before use. Most are organic chemicals, a few are inorganic chemicals. With all drugs, the emphasis is nowadays on a pure active constituent, with carefully controlled properties, rather than on a mysterious concoction of unknown potency and constitution.

1.2 Drug nomenclature and classification

Drug names can lead to confusion. Generally a drug will have at least three names, a full chemical name, a proprietary name, i.e. a trade name registered to a pharmaceutical company, and a non-proprietary name (INN) and/or an approved name. Names that may be encountered include the British Approved Name (BAN), the European Pharmacopoeia (EuP) name, the United States Adopted Name (USAN), the United States Pharmacopoeia (USP) name and the Japanese Approved Name (JAN). The WHO has been introducing a system of recommended INNs (rINN) and it is hoped that this will become the norm for naming drugs, replacing alternative systems. For example, lidocaine is classed as a rINN, USAN and JAN, replacing the name lignocaine that was once a BAN. Often 'ph' is replaced by 'f', as in cefadroxil, even though the group name is cephalosporins. We have elected to use amphetamine rather than amfetamine. Generally, the alternatives obviously refer to the same drug, such as ciclosporin, cyclosporin and cyclosporine. There are some notable exceptions, pethidine is known as meperidine in the United States and paracetamol as acetaminophen. Even a simple molecule like paracetamol may have several chemical names but the number of proprietary names or products containing paracetamol is even greater, including Panadol, Calpol, Tylenol, Anadin Extra. Spelling can also lead to apparent anomolies. For example, cefadroxine is a cephalosporin (Table 1.2). Therefore it is necessary to use an unequivocal approved name whenever possible.

A rigid system for the classification of drugs will never be devised. Increasingly, it is found that drugs possess actions which would permit their categorization in several groups in any one particular classification system. This is shown most strikingly by the use of lidocaine for both local anaesthetic and cardiac effects. Additionally, with constant changes in drug usage, it is not uncommon to find drugs of several different types in use for the same purpose. The number of examples within each type is of course very large. However, drugs are commonly grouped according to one of two major systems. These are on the basis of action or effect, and on the basis of chemistry. It is not possible to include all drugs in either of these groupings, and so a hybrid classification is necessary if all possibilities are to be considered. Table 1.1 shows an abbreviated pharmacological listing. The interpretation of this is quite straightforward, and it is presented as a general aid to the reader of later chapters of this book. Most of the examples quoted in later chapters are mentioned. Not so straightforward is the chemical listing shown in Table 1.2. It will be immediately noticed that while all of the groups of drugs in Table 1.2 are represented in Table 1.1, all of the types in Table 1.1 are not represented in Table 1.2, as a great many drugs are of chemical types of which there is only a single example, and Table 1.1 is only concerned with those chemical groups of drugs which are commonly known by their chemical names. Commonly encountered chemical groups are exemplified in Table 1.3.

1.3 Properties of molecules

Drug molecules may be converted to other molecules either by spontaneous change (i.e. decomposition) or by enzymatic transformation. Enzymes are such efficient catalysts that the rate of a reaction may be increased

Table 1.1   Abbreviated listing of drug groups categorized on the basis of pharmacological use or clinical effect, with examples, or cross-referenced to the chemical types of Table 1.2

THE CENTRAL NERVOUS SYSTEM

General anaesthetics
  I Gases – e.g. nitrous oxide
  II Volatile liquids – e.g. halothane
  III Intravenous anaesthetics, including some barbiturates

Hypnotics including some barbiturates and some benzodiazepines, and newer examples such as zolpidem

Sedatives including certain barbiturates, phenothiazines and benzodiazepines

Tranquillizers
  I Major, including certain phenothiazines and butyrophenones
  II Minor, including certain benzodiazepines
  III Other, newer, examples, such as olanzepine

Antidepressants
  I Dibenzazepines – e.g. nortriptyline
  II Monoamine oxidase inhibitors – e.g. tranylcypromine
  III Lithium
  IV Other newer examples, such as fluoxetine

Central nervous system stimulants
  I Amphetamine-related compounds – e.g methylphenidate and amphetamine
  II Hallucinogens – e.g. lysergic acid diethylamide
  III Xanthines – e.g. caffeine

Analgesics
  I Narcotics – e.g. morphine and pethidine
  II Mild analgesics, including salicylates

Miscellaneous centrally acting drugs, including respiratory stimulants (analeptics), anticonvulsants, certain muscle relaxants, drugs for Parkinson's disease, antiemetics, emetics and antitussives

CHEMOTHERAPY
Drugs used in the chemotherapy of parasitic diseases, including arsenicals
Drugs used in the chemotherapy of microbial diseases, including penicillins, cephalosporins and sulfonamides
Drugs used in the treatment of viral diseases, such as aciclovir
Drugs used in the treatment of fungal diseases, e.g. miconazole
Drugs used in the treatment of cancer, such as alkylating agents, antimetabolites, anthracycline derivatives, trastuzumab, hormone antagonists

PERIPHERAL SYSTEMS

Drugs acting at synapses and nerve endings
  I Acetylcholine and analogues (parasympathomimetic agents)
  II Anticholinesterase drugs – e.g. physostigmine
  III Inhibitors of acetylcholine at parasympathomimetic nerve endings – e.g. atropine
  IV Drugs acting at ganglia – e.g. nicotine
  V Drugs acting at adrenergic nerve endings, including catecholamines and imidazolines
  VI Neuromuscular blocking drugs – e.g. suxamethonium

Drugs acting on the respiratory system
  I Bronchodilators – e.g. salbutamol
  II Drugs affecting allergic responses – e.g. disodium cromoglycate
  III Oral antiasthmatics e.g. montelukast

Autacoids and their antagonists
  I Histamine and 5-hydroxytryptarnine
  II Antihistamines – e.g. diphenhydramine

Drugs for the treatment of gastrointestinal acidity
e.g. ranitidine and omeprazole

Cardiovascular drugs
  I Digitalis and digoxin
  II Antiarrhythmic drugs – e.g. quinidine
  III Antihypertensive drugs, including angiotensin-converting enzyme (ACE) inhibitors ('prils')
  IV Vasodilators – e.g. glyceryl trinitrate
  V Anticoagulants, including heparin and coumarins.
  VI Diuretics, including thiadiazines
  VII Lipid lowering drugs (e.g. 'statins')
  VIII Thrombolytics (e.g. tissue plasminogen activator)

Local anaesthetics – e.g. lidocaine (lignocaine)

Locally acting drugs
e.g. gastric antacids and cathartics

Endocrinology
Hormones, hormone analogues and hormone antagonists, including steroids, sulfonylureas and biguanides (e.g. glipizide, thyroxine and insulin)

Biological response modifiers
e.g. interferon, adalimumab

Immunosuppressants
e.g. ciclosporin

Table 1.2 Some groups of drugs classified on chemical structure rather than pharmacological properties or uses

Group	Parent structure	Chemical example	Uses and examples
Barbiturates		Phenobarbital $R_1 = C_2H_5$ $R_2 = C_6H_5$ $R_3 = H$ $X = O$	As hypnotics and sedatives (pentobarbital) As anticonvulsants (phenobarbital) As general anaesthetics (thiopental)
Benzodiazepines		Lorazepam $R_1 = H$ $R_2 = O$ $R_3 = OH$ $R_7 = Cl$ $R'_2 = Cl$	As anxiolytics (diazepam) As hypnotics (temazepam)
Biguanides		Metformin (as drawn)	As oral hypoglycaemics
Catecholamines		Adrenaline $R_1 = CH_3$ $R_2 = H$ $R_3 = OH$	Sympathomimetic amines
Cephalosporins		Cefadroxil $R_1 =$ $R_1 = CH_3$	As antimicrobial drugs
Coumarins		Warfarin (as drawn)	As anticoagulants
Dibenazazepines		Nortriptyline $R =$ $=CH(CH)_2NHCH_3$	As antidepressants
Imidazoles		Miconazole $R = Cl$	As antifungal drugs

(*continued*)

Table 1.2 (*Continued*)

Group	Parent structure	Chemical example	Uses and examples
Imidazolines		Clonidine	As antihypertensive drugs
Macromolecules: Polysaccharides, peptides, proteins, enzymes, antibodies		Heparin, insulin, trastuzumab	In control of blood clotting, diabetes, cancer, rheumatoid arthritis and other conditions
Penicillins		Penicillin G	As antimicrobial drugs
Phenothiazines		Thioridazine $R_1 = SCH_3$ $R_2 =$ $-(CH_2)_2-$	As antihistamines (promethazine) As antipsychotics (thioridazine) As antiemetics (trifluoperazine)
Prostaglandins		$PGF_{2\alpha}$ (as drawn)	As uterine stimulants and other procedures
Salicylates		Aspirin $R_1 = H$ $R_2 = COCH_3$	As antipyretic, anti-inflammatory and antipyretic drugs
Steroids		Hydrocortisone (as drawn)	Anti-inflammatory drugs
Sulfonamides		Sulfacetamide $R_1 = H$ $R_2 = COCH_3$	As antimicrobial drugs

Table 1.2 (*Continued*)

Group	Parent structure	Chemical example	Uses and examples
Sulfonylureas	Glipizide (As drawn)		As oral hypoglycaemic drugs

Group	Parent structure	Chemical example	Uses and examples
Thiadiazines		Chlorthiazide $R_1 = H$ $R_2 = H$ $R_3 = Cl$	As diuretics

Group	Parent structure	Chemical example	Uses and examples
Xanthines		Theophylline $R_1 = CH_3$ $R_2 = CH_3$ $R_3 = H$	As respiratory stimulants and bronchodilators

by the order of 10^{13} times – in other words some reactions would not, for all practical purposes, proceed but for the presence of enzymes. The role of enzymes in the metabolism of drugs is considered in Section 3.2.

1.3.1 Decomposition of drugs

Spontaneous decomposition needs to taken into consideration during manufacture, storage and use of drugs as well as during bioanalysis, when the products may be mistakenly thought to be metabolites. Although the same compounds may be produced by metabolism, there are occasions, for example, when substances identified in biological fluids arise from decomposition rather than metabolism. Decomposition may result in visible changes and odours when drugs are stored. The reactions tend to be accelerated by the presence of one or more of the following: catalysts, light, heat and moisture.

1.3.1.1 Hydrolysis

Esters and, to a lesser extent, amides are hydrolysed, particularly if catalysed by the presence of acids or bases. Aspirin (acetylsalicylic acid) is hydrolysed to salicylic acid and acetic acid, giving bottles of aspirin a smell of vinegar. Procaine is hydrolysed *p*-aminobenzoic acid and *N*-dimethyl-2-aminoethanol, whilst cocaine is hydrolysed to benzoylecgonine.

Table 1.3 Some important functional groups found in drug molecules

Type of compound	Functional group	Specific example	
		Name	Formula
Acids e.g. carboxylic acids	$-COOH$	Aspirin (also an ester)	**COOH** OCOCH$_3$
Alcohols	$-OH$	Choral hydrate	$CCl_3C(\mathbf{OH})_2$
Amides	$-CONH-$	Lidocaine (also an amine)	**NHCOCH$_2$N(C$_2$H$_5$)$_2$** H_3C CH_3
Bases e.g. amines	$-NRR'$	Amphetamine	CH$_3$ $-CH_2$CH**NH$_2$**
Esters	$-COO-$	Suxamethonium chloride (also a quaternary ammonium compound)	$CH_2\mathbf{COO}CH_2CH_2\overset{\oplus}{N}(CH_3)_3$ $CH_2\mathbf{COO}CH_2CH_2\underset{\oplus}{N}(CH_3)_3$ 2Cl$^{\ominus}$
Ethers	$-C-O-C-$	Enflurane	F F F H$-$C$-$**C**$-$**O**$-$**C**$-$H F F F
Imides	$-CONHCO-$	Thalidomide	
Ketones	$-CO-$	Haloperidol (also an amine and an alcohol)	
Sulfonamides	$-SO_2NH-$	Sulfadiazine (also an amine)	H$_2$N$-$ $-$**SO$_2$NH**$-$
Sulfones	$-SO_2-$	Dapsone (also an amine)	H$_2$N$-$ $-$**S**$-$ $-$NH$_2$
Small neutral molecules		Nitrous oxide	N_2O
Inorganic salts		Sodium bicarbonate Lithium carbonate	$NaHCO_3$ Li_2CO_3

1.3.1.2 Oxidation

Several drugs are readily oxidized, including phenothiazines, which form the corresponding 5-sulfoxides, via coloured semiquinone free radials. Phenothiazines with an electron-withdrawing group in the 2-position, tend to be more stable, thus promethazine (2-H) is more readily oxidized (to a blue product) than chlorpromazine (2-Cl) which gives a red semiquinone radical. Methylene blue is a phenothiazine (Figure 1.1)

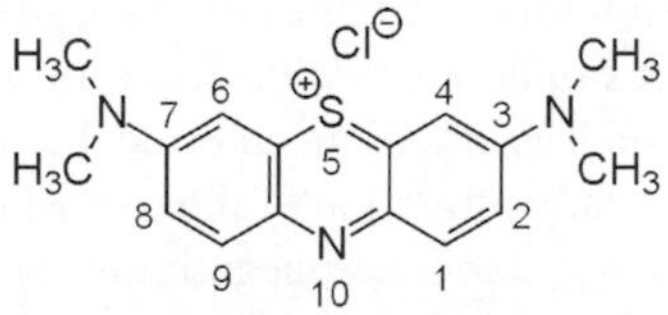

Figure 1.1 Formula of methylene blue (methylthioninium chloride). Other therapeutic phenothiazines usually have 10-substitents and often substituents at position 2.

that is used in medicine as a contrast agent, and as a treatment for methaemoglobinaemia, which is both a congenital and a drug-induced disorder. In this book it appears as an example of tri-exponential plasma concentration decay after intravenous doses (Figure 5.2), and as a valuable tool for laboratory modelling of drug disposition.

Physostigmine is oxidized to rubreserine, which is a deep brown-red colour. The first stage is probably hydrolysis to eseroline:

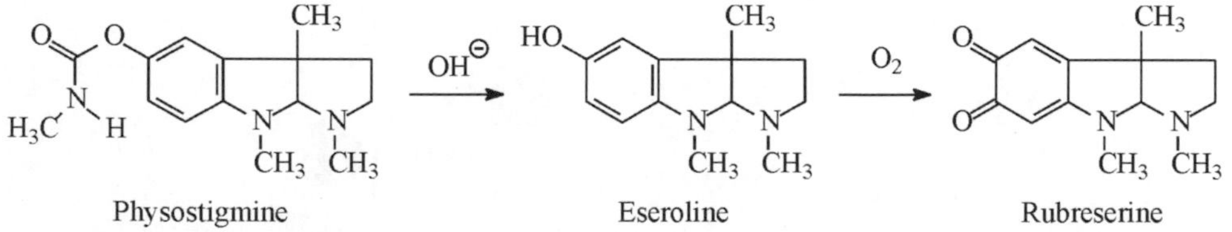

Physostigmine   Eseroline   Rubreserine

Adrenaline oxidizes in a similar manner to a brown–red material, adrenochrome.

1.3.1.3 Photodecomposition

Most compounds are photosensitive if irradiated with intense light of the appropriate wavelength. Some drugs are unstable in natural light, notably the 1,4-dihydropyridine calcium channel blocking drugs such as nifedipine. These drugs have to be formulated and handled in a darkroom under sodium light. Similarly, it is recommended that blood samples taken for nitrazepam or clonazepam analysis are protected from light to avoid photodecomposition.

1.3.1.4 Racemization

Optically active drugs (Section 1.8.2) may undergo racemization. For example during extraction from belladonna (−)-hyoscyamine may be converted to atropine [(±)-hyoscyamine].

1.4 Physicochemical interactions between drugs and other chemicals

In the present context we are principally concerned with interactions between relatively small drug molecules and relatively large endogenous molecules such as proteins e.g. enzymes, receptors and ion channels. The majority of drug–receptor interactions are reversible although some covalent reactions are known, for example non-competitive antagonism.

1.4.1 Chemical bonding and interactions between molecules

The interaction between atoms and molecules is basically electrostatic. The positively charged nuclei of atoms would repel each other if it were not for electrons sharing the space between them such that an electron from one atom is attracted to the nucleus of another. *Ionic bonds* occur when one atom completely donates one or more electrons to another atom, such as in sodium chloride. In the solid the ions are arranged so that the structure is held together by the electrostatic attraction of the oppositely charged ions. Bonds are described as (*non-polar*) *covalent* when atoms of similar electronegativities share electrons more or less equally, or *polar covalent* when the electronegativity of one atom is appreciably greater than the other. The nature of these bonds is somewhere between that of covalent and ionic bonds. The electrons in polar covalent bonds are attracted to the more electronegative atom creating a dipole, i.e. an asymmetric electric charge. The electron-withdrawing effect of oxygen in water molecules results in a strong dipole whereas in methane, CH_4, the electron density is evenly distributed so there is no dipole (Figure 1.2).

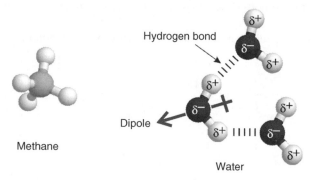

Figure 1.2 Comparison of non-polar methane (CH_4) and water (H_2O) where differences in charge densities between the more electronegative oxygen and hydrogen produces a dipole. Hydrogen bonding occurs between the slightly negatively charged oxygen and the slightly positively charged hydrogen.

Dipoles are responsible for the attractions between molecules. Molecules can align dipole to dipole, the more negative end of one being attracted to the more positive end of another; or a dipole in one molecule can induce a complementary one in an adjacent molecule. The differences in valance electron densities in the more electronegative elements, nitrogen, oxygen and fluorine, leads to *hydrogen bonding* (Figure 1.2) which in some instances can be as strong as some covalent bonds. *Van der Waals* forces arise because the density of the valence electron cloud around an atom can fluctuate causing a temporary dipole which may then induce a dipole in a neighboring atom. These are the weakest but most common forms of attractions between atoms. Another form of bonding is the *hydrophobic bonding* seen in proteins, where hydrophobic regions come together with the exclusion of water.

All the forms of bonding described above are encountered in pharmacology. Covalent binding of groups to enzymes and receptors occurs, such as acetylation of the serine groups in cyclooxgenase by aspirin. Ionic interaction is responsible for the binding of acidic drugs to albumin and of bases to α_1-acid glycoprotein. Hydrophobic binding is also important for binding of molecules to proteins. Similarly, these interactions occur when drugs bind to the active sites of enzymes and to receptors.

1.4.2 Solubility

The physicochemical interactions described above affect the solubility of molecules (solutes) in solvents. The dipole moment and hydrogen bonding in water make this a polar solvent and polar solutes readily dissolve in water, including salts. These solutes are described as hydrophilic – water loving. Organic solvents such as heptane are apolar because they have neither dipoles nor hydrogen bonding. Apolar solvents are very poorly soluble in water and vice versa, water being essentially insoluble in the organic solvent. The two liquids are said to be *immiscible*. Non-polar, non-ionized molecules tend to dissolve readily in organic solvents and lipids, and are referred to as hydrophobic or lipophilic – lipid loving.

To be transferred across lipid membranes drugs must be soluble in the barrier layer of fluid bathing the membrane. Consequently, drugs with low aqueous solubility may be poorly absorbed from the gastrointestinal tract. This can be exploited. An example is sulfasalazine, which is minimally absorbed after oral administration, and is used to treat ulcerative colitis.

1.5 Law of mass action

The Law of Mass Action states: 'the rate at which a chemical reaction proceeds is proportional to the active masses (usually molar concentrations) of the reacting substances'. This means that a non-reversible reaction proceeds at an ever-decreasing rate as the quantity of the reacting substances declines. The Law of Mass Action is easily understood if the assumption is made that, for the reaction to occur, collision between the reacting molecules must take place. It follows that the rate of reaction will be proportional to the number of collisions. The number of collisions will be proportional to the molar concentrations of the reacting molecules.

If a single substance X is in process of transformation into another substance Y, and if at any moment the active mass of X is represented by [X] (usually expressed in moles per litre) then we have:

$$X \rightarrow Y$$

and the rate of reaction at any time point $= k[X]$ where k is the velocity, or rate, constant. This constant varies with temperature and the nature of the reacting substance.

If two substances A and B are reacting to form two other substances C and D, and if the concentrations of the reactants at any particular moment are [A] and [B] then:

$$A + B \rightarrow C + D$$

and the rate of reaction $= k[A][B]$.

1.5.1 Reversible reactions and equilibrium constants

Consider the reaction:

$$A + B \rightleftharpoons C + D$$

The rate of the forward reaction is:

$$\text{forward rate} = k_1[A][B] \tag{1.1}$$

whilst the backward rate is:

$$\text{backward rate} = k_{-1}[C][D] \tag{1.2}$$

where k_1 and k_{-1}, are the rate constants of the forward and backward reactions, respectively. As the reaction proceeds, the concentrations of the original substances A and B diminish and the rate of the forward reaction decreases. At the same time, the substances C and D are produced in ever-increasing quantities so that the rate at which they form A and B increases. Eventually equilibrium is reached when the forward and backward rates are equal:

$$k_1[\text{A}][\text{B}] = k_{-1}[\text{C}][\text{D}] \tag{1.3}$$

The equilibrium constant, K, is the ratio of the forward and backward rate constants, so rearranging Equation 1.3 gives:

$$K = \frac{k_1}{k_{-1}} = \frac{[\text{C}][\text{D}]}{[\text{A}][\text{B}]} \tag{1.4}$$

The term *dissociation constant* is used when describing the equilibrium of a substance which dissociates into smaller units, as in the case, for example, of an acid (Section 4.6). The term is also applied to the binding of a drug, D, to a macromolecule such as a receptor, R, or plasma protein. The complex DR dissociates:

$$\text{DR} \rightleftharpoons \text{D} + \text{R}$$

So:

$$K = \frac{[\text{D}][\text{R}]}{[\text{DR}]} \tag{1.5}$$

An association constant is the inverse of a dissociation constant.

1.5.1.1   Sequential reactions

When a product, D, arises as a result of several, sequential reactions:

$$\text{A} \xrightarrow{k_1} \text{B} \xrightarrow{k_2} \text{C} \xrightarrow{k_3} \text{D}$$

it cannot be formed any faster than the rate of at which its precursor, C, is formed, which in turn cannot be formed any faster than its precursor, B. The rates of each of these steps are determined by the rate constants, k_1, k_2 and k_3. Therefore, the rate at which D is formed will be the rate of the slowest step, i.e. the reaction with the lowest value of rate constant. Say for example, k_2 is the lowest rate constant, then the rate of formation of D is determined by k_2 and the reaction $\text{B} \rightarrow \text{C}$ is said to be the *rate-limiting* step.

1.5.2   *Reaction order and molecularity*

The order of a reaction is the number, n, of concentration terms affecting the rate of the reaction, whereas molecularity is the number of molecules taking part in the reaction. The order of a reaction is measured experimentally and because it is often close to an integer, 0, 1, or 2, reactions are referred to as zero-, first- or second-order, respectively. The reaction

$$\text{X} \rightarrow \text{Y}$$

is clearly monomolecular, and may be either zero- or first-order depending on whether the rate is proportional to X^0 or X^1. The reactions

$$2X \rightarrow Y$$

and

$$A + B \rightarrow C + D$$

are both bimolecular and second-order providing the rate is proportional to $[X]^2$, in the first case, and to $[A][B]$, in the second. Note how the total reaction order is the sum of the indices of each reactant: rate $\propto$ $[A]^1[B]^1$, so $n = 2$. However, if one of the reactants, say A, is present in such a large excess that there is no detectable change in its concentration, then the rate will be dependent only on the concentration of the other reactant, B, that is, the rate is proportional to $[A]^0[B]^1$. The reaction is first-order (rate $\propto [B]$) but still described as 'bimolecular'. Hydrolysis of an ester in dilute aqueous solution is a commonly encountered example of a bimolecular reaction which is first-order with respect to the concentration of ester and zero-order with respect to the concentration of water, giving an overall reaction order of 1.

Enzyme-catalysed reactions have reaction orders between 1 and 0 with respect to the drug concentration. This is because the Michaelis–Menten equation (Section 3.2.5) limits to zero-order when the substrate is in excess and the enzyme is saturated so that increasing the drug concentration will have no effect on the reaction rate. When the concentration of enzyme is in vast excess compared with the substrate concentration, the enzyme concentration is not rate limiting and the reaction is first-order. Thus, the reaction order of an enzyme-catalysed reaction changes as the reaction proceeds and substrate is consumed.

1.5.3   *Decay curves and half-lives*

As discussed above the rate of a chemical reaction is determined by the concentrations of the reactants and from the foregoing it is clear that a general equation relating rate of decline in concentration ($-dC/dt$), rate constant (λ), and concentration (C) can be written:

$$-\frac{dC}{dt} = \lambda C^n \tag{1.6}$$

Note the use of λ, to denote the rate constant when it refers to decay; the symbol is used for radioactive decay, when it is known as the decay constant. Use of λ for elimination rate constants is now the standard in pharmacokinetic equations.

1.5.3.1   *First-order decay*

Because first-order kinetics are of prime importance in pharmacokinetics, we shall deal with these first. For a first-order reaction, substituting $n = 1$ in Equation 1.6 gives:

$$-\frac{dC}{dt} = \lambda C \tag{1.7}$$

that is the rate of the reaction is directly proportional to the concentration of substance present. As the reaction proceeds and the concentration of the substance falls, the rate of the reaction decreases. This is exponential decay, analogous to radioactive decay where the probability of disintegration is proportional to the number of unstable nuclei present. The first-order rate constant has units of reciprocal time (e.g. h^{-1}). Integrating Equation 1.7 gives:

$$C = C_0 \exp(-\lambda t) \tag{1.8}$$

which is the equation of a curve that asymptotes to 0 from the initial concentration, C_0 [Figure 1.3(a)]. Taking natural logarithms of Equation 1.8:

$$\ln C = \ln C_0 - \lambda t \tag{1.9}$$

gives the equation of a straight line of slope, $-\lambda$ [Figure 1.3(b)]. If common logarithms are used (log C versus t) the slope is $-\lambda/2.303$. Another way of presenting the data is to plot C on a logarithmic scale (using 'semilog.' graph paper — not shown). This approach was often used when computers were not readily available and is still frequently used to present data, for example Figure 2.10. Such plots allow computation of C_0 (read directly from the intercept) and the elimination half-life. However, a common misconception is to believe that the slope of this plot is $-\lambda/2.303$. The slope is the same as that of a C versus t plot, but it only appears to be linear — the graph is of the type shown in Figure 1.3(a), which could be viewed as a series of slopes of ever-decreasing magnitude. The slope that matters, permitting calculation of the rate constant and the half-life using linear regression, is that of a graph of $\ln C$ versus t [Figure 1.3(b)].

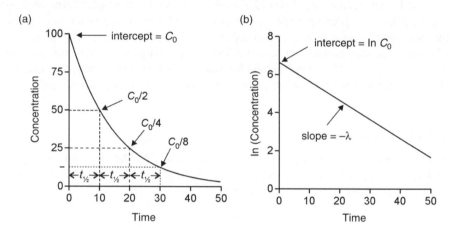

Figure 1.3   Curves for first-order decay plotted as (a) C versus t and (b) $\ln C$ versus t.

The half-life ($t_{1/2}$) is the time for the initial concentration (C_0) to fall to $C_0/2$, and substitution in Equation 1.9 gives:

$$t_{1/2} = \frac{\ln 2}{\lambda} = \frac{0.693}{\lambda} \tag{1.10}$$

as $\ln 2 = 0.693$. This important relationship, where $t_{1/2}$ is constant (independent of the initial concentration) and inversely proportional to λ, is *unique* to first-order reactions. Because $t_{1/2}$ is constant, 50% is eliminated in $1 \times t_{1/2}$, 75% in $2 \times t_{1/2}$, and so on. Thus, when five half-lives have elapsed less than 95% of the analyte remains, and after seven half-lives less than 99% remains.

1.5.3.2   Zero-order decay

For a zero-order reaction, $n = 0$, and:

$$-\frac{dC}{dt} = \lambda C^0 = \lambda \tag{1.11}$$

Thus, a zero-order reaction proceeds at a *constant rate*, and the zero-order rate constant must have units of rate (e.g. $g\,L^{-1}\,h^{-1}$). Integrating Equation 1.11:

$$C = C_0 - \lambda t \tag{1.12}$$

gives the equation of a straight line of slope, $-\lambda$, when concentration is plotted against time [Figure 1.4(a)]. The half-life can be obtained as before, substituting $t = t_{1/2}$ and $C = C_0/2$, gives:

$$t_{1/2} = \frac{C_0}{2\lambda} \tag{1.13}$$

The zero-order half-life is inversely proportional to λ, but $t_{1/2}$ is also directly proportional to the initial concentration. In other words, the greater the amount of drug present initially, the longer the time taken to reduce the amount present by 50% [Figure 1.4(b)]. The term 'dose dependent half-life' has been applied to this situation as well as to Michaelis–Menten kinetics cases.

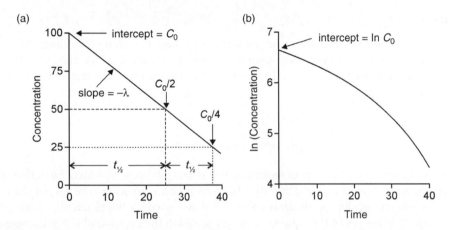

Figure 1.4 Curves for zero-order decay plotted as (a) C versus t and (b) $\ln C$ versus t.

1.5.3.3 Second-order decay

When $n \geq 2$, the integral of Equation 1.6 has a general solution, which when written in terms of λ is:

$$\lambda = \frac{1}{(n-1)t}\left(\frac{1}{C^{(n-1)}} - \frac{1}{C_0^{(n-1)}}\right) \tag{1.14}$$

So, when $n = 2$,

$$\lambda = \frac{1}{t}\left(\frac{1}{C} - \frac{1}{C_0}\right) \tag{1.15}$$

Substituting $C = C_0/2$ when $t = t_{1/2}$ and rearranging gives:

$$t_{1/2} = \frac{1}{\lambda C_0} \tag{1.16}$$

This is to be expected of second-order reactions because the probability of molecules colliding and reacting is much greater at higher concentrations.

Equations such as Equations 1.8 and 1.12 are referred to as *linear* equations. Note that in this context it is important not to confuse 'linear' with 'straight-line'. While it is true that the equation of a straight-line is a linear equation, exponential equations are also linear. On the other hand, nonlinear equations are those where the variable to be solved for cannot be written as a linear combination of independent variables. The Michaelis–Menten equation is such an example.

Despite the importance of the elimination half-life of a drug in pharmacokinetics, it is, in fact, dependent on two other pharmacokinetic parameters, apparent volume of distribution (V) and clearance (CL). The apparent volume of distribution, as its name implies, is a quantitative measure of the extent to which a drug is distributed in the body (Section 2.4.1.1) whilst clearance can be thought of as an indicator of how efficiently the body's eliminating organs remove the drug. Therefore the larger the value of CL, the shorter will be $t_{1/2}$. Changes in half-life are a result of changes in either V or CL or both (Section 4.2.1).

1.6 Ionization

Ionization is a property of all electrolytes, whether weak and strong. For example, sodium chloride (NaCl) is essentially completely ionized in aqueous solutions (forming sodium and chloride ions, Na^+ and Cl^-). Amines and carboxylic acids are only partially ionized in aqueous solutions. Their ionization reactions can be represented as follows:

$$R-NH_2 + H^+ \rightleftharpoons R-NH_3^+$$

$$R-COOH \rightleftharpoons R-COO^- + H^+$$

It will be noted that these ionization reactions are reversible, and the extent to which ionization takes place is determined by the pK_a of the compound and the pH of the aqueous solution. The pK_a of the compound is a measure of its inherent acidity or alkalinity, and it is determined by the molecular arrangement of the constituent atoms. It is the pH of the aqueous solution in which the compound is 50% ionized.

According to the Brønsted–Lowry theory, an acid is a species that tends to lose protons, and a base is a species that tends to accept protons. Acids and bases ionize in solution; acids donating hydrogen ions and bases accepting them. Thus in the examples above, $R\text{-}NH_3^+$ and R-COOH are acids, while $R\text{-}NH_2$ and $R-COO^-$ are bases. Also, $R\text{-}NH_3^+$ and $R\text{-}NH_2$, and R-COOH and $R\text{-}COO^-$, are termed 'conjugate acid–base pairs'.

The term *strength* when applied to an acid or base refers to its tendency to ionize. If an acid, AH, is dissolved in water, the following equilibrium occurs:

$$AH \rightleftharpoons H^+ + A^-$$

The acid dissociation constant is:

$$K_a = \frac{[H^+][A^-]}{[AH]} \tag{1.17}$$

Clearly the more the equilibrium is to the right, the greater is the hydrogen ion concentration, with a subsequent reduction in the concentration of non-ionized acid, so the larger will be the value of K_a. Taking logarithms (see Appendix 1 for details) of Equation 1.14, gives:

$$\log K_a = \log[H^+] + \log[A^-] - \log[AH] \tag{1.18}$$

and on rearrangement:

$$-\log[H^+] = -\log K_a + \log\frac{[A^-]}{[AH]} \qquad (1.19)$$

Because, $-\log[H^+]$ is the pH of the solution:

$$pH = pK_a + \log\frac{[A^-]}{[AH]} = pK_a + \log\frac{[\text{base}]}{[\text{acid}]} \qquad (1.20)$$

where $pK_a = -\log K_a$, by analogy with pH. Note that when $[A^-] = [AH]$ the ratio is 1 and because $\log(1) = 0$, the $pK_a = pH$, as stated earlier.

It is possible to calculate the equilibrium constant, K_b, for a base, B, ionizing in water:

$$B + H_2O \rightleftharpoons BH^+ + OH^-$$

However, one can consider the ionization of the conjugate acid, BH^+ and derive a pK_a for it as above.

$$K_a = \frac{[H^+][B]}{[BH^+]} \qquad (1.21)$$

Note how for a strong base, the concentration of BH^+ is high (high tendency to ionize) and so K_a is small.

$$pH = pK_a + \log\frac{[B]}{[BH^+]} = pK_a + \log\frac{[\text{base}]}{[\text{acid}]} \qquad (1.22)$$

The use of terms such as weak and strong is fraught with danger. However, because the term 'weak electrolyte' is used for all partially ionized materials, the term weak should probably be applied to all acids and bases used as drugs. However, it should not be forgotten that high concentrations of organic acids and bases, in spite of the compounds being weak electrolytes, can appear 'strong' in the sense of being corrosive, removing rust, precipitating protein etc. These compounds, however, are obviously distinct from most inorganic acids (nitric, hydrochloric and perchloric acids) which have pK_a values in the range -1 to -7, and are effectively 100% dissociated at any pH. There are, also, certain weak electrolyte inorganic acids; carbonic acid ($pK_{a_1} = 6.35$, $pK_{a_2} = 10.25$) for example. The bicarbonate to CO_2 ratio is of major significance for buffering the pH of blood.

It should be noted that it is not possible from knowledge of the pK_a alone to say whether a substance is an acid or a base. It is necessary to know how the molecule ionizes. Thiopental, $pK_a = 7.8$, forms sodium salts and so must be an acid, albeit a rather weak one. Diazepam, $pK_a = 3.3$, must be a base as it can be extracted from organic solvents into hydrochloric acid. Molecules can have more than one ionizable group; salicylic acid for example, has a carboxylic acid ($pK_a = 3.0$) and a weaker acidic phenol group ($pK_a = 13.4$). Morphine is amphoteric, that is it is both basic (tertiary amine, $pK_{a_1} = 8.0$) and acidic (phenol $pK_{a_2} = 9.9$) (Figure 1.5).

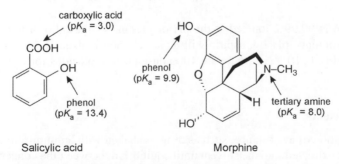

Figure 1.5   Ionizable groups of salicylic acid and morphine.

1.6.1   Henderson–Hasselbalch equation

Equation 1.20 is a form of the Henderson–Hasselbalch equation, which is important in determining the degree of ionization of weak electrolytes and calculating the pH of buffer solutions. If the degree of ionization is α, then the degree non-ionized is $(1 - \alpha)$ and, for an acid:

$$pH = pK_a + \log \frac{\alpha}{1-\alpha} \qquad (1.23)$$

or

$$\log \frac{\alpha}{1-\alpha} = pH - pK_a \qquad (1.24)$$

taking antilogarithms gives:

$$\frac{\alpha}{1-\alpha} = 10^{(pH-pK_a)} \qquad (1.25)$$

on rearrangement:

$$\alpha = \frac{10^{(pH-pK_a)}}{1 + 10^{(pH-pK_a)}} \qquad (1.26)$$

The equivalent equation for a base is:

$$\alpha = \frac{10^{(pK_a-pH)}}{1 + 10^{(pK_a-pH)}} \qquad (1.27)$$

Although Equations 1.26 and 1.27 look complex, they are easy to use. Using the ionization of aspirin as an example: the pK_a of aspirin is $\sim$3.4, so at the pH of plasma (7.4),

$$pH - pK_a = 7.4 - 3.4 = 4$$

$$\alpha = \frac{10^4}{1 + 10^4} = \frac{10000}{10001} = 0.9999$$

In other words aspirin is 99.99% ionized at the pH of plasma, or the ratio of ionized to non-ionized is 10,000 : 1. In gastric contents, pH 1.4, aspirin will be largely non-ionized; $1.4 - 3.4 = -2$, so the ratio of ionized to non-ionized is $1 : 10^{-2}$, i.e. there are 100 non-ionized molecules for every ionized one.

1.7   Partition coefficients

When an aqueous solution of a substance, such as drug, is shaken with an immiscible solvent (e.g. diethyl ether) the substance is extracted into the solvent until equilibrium between the concentration in the organic phase and the aqueous phase is established. Usually equilibration only takes a few seconds. For dilute solutions the ratio of concentrations is known as the distribution, or partition coefficient, P:

$$P = \frac{\text{concentration in organic phase}}{\text{concentration in aqueous phase}} \tag{1.28}$$

Organic molecules with large numbers of paraffin chains, aromatic rings and halogens tend to have large values of P, whilst the introduction of polar groups such as hydroxyl or carbonyl groups generally reduces the partition coefficient. Drugs with high partition coefficients are lipophilic or hydrophobic, whereas those that are very water soluble and are poorly extracted by organic solvents are hydrophilic. Lipophilicity can have a major influence on how a drug is distributed in the body, its tendency to bind to macromolecules such as proteins and, as a consequence, drug activity. A relationship between partition coefficient and pharmacological activity was demonstrated as early as 1901, but it was Corwin Hansch in the 1960s who used regression analysis to correlate biological activity with partition coefficient. He chose *n*-octanol as the organic phase and this has become the standard for such studies (Figure 1.6). Because P can vary between <1 (poorly extracted by the organic phase) to several hundred thousand, values are usually converted to log P, to encompass the large range (see Appendix).

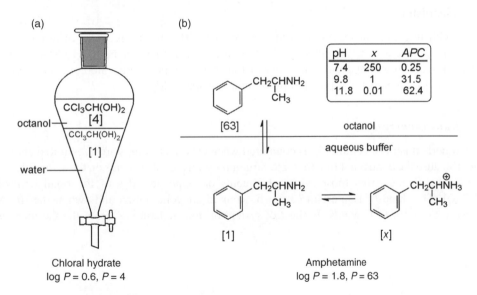

Figure 1.6 presentation of chloral hydrate and amphetamine partitioning.

pH	x	APC
7.4	250	0.25
9.8	1	31.5
11.8	0.01	62.4

Chloral hydrate
log P = 0.6, P = 4

Amphetamine
log P = 1.8, P = 63

Figure 1.6 (a) Partitioning of chloral hydrate is unaffected by buffer pH. (b) Partitioning of non-ionized amphetamine remains constant, 63 : 1. However the ratio of ionized to non-ionized is affected by buffer pH and as a consequent affects the apparent partition coefficient (*APC*) and the proportion extracted (inset).

1.7.1 Effect of ionization on partitioning

Generally, ionized molecules cannot be extracted into organic solvents, or at least not appreciably. The most notable exception to this is the extraction of ion-pairs into solvents such as chloroform. Thus, for weak electrolytes the amount extracted will usually be dependent on the degree of ionization, which of course is a function of the pH of the aqueous solution and the pK_a of the ionizing group as discussed above (Section 1.5.1), and the partition coefficient. If the total concentration (ionized + non-ionized) of solute in the aqueous phase is measured and used to calculate an apparent partition coefficient, D, then the partition coefficient, P, can be calculated. For an acid:

$$P = D[1 + 10^{(pH - pK_a)}] \qquad (1.29)$$

and for a base:

$$P = D[1 + 10^{(pK_a - pH)}] \qquad (1.30)$$

When the pH $= pK_a$ then, because $10^0 = 1$, $P = 2D$. When the pH is very much less than the pK_a, in the case of acids, or very much larger than the pK_a, in the case of bases, there will be no appreciable ionization and then D will be a good estimate of P [Figure 1.6(b)].

Unless stated otherwise, log P is taken to represent the logarithm of the true partition coefficient, i.e. when there is no ionization of the drug. However, for some weak electrolytes, biological activity may correlate better with the partition coefficient between octanol and pH 7.4 buffer solution. These values are referred to as log D.

Differences in the pH of different physiological environments, e.g. plasma and gastric contents can have a major influence on the way drugs are absorbed and distributed. (Section 2.2.1.1).

1.8 Stereochemistry

Compounds with the same molecular formula, but with a different arrangement of atoms are isomers. Structural isomers have different arrangements of atoms, for example, ethanol (CH_3CH_2OH) and diethyl ether (CH_3OCH_3) are structural isomers and have distinct chemical properties. Stereoisomers have the same bond structure but the geometrical positioning of atoms in space differs.

1.8.1 Cis–trans isomerism

Cis–trans isomerism is most commonly encountered when chemical groups are substituted about a double C=C bond. Because the bond is not free to rotate, structures with the substituents on the same side of the bond (*cis*-isomers) are distinct from those with substituents on opposite sides of the bond (*trans*-isomers) [Figure 1.7(a)]. This kind of isomerism can also occur in alicyclic compounds when the ring structure prevents free rotation of C–C bonds. In the *E/Z* system of nomenclature, Z, from the German *zusammen*

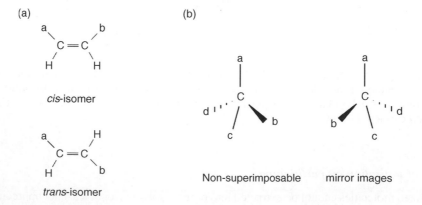

Figure 1.7 (a) *Cis–trans* isomerism occurs because the C=C bond cannot rotate and so the molecules depicted are different. (b) Asymmetric substitution produces isomers that are mirror images. The wedge represents a bond coming out of the page, the broken line represents a bond that recedes behind the page and the carbon atom and other two bonds are in the plane of the page.

meaning together, corresponds to *cis*; whilst *E*, from the German *entgegen* meaning opposite, to *trans*. The *E/Z* system must be used when there are more than two different substituents. The groups are assigned a rank according to the Cahn–Ingold–Prelog (CIP) rules (Section 1.8.2). If substituents of higher priority are on the same side, the isomer is designated *Z*; if they are on opposite sides, it is the *E*-isomer. *Cis–trans* isomerism is encountered in pharmacology. For example, clopenthixol is a mixture of *cis/trans* isomers whereas zuclopenthixol is the purified *Z*-isomer.

1.8.2 Optical isomerism

When a molecule and its mirror image cannot be superimposed, the substance is said to be *chiral*, from the Greek, *cheir*, meaning hand. The distinctive feature of such molecules is that they rotate the plane of plane-polarized light. Asymmetric substitution about carbon produces optical isomers [Figure 1.7(b)]. Other elements that show optical isomerism include sulfur, phosphorus and nitrogen. Individual isomers are referred to as enantiomers. These can be identified by whether they rotate the light to the right (dextrorotatory) or to the left (laevorotatory). The symbols *d*- and *l*- may be used to indicate the direction of rotation but (+)- and (−)- are preferred. A racemic mixture, or racemate, a 50:50 mixture of each enantiomer, is identified by *dl*-, (±)-, or *rac*-. Sometimes it is clear when a drug is an enantiomer from its name, for example the cough suppressant, dextrorphan, and its enantiomer, the analgesic, levorphanol. Often there is no indication, particularly with naturally occurring drugs. Morphine, hyoscine, cocaine and physostigmine are enantiomers. Similarly, many synthetic drugs are marketed as racemates without any indication, but there are examples of deliberate marketing of single isomers, sometimes for reasons connected with patents.

1.8.2.1 Absolute configuration

Although the rotation of plane-polarized light unequivocally defines a compound as one enantiomer or the other, it gives no indication of the spatial arrangement of the groups around the chiral centre – the *configuration*. Until the advent of X-ray crystallography, the absolute configurations of enantiomers were unknown. D-(+)-glyceraldehyde was arbitrarily defined as the D-configuration and all compounds derived from it were designated D-, irrespective of the optical activity, provided that the bonds to the asymmetric carbon remained intact. The naturally-occurring mammalian amino acids in proteins can be related to L-(−)-glyceraldehyde and form an L-series.

The use of D- or L- to define absolute configuration, is not readily applicable to all chiral molecules, and the CIP convention is generally used. The groups are assigned to a priority order according to sequence rules. Simply, the order is determined by the atomic numbers of the atoms attached to the chiral centre, priority being given to the higher numbers; for example O > N > C > H. When groups are attached by the same atom, then the next atom is considered, and so on in sequence until the order has been determined. The arrangement of groups around the chiral centre is 'viewed' with the group of least priority to the rear. Then the spatial arrangement of the groups is determined in decreasing priority order. If the direction is clockwise (i.e. to the right) the configuration is designated *R* (from the Latin, *rectus*, right). If the direction is anticlockwise the configuration is *S* (Latin, *sinister*, left). It must be noted that the D/L and *R/S* notations are not interchangeable. Using the CIP system, all L-amino acids are *S*-, with the exception of cysteine and cystine which, because they contain sulfur atoms (higher priority) that are connected to the chiral carbon, are designated *R*-.

1.8.3 Importance of stereochemistry in pharmacology

Obviously, the physical shape of a drug is important for it to bind to its receptors and so elicit a response. Because receptors are proteins, comprised of chiral amino acids, the spatial arrangement of the atoms in the

interacting drug will be crucial. There are numerous examples where stereoisomers show marked differences in their pharmacology. *R*-Thalidomide is sedative whereas the *S*-isomer inhibits angiogenesis, which is probably part of the mechanism of its teratogenic effect. However, because enzymes and transport systems are proteins, stereochemical differences may be shown in the way in which isomers are metabolized or distributed. With the advent of more convenient methods of measuring enantiomers, such as chiral high performance liquid chromatography phases, we are beginning to understand the full extent of differences in the pharmacokinetics of stereoisomers. The interaction between warfarin and phenylbutazone was not fully understood until it was shown that phenylbutazone selectively inhibits the metabolism of the more active *S*-isomer of warfarin.

Further reading and references

Atkins P, de Paula J. *Atkins' Physical Chemistry*, 8th edn. Oxford: Oxford University Press, 2006.

Brunton L, Lazo J, Parker K. *Goodman and Gilman's the Pharmacological Basis of Therapeutics*, 11th edn. New York: McGraw-Hill, 2005.

Clayden J, Greeves N, Warren S, Wothers P. *Organic Chemistry*. Oxford: Oxford University Press, 2000.

Ganong WF. *Review of Medical Physiology*, 22nd edn. New York: Lang Medical, 2005.

Leo A, Hansch C, Edkins D. Partition coefficients and their uses. *Chem Rev*. 1971; 36: 1539–44.

Nursing 2008: Drug Handbook. Philadelphia: Lippincott Williams & Wilkins, 2008.

Patrick GP. *An Introduction to Medicinal Chemistry*, 4th edn. Oxford: Oxford University Press, 2009.

Troy DB. (ed.) *Remington: The Science and Practice of Pharmacy*, 21st edn. Philadelphia: Lippincott Williams & Wilkins, 2006.

World Health Organization. Guidelines on the use of international non-proprietary names (INNs) for pharmaceutical substances. 1997 Geneva. Available one line http://whqlibdoc.who.int/hq/1997/WHO_PHARM_S_NOM_1570.pdf (accessed 15 October 2009).

2

Drug Administration and Distribution

2.1 Introduction

In order to achieve its effect, a drug must first be presented in a suitable form at an appropriate site of *administration*. It must then be *absorbed* from the site of administration and *distributed* through the body to its site of action. For the effect to wear off the drug must nearly always be *metabolized* and/or *excreted*. These processes are often given the acronym, ADME, and occasionally, LADME, where L stands for liberation of drug (from its dosage form). Finally, drug residues are *removed* from the body (Figure 2.1). Removal refers to loss of material, unchanged drug and/or metabolic products, in urine and/or faeces, once this material has been excreted into the bladder or bowel by the kidneys and liver. Absorption and distribution comprise the *disposition* (placement around the body) of a compound. Metabolism and excretion comprise the *fate* of a compound. It should however be noted that pharmacokineticists sometimes use the word disposition in a slightly different context, invoking concepts of 'disposal'.

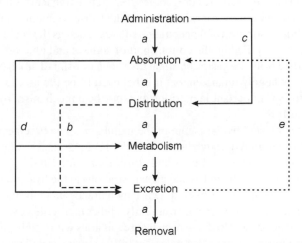

Figure 2.1 General scheme showing the relationship between the various events involved in drug disposition and fate. See text for explanation of letters *a–e*.

The most common pathway for an orally administered drug as indicated in Figure 2.1 is by route *a*. This pathway involves metabolism, and excretion of both unchanged drug and metabolites. A drug that is excreted in its unmetabolized form will by-pass metabolism (pathway *b*). An intravenously administered drug undergoes no absorption (pathway *c*). An oral dose can be rapidly converted to its metabolites in the intestinal

Drug Disposition and Pharmacokinetics, By Stephen H. Curry and Robin Whelpton
© 2011 John Wiley & Sons, Ltd

mucosa and/or the portal circulation and then excreted in bile before distribution through the body can occur (pathway *d*) ('pre-systemic metabolism or elimination' or the so-called 'first-pass' effect). Excretion products in the intestine can be reabsorbed (pathway *e*).

2.2   Drug transfer across biological membranes

Absorption, distribution and excretion of drugs involves transfer of drug molecules across various membranes, such as the gastrointestinal epithelium, the renal tubular epithelium, the blood-brain barrier and the placental membrane. Transfer of substances across biological membranes can occur by one or more of five possible mechanisms:

- *Passive diffusion.* Through the membrane, down a concentration gradient.
- *Filtration.* Through pores in the membrane.
- *Active transport.* Involving carrier molecules (transporters), requiring energy and occurring against a concentration gradient.
- *Facilitated diffusion.* A carrier-mediated process that does not require energy and where the net flow is down a concentration gradient.
- *Pinocytosis.* Microscopic invaginations of the cell wall engulf drops of extracellular fluid and solutes are carried through in the resulting vacuoles of water.

Historically, passive diffusion has been viewed as by far the most important for foreign molecules. Filtration is important in the transfer of small molecules into interstitial fluid via the fenestrations in peripheral capillaries and it plays a major part in urinary excretion of drug molecules. Active transport is important for the absorption of a few drugs, but it is involved to a considerable extent in drug excretion processes. Transporters as mechanisms of reducing drug absorption and penetration of areas such as the brain are considered in Section 2.3.1.2. Facilitated diffusion is typified by the co-transport of sodium and glucose, and is the mechanism by which vitamin B_{12} is absorbed from the gastrointestinal tract. The transfer of small foreign compounds across membranes by pinocytosis is largely unknown, but it is believed to be the mechanism by which botulinum toxin is absorbed. Pinocytosis is important in nutrition and may be a mechanism by which small amounts of ionized molecules are absorbed.

Drug transfer will also be affected by the nature of the membrane. The basement membrane of peripheral capillaries has gaps or *fenestrations* through which small molecular mass drugs readily filter. Albumin ($M_r \sim 69,000$) is too large to be filtered and is excluded from interstitial fluid. At the glomerulus drugs up to about the size of albumin are freely filtered, while the fenestrations in the liver sinusoids are large enough for macromolecules such as lipoproteins to be filtered. Other membranes, notably those of the GI tract and the placenta do not have fenestrations. The observation from early studies that certain dyes did not enter the brain, led to the concept of the blood–brain barrier (BBB) – specialized capillaries with tightly packed endothelial cells. For these membranes drugs must traverse them either by passive diffusion or specialized carrier-mediated transport.

2.2.1   Passive diffusion

The rate of diffusion of a drug, dQ/dt, is a function of the concentration gradient across the membrane, ΔC, the surface area over which transfer occurs, A, the thickness of the membrane, Δx, and the diffusion coefficient, D, which is characteristic of the particular drug. The diffusion coefficient of a compound is a function of its solubility in the membrane, its relative molecular mass, and its steric configuration. Thus, the rate of diffusion is governed by Fick's law:

$$\frac{dQ}{dt} = RDA\frac{\Delta C}{\Delta x} \tag{2.1}$$

where R is the partition coefficient of the drug between the membrane and the aqueous phase. Because A, D, R and the distance over which diffusion is occurring are constants these can be combined into one constant, permeability, P, so that:

$$\text{rate of diffusion} = P\Delta C = P(C_1 - C_2) \tag{2.2}$$

2.2.1.1 pH-partition hypothesis

As far as foreign molecules are concerned, biological membranes behave as if they are simple lipid barriers. This is remarkable, in view of the obvious complexity of such membranes, but it should be remembered that they exist primarily to transfer nutrients, not to transfer drugs. Their function, as far as foreign molecules are concerned, is more likely to be one of exclusion than one of transfer, so that in a sense drugs are absorbed against the odds. The fact that drugs pass through biological membranes mostly by simple diffusion has been repeatedly verified in experimental work. In this transfer, drug molecules have to dissolve in the membrane and so lipophilic species diffuse freely, but polar, particularly ionized, molecules do not. Therefore whether a weak electrolyte is ionized or not, will be a major determinant of whether it will diffuse through a biological membrane. The ratio of ionized to non-ionized forms is a function of the pH of the aqueous environment and the acid dissociation constant, K_a, as explained in Section 1.6. This means that in any aqueous solution both ionized and non-ionized forms are present.

Calculation of the equilibrium distribution of aspirin between stomach contents and plasma water illustrates the importance of pH-partitioning (Figure 2.2). It will be readily appreciated that, at equilibrium,

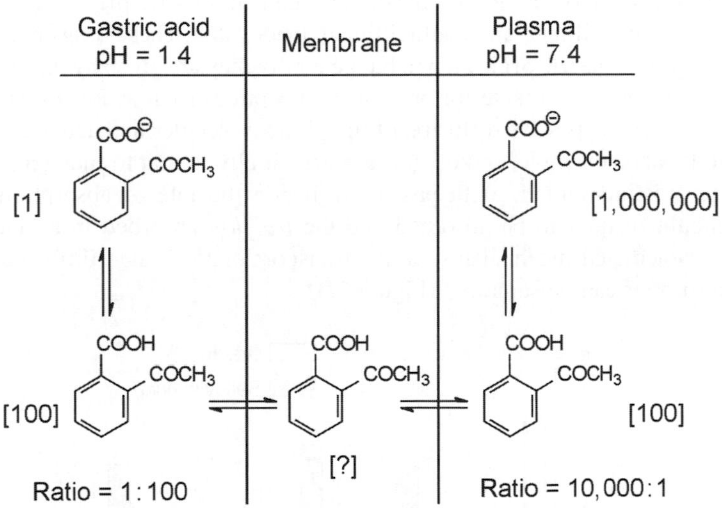

Figure 2.2 Equilibrium distribution of aspirin ($pK_a = 3.4$) between two solutions of pH 1.4 and pH 7.4 separated by a simple membrane. Relative concentrations are shown in [ ].

for the *non-ionized* species (which alone passes the lipid membrane) the concentration ratios will be the same:

$$\frac{\text{Concentration in membrane}}{\text{Concentration in gastric acid}} = \frac{\text{Concentration in membrane}}{\text{Concentration in plasma water}}$$

Thus for the non-ionized species the concentrations on each side of the membrane will be identical. For convenience we have designated this concentration [100]. However, from the Henderson–Hasselbalch equation (Section 1.6.1) the ratio of ionized to non-ionized aspirin at pH 1.4 is 1 : 100. In other words, for every [100] non-ionized molecules of aspirin there is [1] ionized molecule. In the plasma (pH 7.4), aspirin will be highly ionized, the ratio being 10,000 : 1, i.e. for every 100 molecules of non-ionized aspirin there will be $10,000 \times 100 = 1,000,000$ molecules ionized. Therefore at equilibrium, the ratio of total material on the two sides will be:

$$\frac{\text{Total concentration in acid}}{\text{Total concentration in plasma}} = \frac{100 + 1}{1\,000\,000 + 100} = \frac{101}{1\,000\,100} \approx \frac{1}{9900}$$

or almost 10,000 to 1 in favour of plasma. If drug molecules are introduced into any part of the system, they will transfer between the various media, including the membrane, until this concentration ratio is achieved.

2.3   Drug administration

The route of drug administration will be determined by the nature of the drug and the indication for its use.

2.3.1   Oral administration

This route is popular as it is generally convenient, and requires no medical skill or sterile conditions. Thus it is appropriate for outpatient use and medicines bought over the counter (OTC). For those with difficulty swallowing tablets or capsules, the elderly or infants for example, the drug may be may be given as a solution or suspension in liquid.

However, the gastrointestinal (GI) tract is a harsh environment. Gastric pH is low and acid-labile drugs such as benzylpenicillin (penicillin G) and methicillin are inactivated. These drugs may be used in the very young or the elderly in whom gastric pH is higher, but generally they are given by injection. The presence of proteases makes the oral route unsuitable for proteins and peptides such as insulin and oxytocin.

Absorption occurs chiefly by passive diffusion of lipophilic molecules and active transport of drugs that are endogenous, levodopa, for example, or those that are structurally similar to endogenous compounds, such as the cytotoxic agent 5-fluorouracil. With passive diffusion the rate of absorption is proportional to concentration or amount of drug to be absorbed and the *fraction* absorbed in a given interval remains constant. With carrier-mediated mechanisms, active transport or facilitated diffusion, there is a limited capacity and the transporter can be saturated (Figure 2.3).

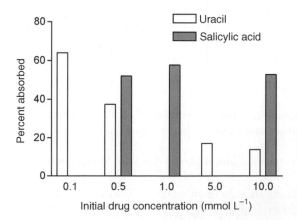

Figure 2.3   Absorption of uracil and salicylic acid from the rat small intestine as a function of initial drug concentration. Increasing concentrations of uracil saturate the active carrier so the proportion absorbed decreases, whereas for salicylic acid, which is absorbed by passive diffusion, the proportion absorbed is independent of the concentration. (Redrawn from the data of Brodie, in Binns, 1964.)

Being largely non-ionized in acid, aspirin can be absorbed from the stomach (Section 2.2.1.1), however most of the absorption occurs in the small intestine where the large surface area compensates for the less favourable degree of ionization. Weak bases cannot be absorbed until they have left the stomach, so delayed gastric emptying can delay the effect of such drugs. Quaternary ammonium compounds (Figure 2.4), which

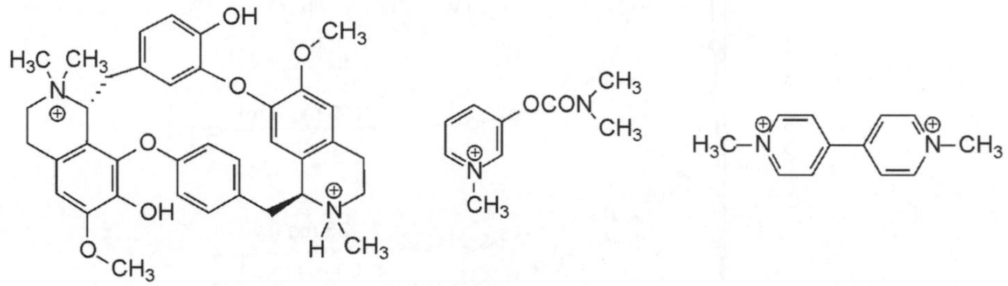

| (+)-Tubocurarine | Pyridostigmine | Paraquat |

Figure 2.4 Some quaternary ammonium compounds.

are permanently ionized, would not be expected to be absorbed to an appreciable extent by passive diffusion and this appears to be the case. (+)-Tubocurarine, the purified alkaloid from the arrowhead poison, curare, is not absorbed and the oral absorption of pyridostigmine is very low and erratic, such that an oral dose is typically 30 to 60 times larger than an equivalent intravenous one. Of toxicological importance is the absorption of the diquaternary ammonium herbicide, paraquat. Its absorption is saturable which supports the hypothesis that a carrier (probably for choline) is involved.

2.3.1.1 Presystemic metabolism

Materials absorbed from the stomach and intestine are carried, via the mesenteric capillary network and the hepatic portal vein, to the liver. If the drug is largely metabolized as it passes through the liver, then little of it will reach the systemic circulation. The consequences of this *first-pass metabolism*, as it is sometimes known, will depend on whether or not the metabolites are pharmacologically active. In the case of glyceryl trinitrate (GTN, nitroglycerin), which is almost totally metabolized, the di- and mono-nitrate metabolites have very reduced activity and generally this drug is considered to be inactive when taken orally, although some clinical trials have confirmed at least some clinical value of high dose sustained-release oral preparations. GTN is generally given by more suitable, alternative routes. The extent to which a drug undergoes presystemic metabolism can be obtained by comparing the plasma concentration–time curves and areas under the curves (AUC), after an oral and an intravenous dose of the drug (Section 8.3, Equation 8.2). This is illustrated for chlorpromazine in Figure 2.5, where the systemic availability of chlorpromazine is markedly reduced after oral administration, as evidenced by the reduced $AUC_{p.o.}$ compared to the area under the intravenous curve, $AUC_{i.v.}$. The low oral bioavailability is known to be largely due to presystemic metabolism, because studies with radioactive drug and related phenothiazines show that radioactive metabolites appear in the systemic circulation.

First-pass metabolism may influence the relative proportions of metabolites produced, as with propranolol, for example. When given orally, pharmacologically active 4'-hydroxypropranolol is produced, however little of this metabolite is measurable when propranolol is given intravenously. It has been suggested that this

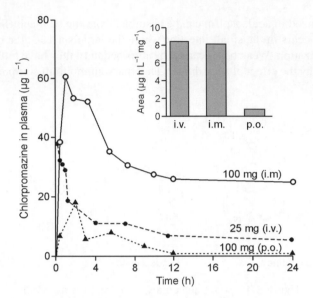

Figure 2.5   Plasma concentrations of chlorpromazine after three routes of administration and the areas under the curves after normalization for dosage (inset).

is because the high concentrations of propranolol reaching the liver after oral administration saturate the pathway that produces naphthoxylacetic acid, a major metabolite after intravenous (i.v.) injection. Comparison of the *AUC* values after oral and intravenous doses support this supposition (Figure 2.6). The

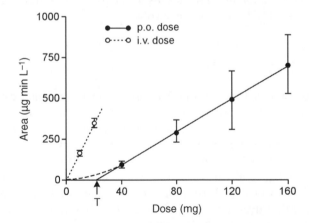

Figure 2.6   *AUC* values as a function of propranolol dose. The data provide evidence of saturable first-pass metabolism, with an apparent threshold dose (T) being required before any propranolol is measurable in plasma. (Redrawn from Shand & Rangno, 1972).

amount of drug reaching the systemic circulation is very reduced when given orally and, furthermore, there appears to be a threshold dose of approximately 20 mg below which little or no propranolol is measurable in the plasma, suggesting that at low doses all of the dose may be metabolized and that higher doses saturate some of the drug metabolizing pathways. When given to a patient with portocaval anastomosis the drug bypassed the liver and was completely available to the systemic circulation and no threshold was observed.

Saturation of first-pass metabolism may apply to other drugs that are normally extensively metabolized as they flow through the liver.

Drugs may be metabolized before they even reach the liver. GI tract mucosal cells contain several drug-metabolizing enzymes, probably the most important being cytochrome P450 3A4 (CYP3A4; Section 3.2.1.1). As described above, presystemic metabolism may render a drug inactive, or larger doses of a drug may have to be administered when given orally. On the other hand, prodrugs, that is drugs that are inactive until they have been metabolized (Section 3.2.1.1), will be activated by presystemic metabolism. Dose for dose such drugs may be more pharmacologically active when given orally than by other routes.

Inhibition of presystemic metabolism can have important consequences, including dangerous drug–drug or drug–food interactions (Section 17.6.2).

2.3.1.2 P-Glycoprotein

Not only are drugs metabolized by intestinal cells, drug that is not metabolized may be returned to the gut lumen by efflux. P-Glycoprotein (P-gp) transports substrates from the intracellular to the extracellular side of cell membranes. The gene expressing P-gp was first recognized in tumour cells and it was originally referred to as the multi-drug resistance (*MDR1*) gene. It is now known that the gene is highly expressed in the apical membrane of enterocytes lining the GI tract, renal proximal tubular cells, the canalicular membrane of hepatocytes, and other important blood–tissue barriers such as those of the brain, testes and placenta. Its location suggests that it has evolved to transport potentially toxic substances out of cells. P-gp is inducible and many of the observations that were once ascribed to enzyme induction or inhibition (Section 17.4) may in fact be due to changes in P-gp activity.

2.3.1.3 Gastrointestinal motility and splanchnic blood flow

Absorption is facilitated by thorough mixing of the drug within the gastrointestinal tract. Mixing increases the efficiency with which the drug is brought into contact with surfaces available for absorption. Therefore, in some instances, a reduction in gastrointestinal motility, for example by opiates or antimuscarinic drugs, may reduce absorption, including their own. On the other hand, excessive motility and peristalsis will reduce the transit time and this may be critical for drugs that are slowly absorbed or are only absorbed from particular regions of GI tract. The most favourable site for the majority of drugs is the small intestine with its large surface area due the presence of microvilli – sometimes referred to as the 'brush-border' because of its appearance. The transit through this part of the GI tract is usually about 3 to 4 hours. The surface area for absorption decreases from the duodenum to the rectum although the transit time in the large bowel is generally longer, 12–24 hours, possibly more. It is thought that some drugs are absorbed from particular regions, for example part of an oral dose of drug may be absorbed for 3 to 4 hours with little more appearing in the blood after that time. The remaining portion of the dose is expelled in the faeces. Increased intestinal motility may reduce the absorption of such a drug because the time it is in the optimal region for absorption is reduced. However, a drug that is rapidly and extensively absorbed from the duodenum will be less affected. Therefore it is sometimes difficult to predict how changes in GI motility will affect the oral availability of a drug.

Splanchnic blood flow will affect the rate of removal of the drug from the site of absorption, as blood flow and transport of drugs by plasma proteins is the major mechanism by which drugs are carried away from their sites of absorption and around the body. The blood flow to the GI tract, which represents approximately 30% of the cardiac output, is lower during the fasting state than after feeding. Weight for weight, the mucosa of the small intestine receives the largest proportion of the flow, followed by that of the colon and then the stomach.

Gastric emptying, or rather the lack of it, such as with the pyloric stenosis that often follows surgery, can have a major influence on oral availability. Tablets and capsules, particularly those designed not to release their contents until they reach the intestine will be trapped.

2.3.1.4   Food and drugs

Food may have a major, but not always predictable, effect on oral availability. Food generally delays gastric emptying but, as discussed above, increases splanchnic blood flow. It is often assumed that food delays absorption without necessarily reducing it. The absorption of griseofulvin, a very poorly soluble antifungal drug, is increased when taken with a 'fatty' meal. The constituents of a meal may interfere with the processes of absorption, such as the components of grapefruit juice (Section 17.6.2), or they may interact directly with the drug. Tetracylines chelate divalent metal ions to form unabsorbable complexes, and so their absorption is reduced by milk (high in calcium), magnesium containing antacids, and ferrous sulfate.

From the foregoing, it should be obvious that concomitant use of other drugs can affect oral absorption in a number of ways, including changing gastric pH or gut motility, or by forming unabsorbable complexes. Yet despite these problems (Figure 2.7), the oral route is the preferred one for outpatient and OTC medicines.

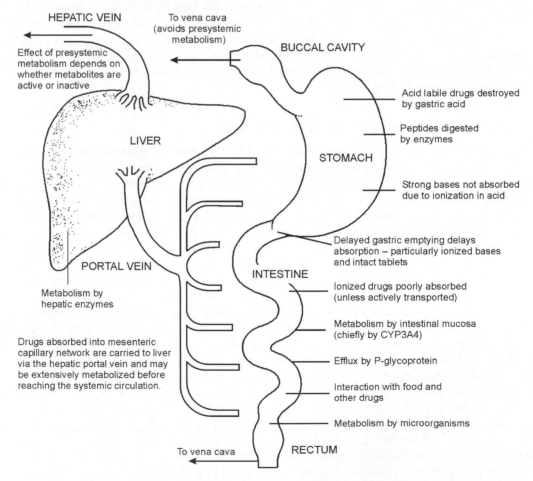

Figure 2.7   Some of the factors that may reduce oral availability. Other factors include GI motility, absorptive surface area and splanchnic blood flow.

2.3.2 Sublingual administration

Tablets are chewed or crushed and held under the tongue for the drug to be absorbed from the mouth. This occurs by diffusion and although the surface area for absorption is small, absorption occurs rapidly. Furthermore, materials absorbed across the buccal membrane avoid the hepatic portal system so this route is useful for drugs such as GTN with which a rapid effect is desired and which can be destroyed in the gastrointestinal tract before absorption or by presystemic metabolism (Section 2.3.1.1, Figure 2.7). Because GTN is volatile and may be lost from tablets, limiting its shelf-life, it is available as an aerosol for spraying into the mouth as a metered dose. Buprenorphine is available as a lozenge that can be held in the mouth to give sustained release of this potent opioid. Most orally administered drugs are deliberately formulated as tablets and capsules that are swallowed, thereby avoiding contact with the oral mucosa.

2.3.3 Rectal administration

Substances absorbed from the lower part of the rectum are carried to the vena cava and so avoid first-pass metabolism. However, this is not the chief reason for choosing this route of administration. The surface area for absorption is not especially large, but the blood supply is extremely efficient, and absorption can be quite rapid. This is considered sufficiently advantageous to warrant extensive use of this route of administration for a wide range of drugs in certain countries. Apart from situations in which a local effect is sought, in the United Kingdom suppositories tend to be used for more specialized purposes, such as the antiemetic, prochlorperazine (there is little point in taking it orally when one is vomiting). Diazepam suppositories are available for use in epileptic infants, when insertion of an intravenous cannula might be considered dangerous. Slow-release aminophylline is given rectally, often at night, to ease breathing in asthmatic children. Disadvantages of this route, apart from patient acceptability, include local irritation and inflammation and the possibility that the patient may need to defecate shortly after insertion of the suppository.

2.3.4 Intravenous and intra-arterial injections

Injecting a drug directly into the circulation avoids any problems of absorption and usually produces the most rapid onset of effects of any route. Peak blood concentrations occur immediately after a rapid 'bolus' injection, but if this presents a problem or a sustained effect is required, then the drug can be given as a slow intravenous infusion, over several minutes, hours or even days. For the longer periods, portable (ambulatory) pumps are available. These have been used to deliver insulin, opiates (usually in terminally ill patients), for treatment of iron poisoning and for the treatment of certain cancers, amongst other applications.

Intravenous injections are not used simply to overcome problems that may be encountered using other routes, but when the i.v. route is the most appropriate. Examples include the i.v. general anaesthetics (thiopental, propofol), muscle relaxants (tubocurarine, suxamethonium) and neostigmine (to reverse the effects of tubocurarine-like drugs); all these drug examples are used during surgery.

Intra-arterial injections are more specialized and are typically used in the treatment of certain tumours. By injecting a cytotoxic drug into an appropriate artery the drug is carried directly to the tumour.

Disadvantages of i.v. injections include the fact that sterile preparations and equipment are required. A high degree of skill is required, particularly to prevent extravasation (i.e. injection near to the vein or leakage from it), which can lead to serious tissue damage.

2.3.5 Intramuscular and subcutaneous injections

Some of the problems of low oral bioavailability can be avoided by intramuscular (i.m.) or subcutaneous (s.c.) injections. Unlike bolus i.v. injections there is no immediate peak plasma concentration as the drug has to be absorbed from the injection site. Drugs, including ionized ones, enter the systemic circulation via the fenestrations in the capillary walls. Size does not appear to be a limiting factor up to approximately $M_r \sim 5000$, although the absorption is flow dependant and increasing local blood flow, for example by warming and massaging the injection site can increase the rate of absorption. The rate of absorption can be delayed by co-injection of a vasoconstrictor to reduce blood flow to the area; for example the use of adrenaline to prolong the effect of a local anaesthetic.

The rate of absorption may be faster or slower than that following oral administration. Absorption after i.m. injection of chlordiazepoxide or diazepam may be delayed because the drugs precipitate at the injection site. Some i.m. preparations may be formulated to provide sustained-release from the injection site; for example microcrystalline salts of penicillin G (i.m.) and various insulin preparations (s.c.).

Sterile preparations and equipment are required but as they require less skill, patients or their carers can be trained to perform i.m or s.c. injections.

2.3.6   *Transdermal application*

The epidermis behaves as a lipoprotein barrier while the dermis is porous and permeable to almost anything. Consequently, lipophilic molecules penetrate the skin easily and rapidly, whilst polar, ionized molecules penetrate poorly and slowly. For many years it has been known that chlorinated solvents and some organic nitro compounds were potentially toxic because they are rapidly absorbed through the skin. More recently, several pharmacological preparations, designed for absorption across the skin have been introduced. These may be in the form of ointments and creams to be rubbed on to the skin or patches to be stuck on. Some of the patches incorporate a rate-limiting membrane to ensure a steady, sustained-release of the drug. Generally the drugs have to be potent as well as lipophilic as large doses would be problematic. Examples of drugs applied to the skin for systemic effects include glyceryl trinitrate, hyoscine, buprenorphine, steroids (contraceptives and hormone replacement) and nicotine.

2.3.7   *Insufflation*

The nose with its rich blood supply, highly fenestrated capillaries and an epithelium with gaps around the goblet cells allows absorption of drugs that cannot be given orally. Peptide hormones, insulin, calcitonin and desmopressin can be given as nasal sprays. Furthermore, it has been suggested that some drugs can enter the central nervous system (CNS) directly via the nose and this route is being investigated for molecules which do not normally enter the brain.

2.3.8   *Inhalation*

Volatile and gaseous general anaesthetics are given via the lungs; their large surface area gives rapid absorption and onset of effect. Because most of these anaesthetics are also excreted via the lungs, the level of anaesthesia can be controlled by adjusting the partial pressure of the drug in the apparatus used to give it.

Other drugs frequently given by inhalation are those to relieve bronchiolar constriction in asthmatic patients. The β-adrenoceptor agonists, salbutamol and terbutaline, and the antimuscarinic drug, ipratropium, are examples. Applying these agents directly to the lungs gives rapid relief, reduces the dose of drug required and so lessens the severity of any systemic adverse effects. Disodium cromoglycate is very poorly absorbed when given orally but is absorbed into lung tissue after inhalation as a fine dry powder.

2.3.9   *Other routes of administration*

Drugs may be applied to various other sites, usually for a local effect. Intravaginal applications of antifungal creams are usually for local effects against *Candida albicans*. Prostaglandin pessaries may used to induce labour. Lipophilic drugs such as physostigmine, pilocarpine and timolol are absorbed across the cornea and used to treat glaucoma. Tropicamide eye-drops may be used to dilate the pupil to aid ophthalmic examinations.

Drugs are generally well absorbed from the peritoneal cavity and although this route is rarely used in human beings, intraperitoneal injection (i.p.) is a convenient method for dosing laboratory animals.

2.4 Drug distribution

The majority of drugs have to be distributed to their site(s) of action. Only rare examples such as anticoagulants, heparin and the like, which have their effects in the bloodstream, do not. Drugs are carried by the circulation, often bound to plasma proteins from where they equilibrate with their sites of action or other storage sites (Figure 2.8).

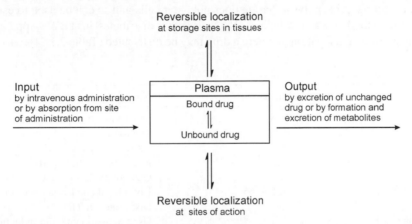

Figure 2.8 Scheme illustrating the key role of plasma in transporting drugs to sites of action, other storage sites and sites of elimination.

2.4.1 Extent of distribution

The extent and rate at which a drug is distributed is dependent on its physiochemical characteristics. Small lipophilic drugs that readily penetrate membranes are generally widely distributed, whereas polar, ionized ones and macromolecules are often contained in particular anatomical volumes. The body is made up of approximately 60% water, 18% protein, 15% fat and 7% minerals. Body water can be subdivided into that in the cells [intracellular water (ICF, 40% of total)] and the remaining extracellular water (ECF, 20%), which can be subdivided further into interstitial fluid (15%) and plasma (5%). The blood volume is 9% of total body water (TBW); the 4% of body water associated with the red cells is part of the intracellular volume.

2.4.1.1 Apparent volume of distribution

The extent of distribution of a drug is quantified by a parameter known as the apparent volume of distribution (V). This can be defined as 'the volume of fluid that is required to dissolve the amount of drug in the body (A) to give the same concentration as that in plasma at that time'. In other words:

$$V = \frac{A}{C} \tag{2.3}$$

where C is the plasma concentration. Immediately following a rapid intravenous injection of drug, the amount in the body is, for all intents and purposes, the dose, D, that has been injected. Therefore, if the plasma concentration was known at this time then the volume could be calculated. The practical situation is complicated because time has to be allowed for the drug to mix within the circulation and be distributed. Therefore, it is usual to measure the plasma concentrations over a period of time and to plot them, or ln C, against time (Figure 1.3) and to extrapolate to $t = 0$ to obtain a value of C_0, from which V can be calculated:

$$V = \frac{D}{C_0} \tag{2.4}$$

Although C_0 may be referred to as the concentration at time zero, it is the theoretical plasma concentration that would occur if it were possible for the drug to be distributed instantaneously, which of course it cannot.

Compounds such as Evans' blue, inulin and isotopically labelled water can be used to measure the volumes of plasma, ECF and TBW, respectively. Evans' blue binds so avidly to albumin that it does not leave the plasma. Inulin, a water-soluble polysaccharide does not enter cells and so can be used to estimate ECF. Of course, isotopically labelled water, whether ^2H_2O or ^3H_2O, distributes in TBW. Apparent volumes of distribution may give some indication of where a drug may be distributed (Table 2.1). Heparin is too large to

Table 2.1 Examples of apparent volumes of distribution

Compound	V (L kg^{-1})^a	Notes
Evans' Blue^b	0.05	Dye to measure plasma volume
Heparin	0.06	Macromolecule – cannot enter interstitial fluid
Inulin	0.21	Used to measure ECF
Penicillin G	0.2	Does not penetrate cells
Tubocurarine	0.2	Quaternary ammonium compound
Deuterium oxide (D$_2$O)	0.55–0.65	Isotopic labelled water – to measure TBW
Ethanol	0.65	Distributes in TBW
Antipyrine (phenazone)	0.6	Used to assess enzyme induction (Chapter 17)
Digoxin	5	Binds to Na$^+$/K$^+$ ATPase
Chlorpromazine	20	
Amiodarone	62	Little found in the CNS
Quinacrine	500	Intercalates in DNA

^{a} Normalized to body weight.
^{b} Compounds in italics used to measure anatomical volumes.

pass through the fenestrations in the peripheral capillaries and so is confined to plasma. Several drugs including the penicillins and tubocurarine do not readily enter cells but are small enough to filter into interstitial fluid. Drugs which can cross lipid cell membranes but are not concentrated (sequestered) in cells nor bound to plasma proteins have volumes approximately equal to those of TBW. Ethanol and the now obsolete antipyretic drug, antipyrine (phenazone), are examples of such compounds. Centrally acting drugs generally have to be lipophilic enough to cross the BBB and often have apparent volumes of distribution $> 1 \, L \, kg^{-1}$. However, the converse is not true, a drug such as digoxin has a large value of V because it binds to cardiac and skeletal muscle Na$^+$/K$^+$ ATPase.

2.4.2 Mechanisms of sequestration

Sequestration of drugs in various parts of the body arises because of differences in pH, binding to macromolecules, dissolution in lipids, transportation (usually against the concentration gradient), and what may be termed 'irreversible' binding.

2.4.2.1 pH differences

Local differences in pH may lead to high concentrations of weak electrolytes in one area relative to another because of differences in the degree of ionization. This is predictable from pH-partition considerations (Section 2.2.1.1.). For example the ionized to non-ionized ratio of salicylic acid ($pK_a = 3.0$) at pH 6.8 (intracellular pH) is 6,300 : 1 whereas at the pH of plasma (7.4) the ratio is 25,000 : 1. This represents a ratio of $\sim$4 : 1 in favour of plasma water and is true for any weak acid. The converse is the case for bases, when the difference in intracellular and plasma pH means that bases will be distributed $\sim$4 : 1 in favour of the more acidic fluid.

2.4.2.2 Binding to macromolecules

Drugs may be concentrated by binding to several types of macromolecules: plasma proteins, tissue proteins, including enzymes, and nucleic acids. Quinacrine and chloroquine are concentrated in tissues because of binding to DNA. Chlorthalidone binds to red cell carbonic anhydrase, whilst anticholinesterases such as neostigmine and pyridostigmine bind to red cell acetylcholinesterase. Binding to plasma proteins, which can lead to important differences in distribution and hence pharmacological activity is discussed in Section 2.5.3. The possible effects of protein binding on drug kinetics are considered in Section 7.4.1.

2.4.2.3 Dissolution in lipids

Lipophilic drugs are often concentrated in lipid cell membranes and fat deposits. This is thought to be simple partitioning, analogous to solvent:water partitioning. The distribution of thiopental in adipose tissue is an important determinant of its duration of action (Section 2.4.3.1).

2.4.2.4 Active transport

As mentioned earlier (Section 2.2) compounds that are structurally similar to endogenous molecules may be substrates for transport proteins. Guanethidine (and probably other adrenergic blocking drugs) is concentrated in cardiac tissue by active uptake. Amphetamine and other indirectly acting sympathomimetic drugs are actively transported into aminergic nerves. Paraquat is concentrated in the lungs because it is a substrate for a putrescine transport protein.

2.4.2.5 Special processes

This refers to binding processes which, for all intents and purposes, can be considered as irreversible. Examples include deposition of tetracyclines in bone and teeth and of drugs in hair. These areas have very poor blood supply so that penetration is slow and loss of drug by diffusion back to the blood is in effect zero. It is probable that drug residues are laid down as hair and teeth are formed, in areas of rich blood supply, and the deposits are carried away from these areas as the tissues grow.

Other examples of irreversible binding include covalent binding in tissues with a good blood supply. Such localization usually involves only a small proportion of the total amount of drug in the body but it is particularly important as a mechanism of drug toxicity (Chapter 18).

2.4.3 Kinetics of distribution

Drugs carried in the bloodstream will penetrate those tissues which they can, net transfer being down the concentration gradient, a process sometimes referred to as 'random walk'. A drug placed at any point within the system will diffuse backwards and forwards until characteristic equilibrium concentration ratios are reached (Figure 2.9). Within equilibrium of course, diffusion continues, but the relative concentrations at the various points do not change. Removal of drug from ECF, by metabolism and excretion, reduces the plasma concentration so that net movement is now from tissues to plasma.

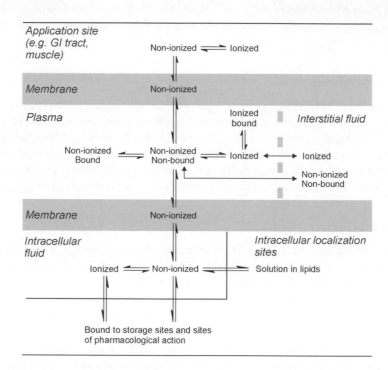

Figure 2.9   Simplified diagram for the equilibrium distribution of an ionizable drug of small relative molecular mass. The drug readily filters into interstitial fluid so the unbound concentrations are equal to those in plasma.

The result of tissue localization is a pattern throughout the body of concentrations in each tissue higher than concentrations in plasma by varying amounts. Highest concentrations are commonly found in liver, lung and spleen, but the significance of this, except perhaps in regard to drug metabolism and excretion by the liver, remains obscure. The speed with which a particular ratio is achieved is governed by the nature of the drug and the tissues in which it is distributed. A highly ionized drug like pyridostigmine rapidly enters interstitial fluid via capillary fenestrations and, because it does not enter cells, the equilibration time is very short. Because lipophilic drugs rapidly diffuse across cell walls, the rate-limiting step for equilibration is delivery of the drug to the tissue, that is, it is flow-limited. Thus the rate of equilibration will be a function of the vascularity of the particular tissue and is rapidly established with well-perfused tissues such as kidney, liver, lung and brain. Muscle is intermediate, in that a rising ratio can often be detected. Poorly perfused adipose tissue can require many hours for equilibrium to be achieved. Presumably drugs of intermediate lipophilicity will show a mixture of flow-limited and diffusion-limited equilibration. A number of drugs has been studied in detail with regard to tissue distribution.

2.4.3.1 Tissue distribution of thiopental

Thiopental is a lipophilic barbiturate which is used as a short-acting general anaesthetic. Brodie and his colleagues investigated the distribution of this drug in the 1950s as part of their studies to explain why repeated, or higher, doses gave disproportionate increases in duration of action (Brodie *et al.*, 1952, 1956; Brodie and Hogben, 1957; Brodie, 1967; Mark *et al.*, 1957). Their work made a major contribution to the understanding of the kinetics of drug distribution. After an intravenous injection in a dog, plasma and liver thiopental concentrations fell rapidly. The log(concentration)–time plots were parallel throughout the sampling period, showing that the liver:plasma concentrations had equilibrated by the time of the first sampling point [Figure 2.10(a)].

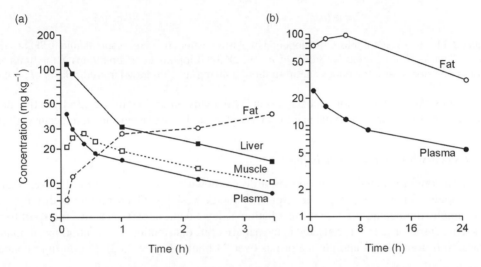

Figure 2.10 Concentrations of thiopental (logarithmic scale) in the plasma and various tissues of a dog after intravenous administration of 25 mg kg^{-1}. (a) 0–3.5 h (b) 0–24 h (Brodie *et al.*, 1952).

The concentration in skeletal muscle rose over the first 30 minutes and had equilibrated by about 1 hour after which the muscle to plasma ratios remained constant. Thiopental concentrations in fat continued to rise for several hours, equilibrated and then declined in parallel with the plasma concentration when plotted on a semi-logarithmic plot. [Figure 2.10(b)]. A slow rate of return from fat to plasma is responsible for the slow decline in plasma concentrations at later times. The peak concentrations in cerebrospinal fluid (CSF) were recorded at 10 minutes after which the concentrations were similar to those in plasma water [Figure 2.11(a)]. Clearly, the short duration of action of thiopental cannot be explained by it being rapidly removed from the body. However, in a separate study it was shown that brain concentrations rapidly equilibrated with those in the plasma [Figure 2.11(b)] and thus the steep decline in plasma concentrations is accompanied by a similar sharp fall in brain concentrations. Hence, the rapid onset and short duration of action of thiopental can be explained in terms of the kinetics of its distribution. After an i.v. injection, this lipophilic drug rapidly crosses the BBB to enter the brain giving an almost immediate loss of consciousness. Over the next few minutes, the plasma concentration declines, not due to elimination of the drug, but because of loss of the drug from the plasma to the less well-perfused tissues. Because of the rapid transfer between brain and plasma, brain concentrations quickly fall and the patient regains consciousness. In other words, the short duration of action of thiopental is due to redistribution of the drug from the brain, via the plasma, to less well-perfused tissues,

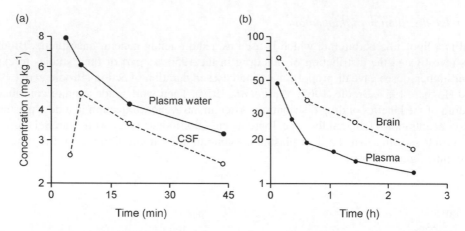

Figure 2.11   (a) Concentrations of thiopental in plasma water and cerebrospinal fluid (CSF) of a dog given 25 mg kg^{-1} intravenously (Brodie *et al.*, 1952). (b) Thiopental concentrations in the brain and plasma of a dog after intravenous administration of 40 mg kg^{-1} (redrawn from Brodie *et al.*, 1956).

probably muscle. This phenomenon probably occurs with many other lipophilic centrally acting drugs such as the related barbiturate, methohexital and the opioid, fentanyl. It has been shown to occur with propofol.

2.4.3.2   *Tissue distribution of guanethidine*

Guanethidine belongs to a class of antihypertensive drugs known as adrenergic neurone blockers that work by displacing noradrenaline (norepinephrine) from its storage vesicles. Guanethidine is different from the previous example, thiopental, because its distribution is dependent, in part, on active transport to maintain high tissue concentrations. This is reflected in the pattern of its distribution. Concentrations of guanethidine were measured in four tissues and plasma of rats over 24 hours (Figure 2.12). Equilibrium concentration

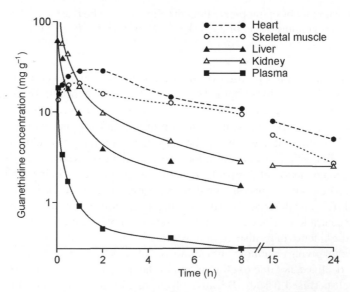

Figure 2.12   Mean concentrations of guanethidine in various rat tissues after 28 mg kg^{-1} (i.v.). Each point is the mean of data from four rats. (Redrawn from the data of Schanker & Morrison, 1965).

ratios were achieved almost instantaneously for liver, kidney and plasma. These ratios were maintained from the first sampling time (5 minutes) to the 8 hour collection time. It is difficult to conclude much for these tissues between 15 and 24 hours because of the paucity of data.

Skeletal muscle and heart showed initial rises in their concentrations, with peak concentrations between 1 and 2 hours. Extensive studies following up this observation have shown that the slow achievement of equilibrium in the case of muscle occurs as a result of a somewhat less efficient blood supply to this tissue. It is believed that localization in liver and kidney results from reversible process as discussed earlier.

Heart, however, is different. Guanethidine concentrations in heart are maintained at high concentrations by an active transport process. This accounts for the slow achievement of heart to plasma equilibrium ratios. The drug is thought to be taken up into noradrenergic nerve terminals by the high affinity neuronal transporter (Uptake 1) that is normally responsible for the reuptake of noradrenaline. The example of guanethidine illustrates how the tissue distribution of one drug may be more complex than another, particularly when active transport is involved and the simple model of passive diffusion does not apply.

2.4.4   Tissue distribution: more modern approaches

At one time tissue distribution studies of the type shown for thiopental and guanethidine were commonplace. The objective was first to determine which tissues showed selective uptake, in the hope of discovering sites of action, and second to aid the understanding of the time course of both the drug in the body and pharmacological effect. The thiopental studies described above provided a fundamental basis for the later concepts of pharmacokinetic compartment modelling, and also influenced dosing practices with intravenous anaesthetics. The guanethidine study, one of the last studies of this type to be published, showed that highly specific uptake at sites of action, using active transport mechanisms, could occur, and probably presaged modern effect-compartment PK/PD modelling (Chapter 14). Studies of this type were time-consuming, and they have largely been superseded by such techniques as microdialysis, and by imaging methods such as whole body autoradiography and positron emission tomography (PET).

2.4.4.1   Microdialysis

This technique involves insertion of very fine probes into the tissues of the living body, mostly into fluid spaces, such as the CSF, and other extracellular fluids where possible. The probes consist of at least two concentric tubes, and a semipermeable membrane separating them, positioned such that an artificial extracellular dialysis fluid can be slowly infused through the probe and past the membrane. Unbound drug molecules in the tissues surrounding the probe diffuse into the flowing dialysate, which is then collected for analysis.

Microdialysis was introduced for the measurement of extracellular concentrations of neurotransmitters in the brain. It has been used in both animals and humans, and is quite commonly used in intensive care units for measuring glucose and lactate. It can be used for both administration of drugs and for sampling fluids for drugs. Technical difficulties include determining the calculation corrections needed in quantification, because of the time over which samples are collected. The greatest challenge is determining the recovery of analyte. This is complicated by dilution of the sample and extracellular fluid by the dialysate, and the fact the analyte is not equilibrated between sample and the flowing dialysate. This is considered to be more important than the invasive nature of the technique. As examples of applications, this technique can be used to study transfer of drugs into the CSF through the choroid plexus, in the search for information on whether drugs then diffuse back into the brain tissue through the monocellular blood–CSF barrier, rather than diffusing out of the brain into the CSF, as a method of removal of drugs from the brain.

Cerebral concentrations of morphine were measured using microdialysis in a patient with a head injury (Figure 2.13). Samples were collected from two sites, an injured one and an uninjured one. Calculations of

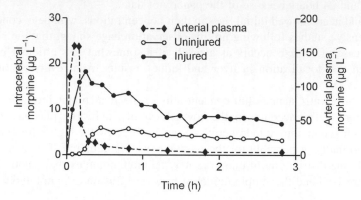

Figure 2.13 Morphine concentrations in a patient with a head injury. (Adapted from Ederoth *et al.*, 2004 and Upton, 2007).

the area under the curve of the unbound drug concentration in uninjured brain (open circles) was less than that in the arterial plasma (diamonds) or the injured part of the brain (closed circles). This was consistent with (i) active efflux of morphine from the brain, and (ii) a damaged blood-brain barrier in the injured area. Passive diffusion of morphine into the injured area was the major mechanism controlling brain exposure to the drug after the injury.

2.4.4.2 *Autoradiography*

In this approach, the radioactively labelled substance under investigation is administered to test animals, usually mice or rats, which are killed at suitable time intervals, and frozen sections of tissues or of the whole body prepared. An image of the radiation is obtained by placing the sections next to a photographic emulsion. Thus, if a whole body section is used, and the radioactivity is specifically localized in highly perfused organs such as the lungs and liver, the image will reveal this. Comparison with a normal photograph of the slice is used to identify which tissues contain the radioactivity. Densitometry can be used to quantify the relative amounts of radioactivity. This technique has been used to show highly-specific localization of endogenous materials, such as iodine in the thyroid gland, and cholecystokinin in the walls of the stomach and intestine. Less specific but no less valuable information is obtained with drugs.

Figure 2.14 shows three whole body autoradiographs of rat sections at different times after administration of hydralazine labelled with two different radioactive isotopes. Quite specific localization of this vasodilator drug can be seen in the arterial walls, the walls of the aorta, and the walls of the vena cava, amongst many other specific deposits, particularly those connected with the disposition of the drug, such as intestinal contents, the bladder, and the heart, kidney and liver. It can be presumed that the blood vessel labelling is connected with specificity for sites of action.

Autoradiography requires the synthesis of labelled drug, careful handling of materials, and long photographic exposures (up to 6 months). Also, there is no differentiation between parent drug and any metabolites, clearly a problem with drugs that are extensively metabolized. In the hydralazine study it can be presumed that the differences between the images obtained with ^{14}C and ^{3}H, which would have been at different locations within the hydralazine molecules, and therefore retained or lost in metabolites in different ways, reflect metabolic degradation and excretion. These problems have been overcome to some extent by using alternative analytical techniques, such as fluorescence imaging and radioluminography.

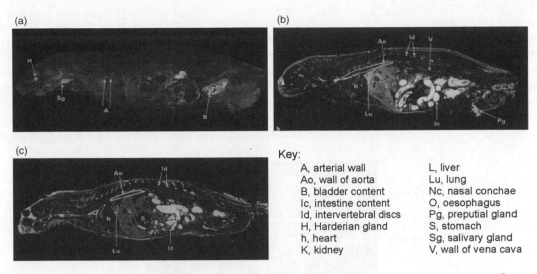

Key:
A, arterial wall   L, liver
Ao, wall of aorta   Lu, lung
B, bladder content   Nc, nasal conchae
Ic, intestine content   O, oesophagus
Id, intervertebral discs   Pg, preputial gland
H, Harderian gland   S, stomach
h, heart   Sg, salivary gland
K, kidney   V, wall of vena cava

Figure 2.14 Whole-body autoradiographs of rats following intravenous administration of [^{3}H]-hydralazine (a and b) or [^{14}C]-hydralazine (c). (a) Five minutes (b and c) 6 h after injection. (From Baker *et al.*, 1985).

2.4.4.3 Positron emission tomography (PET)

Synthetic radioactive isotopes (e.g. ^{11}C, ^{13}N, ^{15}O and ^{18}F) with atomic masses less than the naturally occurring stable isotopes have half-life values of 2–110 minutes and emit positrons that interact with electrons to emit gamma radiation that can be detected outside the body. The isotopes are generated in a cyclotron and incorporated into the drug molecules immediately before the administration of the drug. PET scanning permits the production of images of live organisms including humans. The disposition of [^{11}C]-triamcinolone after intranasal administration is shown in Figure 2.15(a). The image is of a single

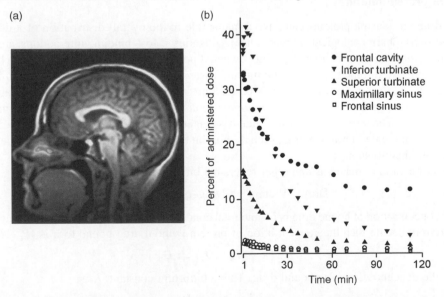

Figure 2.15 (a) PET scan of [^{11}C]- triamcinolone (see text for details).(b) Time course of [^{11}C]-triamcinolone in selected regions. Each point is the mean from three volunteers plotted at the midpoint of each PET scan. (Adapted from Berridge *et al.*, 1998).

sagittal 'slice' through a three-dimensional data set, 8 mm from the central plane of the head. The study showed that the drug was delivered rapidly into the turbinates and frontal regions of the nose and the sinuses. Thereafter, it was found to gradually diffuse away, some of it being swallowed, and also, some being absorbed into the systemic circulation. This technique is quantitative, and a graph of percentage of administered dose, against time up to 2 hours, in the frontal cavity, the turbinates, and the sinuses was generated [Figure 2.15(b)]. Estimates of the half-life of the drug in these locations were possible

Another application of this technique is exemplified by Figure 2.16. Dopamine receptors within the brain were labelled using the precursor, [18F]-fluoroDOPA, to study the effect of drugs that modify dopamine

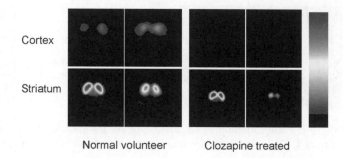

Figure 2.16   PET scans of the cortex and striatum after administration of [18F]-fluoroDOPA, the precursor of [18F]-fluorodopamine.

function in exerting their beneficial effect on psychiatric illness. The figure shows eight images, four generated within a healthy volunteer and four generated in a patient. Two areas of the brain were examined, and the disposition of [18F]-fluorodopamine was studied at two different time points. The images clearly show clozapine reducing dopamine binding in the cortex of the brain.

2.5   Plasma protein binding

Binding of drugs to plasma proteins can have a major role in the overall distribution of a drug and may influence both its pharmacological activity and its kinetics. Most binding interactions are reversible, probably due to ionic and hydrophobic bonding – where lipophilic molecules associate with a hydrophobic part of the protein. Acids tend to bind to albumin and bases to α_1-acid glycoprotein and albumin. Covalent bonding, when it occurs, may result in antibody production and hypersensitivity reactions (Chapter 18). The extent of binding can be described by the fraction or percent of drug bound, β, or less commonly by the unbound fraction, α. The term 'free' is best avoided as the term is used by some to describe the non-ionized form or the non-conjugated form of a drug, such as morphine rather than morphine glucuronide, potentially resulting in untold confusion.

The interaction can be treated as any other reversible binding isotherm:

$$\text{Drug} + \text{Protein} \rightleftharpoons \text{Drug–protein complex}$$

If the molar concentration of bound drug is D_b and total concentration of protein is P_t, assuming one binding site per protein molecule then the concentration of protein without drug bound to it is $(P_t - D_b)$, so:

$$D_f + (P_t - D_b) \rightleftharpoons D_b$$

where D_f is the concentration of unbound drug. The equilibrium constant, K is:

$$K = \frac{k_{-1}}{k_1} = \frac{D_b}{D_f(P_t - D_b)} \tag{2.5}$$

Rearrangement gives:

$$\frac{D_b}{P_t} = \frac{KD_f}{(1+KD_f)} = r \qquad (2.6)$$

where r is the number of moles bound per total number of moles of protein. Equation 2.6 has been rearranged in a number of ways so that K and, usually, n can be estimated. Note that double reciprocal plot of Klotz [Figure 2.17(a)] and the Scatchard plot [Figure 2.17(b)] require the molar concentration of the protein to be

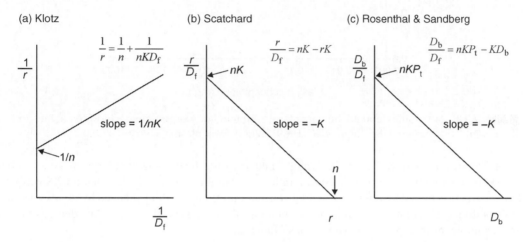

(a) Klotz

(b) Scatchard

(c) Rosenthal & Sandberg

$$\frac{1}{r} = \frac{1}{n} + \frac{1}{nKD_f}$$

$$\frac{r}{D_f} = nK - rK$$

$$\frac{D_b}{D_f} = nKP_t - KD_b$$

slope = 1/nK

slope = -K

slope = -K

Figure 2.17   Diagram showing three methods for solving protein-binding data.

known, which of course means one has to know the characteristics of the protein to which the drug is binding. Plotting D_b/D_f against D_b overcomes this and is more useful when measuring binding in plasma. If P_t is known then n can be derived from the y-intercept [Figure 2.17(c)]. This plot is frequently used for receptor–ligand binding studies, and, although it is often referred to as a Scatchard plot, it was first proposed by Sandberg and Rosenthal (Rosenthal, 1967).

Curvature of binding plots is indicative of the existence of more that one class of site and the number of moles bound is the result of binding to all the sites:

$$r = \sum_{i=1}^{i} \frac{n_i K_i D_f}{1 + K_i D_f} \qquad (2.7)$$

Frequently, protein binding has been evaluated in terms of the fraction bound, β, which for a single class of binding sites:

$$\beta = \left(1 + \frac{D_f}{nP_t} + \frac{1}{nKP_t}\right)^{-1} \qquad (2.8)$$

Often a change in the fraction of drug bound in plasma is barely observable within the therapeutic range and when the concentration is increased over several orders of magnitude the change may be small (Figure 2.18). Experimentally, proteins appear to be able to 'mop-up' some drugs from aqueous solutions or suspensions, even when, in some cases the solubility in water as been exceeded. Scatchard plots are often curved when binding is studied over a large range of concentrations, suggesting that there is specific binding, with a small value of n and relatively large K, and non-specific binding which prevails at high concentrations where K is small but n is large.

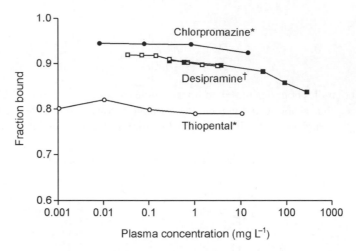

Figure 2.18 Fraction bound to human plasma protein for desipramine, chlorpromazine and thiopental ([†]two individuals, Borga *et al.*, 1969; [*]blood bank plasma, Curry, 1970).

Using a literature value of $K = 2.3 \times 10^4 \, \text{L mol}^{-1}$ and $n = 1$ for the binding of chlorpromazine to albumin ($M_r = 69,000$) and assuming the plasma concentration of albumin to be $40 \, \text{g L}^{-1}$, Equation 2.8 was used to calculate the fraction bound as a function of total concentration. The binding changed by <0.4% over a 100-fold range of drug concentration, so even specific binding can result in an almost constant degree of binding, as in this example, at therapeutic concentrations (Table 2.2).

Table 2.2   Calculated values of β for a range of chlorpromazine concentrations in man

Total concentration		Fraction bound (β)
$\mu\text{g L}^{-1}$	mol L^{-1}	
1.38	3.89×10^{-9}	0.9276
13.8	3.89×10^{-8}	0.9273
131.9	3.72×10^{-7}	0.9242
947.9	2.67×10^{-6}	0.9845
3093.1	8.71×10^{-6}	0.6767

Thus, for many drugs there appears to be excess binding capacity, however this may not be the case for less potent drugs that are used at higher doses. For example, therapeutic concentrations of phenylbutazone (approximately $50–100 \, \text{mg L}^{-1} = 0.16–0.32 \, \text{mmol L}^{-1}$) are not very far removed from the molar concentration of albumin in plasma ($0.58 \, \text{mmol L}^{-1}$). A phenylbutazone concentration of $\sim 180 \, \text{mg L}^{-1}$ represents a 1 : 1 ratio of drug: albumin and the fraction bound shows a marked increase in unbound concentration at concentrations higher than this (Figure 2.19).

2.5.1   Assessing protein binding

An obvious way of determining the extent of binding to plasma proteins is to separate and measure the concentration of the unbound drug. For highly protein-bound drugs, measuring the unbound drug concentration presents an analytical challenge and for drugs that are >99% bound it may not be possible to obtain accurate concentrations using standard laboratory methods, particularly if the analyte and protein

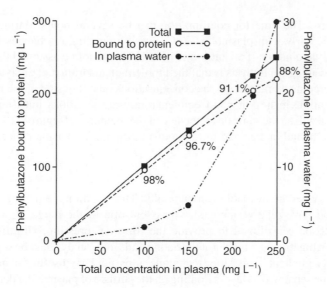

Figure 2.19 Binding of phenylbutazone to plasma protein as a function of phenylbutazone concentration. The percentages indicate the binding at 100, 150, 225 and 250 mg L^{-1} (Brodie & Hogben, 1957).

concentrations are those attained after therapeutic dosing. Under these conditions it may be necessary to use radiolabelled drug. A very small quantity of high specific activity labelled drug is added to plasma, incubated to ensure equilibration with non-labelled analyte, and the free and bound fractions are then separated for radioactive counting, usually by liquid scintillation spectrometry.

2.5.1.1 Equilibrium dialysis

This is the most unequivocal method for assessing the unbound fraction. The sample is placed on one side of a membrane that allows small molecular mass drugs to pass into the dialysate (usually pH 7.4 buffer) on the other side but not the protein [Figure 2.20(a)]. After equilibration, the concentrations on the protein (C_t) and buffer side (C_f) of the membrane are measured, the bound concentration, C_b being $C_t - C_f$. The technique is

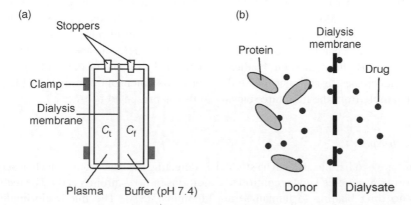

Figure 2.20 (a) Dialysis cell; the stoppers can be removed for sampling. (b) When equilibrated the unbound drug concentrations (C_f) on either side of the membrane are equal. Measuring the total drug concentration (C_t) on the protein side avoids errors due to binding of drug to the membrane.

not without its problems. The time for equilibration can be several hours, during which there may be microbial growth, particularly with plasma at 37 °C, possibly leading to changes in protein and analyte concentrations and analyte binding. If an antibiotic is added it cannot be assumed that it does not interfere with the binding. An advantage of dialysis is that the problem of adsorption of analyte to the membrane and apparatus is largely overcome by measuring the concentrations on either side of the membrane. Adsorption will reduce the concentrations in the donor and recipient (dialysate) solutions, but at equilibrium the unbound concentration in solution will be the same on either side of the membrane [Figure 2.20(b)]. Thus, it is possible to calculate the fraction bound or free and to relate this to the initial plasma concentration of analyte.

2.5.1.2   Ultrafiltration

Several ultrafiltration devices are available commercially. The filtration membranes used are made from a variety of materials and have relative molecular mass cut-offs in the range $M_r = 10,000$–30,000. Most devices are designed to be centrifuged to provide the filtration pressure. Ultrafiltration should not be confused with ultracentrifugation in which protein-bound and 'free' analyte can be separated as layers, often with the aid of a density gradient. This latter technique is particularly useful for investigating binding to lipoproteins which can be separated as layers floating on the surface of plasma, the density of which has been adjusted with potassium bromide. Ultrafiltration is more convenient and quicker than equilibrium dialysis, but the protein concentration in the retentate increases during filtration, potentially increasing the proportion of analyte bound. To minimize this problem as small a volume of ultrafiltrate should be collected as practicable, and the volume of retentate made up with an appropriate buffer solution periodically during centrifugation. Binding of the analyte to the filtration membrane (common with many lipophilic drugs) will reduce the concentration in the filtrate. Control experiments to ascertain the magnitude of this problem should be conducted. It is good practice to collect serial samples of filtrate for analysis to ensure that the sample is representative of the unbound concentration. Some filtration membranes need to be soaked in water or buffer before use and as a result the first ultrafiltrate collection(s) may be unrepresentatively dilute.

2.5.2   Molecular aspects of protein binding

Plasma protein binding is a physical interaction between a small molecule and a large molecule, and as such it is amenable to study using calorimetric and spectroscopic instrumentation, and analysis in thermodynamic terms. The purpose of such studies is to understand the sites and nature of the interaction on the protein surface and the effects, if any, of the binding drug on the protein, and to facilitate prediction of protein binding properties of new molecules, alone and in combination. Thus, among the spectroscopic techniques, changes in UV absorption or fluorescence spectra can be interpreted in terms of polarity of the drug. Also, nuclear magnetic resonance spectra can indicate which parts of the drug molecule are involved in the binding, and circular dichroism yields information on the three-dimensional structure of the binding site of the protein. Electron spin resonance spectroscopy permits study of the involvement of free radicals in the binding, when it occurs. A small sample from the literature relevant to this is presented below.

2.5.2.1   Microcalorimetry

Binding reactions are exothermic and can be studied by detecting the heat produced when separate solutions of the drug and protein are mixed under controlled conditions in flow-through cells. The technique has been applied mainly to drug binding to human serum albumin (HSA). The microcalorimeter is calibrated electrically, and the heat output measured in microvolts. Generally, a constant concentration of HSA and varied concentrations of drug are used, producing a diagram of the type in Figure 2.21. This shows the voltage produced with a range of concentrations of sulfathiazole, and seems to indicate a minor trend towards

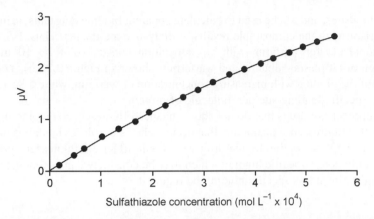

Figure 2.21 Heat produced (assessed as microvolts) as a function of sulfathiazole concentration at a fixed HSA concentration of $3.25 \times 10^{-4}\,mol\,L^{-1}$ (Hardee *et al.*, 1984).

saturation. The voltage output associated with formation of one mole of complex is calculated, and the heat flux is calculated in microwatts – the numbers are quite small, in the range $1–40\,\mu W$.

This technique is particularly applicable to the study of drug interactions, as the heats of reaction for two separate drug–protein combinations should be additive. If they are less than additive it is proposed that a displacement of one drug by the other is occurring. In this case the interaction must be further studied using more traditional techniques such as dialysis and ultrafiltration. For example, it was possible to show both enhancement (greater than additive heat) and displacement (less than additive heat) with various combinations of non-steroidal anti-inflammatory drugs and coumarin anticoagulants (Hardee *et al.*, 1984).

2.5.2.2 *Fluorescence spectroscopy*

Binding of a fluorescent drug to protein may enhance the fluorescence and this can be used to study the interaction. For example, when warfarin binds to HSA, the fluorescent quantum yield of the drug increases eightfold and its emission maximum moves to shorter wavelengths. Rat and dog plasma albumins induce even greater 'blue shifts' in the emission [Figure 2.22(a)]. These changes can be interpreted in terms of the

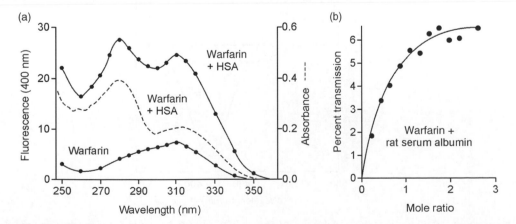

Figure 2.22 (a) Fluorescence emission spectrum of warfarin showing the increased quantum yield in the presence of human serum albumin (HSA). (b) Fluorescence titration of rat serum albumin $(1 \times 10^{-5}\,mol\,L^{-1})$ with warfarin. Mole ratio = wafarin/rat serum albumin. (From Chignell, 1973).

polarity of the binding sites, and can be used to calculate apparent binding constants, using Scatchard plots, facilitating comparison with the comparable results of dialysis experiments. Thus, HSA has one specific binding site per molecule for warfarin, with an association constant of $8.2 \times 10^5 \, mol \, L^{-1}$. A typical fluorimetric titration of rat plasma albumin and warfarin is shown in Figure 2.22(b). The titration appears to show saturation of the binding with increasing concentration of warfarin, while a Scatchard plot showed approximately one specific binding site per molecule of protein.

An alternative approach for drugs that do not show enhanced fluorescence on binding is to monitor the quenching of the native fluorescence of albumin that occurs when some drugs bind. It is important to correct for any effects due to light absorption by the drug which will affect the fluorescent yield. Spectroscopic techniques usually only detect specific binding which may be considered an advantage or a disadvantage, should the contribution from less specific binding be required.

2.5.2.3   Circular dichroism (CD)

This spectroscopic technique takes advantage of the differential absorption properties of left and right-handed circular polarized light. This is particularly applicable to the determination of the secondary structure of proteins; as stated earlier, it is possible to detect binding sites, and to observe changes in the α-helix of proteins, induced by binding of ions and organic molecules, including drugs. This technique is closely related to another approach, optical rotatory dispersion (ORD) (Chignell, 1973).

The power of CD is illustrated by studies conducted with dog serum albumin and copper and nickel (Mohanakrishnan & Chignell, 1984). Copper was described as showing a single positive 'extremum' at 664–667.5 nM; this was maximal at a Cu^{2+}/albumin molar ratio of three. Also, there was an absence of 'extrema' at 560–570 nM and 480–510 nM, said to indicate the absence of involvement of, for example, histidine residues – this aided the identification of the binding sites. In the case of nickel, it was shown that the binding was at the N-terminal tripeptide. Also, there was binding at sulfur-containing residues at relatively high Ni^{2+}/albumin ratios (Figure 2.23). The figure shows difference CD spectra of nickel binding at three different molar ratios. This technology has been recently reviewed by Ascoli *et al.* (2006).

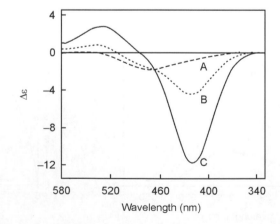

Figure 2.23   Difference circular dichroism spectra of nickel binding to 87.56 μM DSA (0.15 M NaCl pH 10.2). The Ni^{2+}/DSA ratios are 2 for spectrum A, 3 for spectrum B and 6 for spectrum C. (Mohanakrishnan & Chignell, 1984).

2.5.2.4   *Electron paramagnetic resonance spectroscopy (EPR)*

This is a spectroscopic technique for studying chemical species that have one or more unpaired electrons, such as organic and inorganic free radicals or inorganic complexes possessing a transition metal ion. This is similar to nuclear magnetic resonance, but it is the unpaired electrons that spin if excited instead of the atomic nuclei. Most chemical solvents do not give rise to EPR spectra. This technique has been applied to the search for free-radical mechanisms in covalent binding of drugs that oxidize to free radicals, such as phenothiazine drugs and halogenated solvents. Binding of this type would be expected to be associated with long-term persistence in the body, and possibly with an adverse clinical outcome. The technique can also employ 'spin labelling', in which a free radical is attached to a molecule of interest in order to track the disposition and fate of the compound so tagged. It has been proposed that this, and other spectroscopic techniques, would be especially applicable to studies of α_1-glycoprotein, which particularly binds basic drugs and remains relatively unstudied (Israili and Dayton, 2001). It is also of interest that albumin has a major role in the transport of nitric oxide, accounting for most of the antioxidant properties of human plasma (Fasano *et al.*, 2005).

2.5.2.5   *Commentary*

Over the last 35 years or so, these, and other instrumental approaches to the study of plasma protein binding of drugs have undoubtedly enhanced our understanding of the chemistry of the phenomenon, and have certainly facilitated the partial prediction and prevention of drug binding interactions. They have made possible quantitative structure activity relationship (QSAR) studies, in the search for a measure of predictability, especially in relation to new chemical entities. However, an age-old question remains unanswered, whether *in vitro* experiments, even when conducted at 37 °C, reproduce faithfully the *in vivo* events, and studies of this type have not entered the mainstream of the new drug discovery process, or provided an explanation for many unexpected pharmacokinetic phenomena. It is still possible to treat protein binding with a sense of awe, as did Fasano *et al.* (2005), in their review entitled: 'The extraordinary ligand binding properties of human serum albumin'.

2.5.3   *Pharmacological importance of binding to plasma proteins*

Clearly plasma protein binding has a major influence on the distribution of drugs. Extensive binding to plasma proteins reduces the apparent volume of distribution because a larger proportion of the amount of drug in the body will be in the plasma. It is usual to measure the 'total' concentration of drug (i.e. bound + unbound) in plasma.

Binding to plasma proteins, provides an efficient way of transporting drugs in the circulation, sometimes at concentrations that exceed their solubility in plasma water. Binding has an important role in absorption, as it maintains a favourable concentration gradient for the unbound drug. It is generally assumed that plasma protein binding reduces the proportion of a dose of drug available to its receptors and so it can have a major influence on drug activity. Changes in the faction bound may occur because of:

- Displacement by a second drug. This may be clinically important with salicylate and valproic acid, displacing drugs that attain molar concentrations similar to that of the binding protein.
- Changes in protein concentration, often as result of disease (Chapter 12).
- Concentration dependent binding.

Many *in vitro* studies have demonstrated displacement of one drug by another, but *in vivo* the situation is more complex. The 'total' concentration of a displaced drug in plasma will be reduced as some of the liberated drug diffuses into tissues as new equilibria are established. The increased concentration of unbound drug may lead to greater, possibly toxic, effects. Hence, measurement of the 'total' (bound + unbound) concentration of a drug in plasma may be misleading under certain circumstances. When phenytoin was displaced by salicylate, for example, the percentage unbound increased from 7.14 to 10.66%, and this was accompanied by a significant decrease in total serum phenytoin concentration from 13.5 to 10.3 mg L^{-1}. The salivary phenytoin concentration rose from 0.97 to 1.13 mg L^{-1} (Leonard *et al.*, 1981).

The effect of protein binding on drug elimination is more complicated than it might at first appear. It is reasonable to assume that binding will reduce glomerular filtration, as the composition of the filtrate is in essence that of plasma water. However, other factors such as urine flow rate and reabsorption of drug from renal tubular fluid need to be considered (Section 3.3.1). The effects on drug metabolism are more complex. Briefly, binding reduces the rate of elimination of those drugs that are poorly extracted by the liver but not those that are extensively metabolized. In fact, for these drugs plasma protein binding can be considered as an efficient mechanism for delivering the drug to its site of metabolism. This is discussed in more detail in Section 7.4.1.

2.6   Summary

It is interesting to note that twentieth century man is fortunate that so many therapeutically useful compounds are successfully absorbed from the GI tract. The intestine is undoubtedly a very efficient organ for absorbing essential nutrients by active transport, but evolution appears to have designed it to exclude rather than transfer drugs, in that it is a lipid barrier, keeping out all but the most lipid-soluble foreign molecules. In addition, the intestine and the portal circulation are well endowed with enzymes which very effectively destroy many of the foreign molecules which penetrate the barrier. This design feature makes possible efficient transfer of food and virtual exclusion of non-food chemicals. Out of the minority of non-nutritive materials which do pass successfully from the gastrointestinal tract to the blood, a number have emerged as therapeutically useful drugs.

References and further reading

Ascenzi P, Bocedi A, Notari S, Fanali G, Fesce R, Fasano M. Allosteric modulation of drug binding to human serum albumin. *Mini Rev Med Chem* 2006; 6: 483–9.

Ascoli GA, Domenici E, Bertucci C. Drug binding to human serum albumin: abridged review of results obtained with high-performance liquid chromatography and circular dichroism. *Chirality* 2006; 18: 667–79.

Baker JR, Bullock GR, Williamson IH. Autoradiographic study of the distribution of [3H]- and [14C]-hydrallazine in the rat. *Br J Pharmacol* 1985; 84: 107–20.

Bauer M, Langer O, Dal-Bianco P, Karch R, Brunner M, Abrahim A,*et al.* A positron emission tomography microdosing study with a potential antiamyloid drug in healthy volunteers and patients with Alzheimer's disease. *Clin Pharmacol Ther* 2006; 80: 216–27.

Bauer M, Langer O, Dal-Bianco P, Karch R, Brunner M, Abrahim A,*et al.* A positron emission tomography microdosing study with a potential antiamyloid drug in healthy volunteers and patients with Alzheimer's disease. *Clin Pharmacol Ther* 2006; 80: 216–27.

Berridge MS, Heald DL, Muswick GJ, Leisure GP, Voelker KW, Miraldi F. Biodistribution and kinetics of nasal carbon-11-triamcinolone acetonide. *J Nucl Med* 1998; 39: 1972–7.

Binns TB (ed.). *Absorption and Distribution of Drugs*. Edinburgh: Churchill Livingstone, 1964.

Borga O, Azarnoff DL, Forshell GP, Sjoqvist F. Plasma protein binding of tricyclic anti-depressants in man. *Biochem Pharmacol* 1969; 18: 2135–43.

Brodie BB. Physicochemical and biochemical aspects of pharmacology. *JAMA* 1967; 202: 600–9.

Brodie BB, Hogben CA. Some physico-chemical factors in drug action. *J Pharm Pharmacol* 1957; 9: 345–80.

Brodie BB, Bernstein E, Mark LC. The role of body fat in limiting the duration of action of thiopental. *J Pharmacol Exp Ther* 1952; 105: 421–6.

Brodie BB, Burns JJ, Mark LC, Papper EM. Clinical application of studies of the physiologic disposition of thiopental. *N Y State J Med* 1956; 56: 2819–22.

Chignell CF. Drug-protein binding: recent advances in methodology: spectroscopic techniques. *Ann N Y Acad Sci* 1973; 226: 44–59.

Curry SH. Plasma protein binding of chlorpromazine. *J Pharm Pharmacol* 1970; 22: 193–7.

Curry SH. Theoretical changes in drug distribution resulting from changes in binding to plasma proteins and to tissues. *J Pharm Pharmacol* 1970; 22: 753–7.

Curry SH. *Drug Disposition and Pharmacokinetics*, 3rd edn. Oxford: Blackwell Scientific, 1980.

Curry SH, McCarthy D, Morris CF, Simpson-Heren L. Whole body autoradiography of CCK-8 in rats. *Regul Pept* 1995; 55: 179–88.

Dimitrakopoulou-Strauss A, Strauss LG, Gutzler F, Irngartinger G, Kontaxakis G, Kim DK, *et al.* Pharmacokinetic imaging of 11C ethanol with PET in eight patients with hepatocellular carcinomas who were scheduled for treatment with percutaneous ethanol injection. *Radiology* 1999; 211: 681–6.

Ederoth P, Tunblad K, Bouw R, Lundberg CJ, Ungerstedt U, Nordstrom CH, Hammarlund-Udenaes M. Blood–brain barrier transport of morphine in patients with severe brain trauma. *Br J Clin Pharmacol* 2004; 57: 427–35.

Fasano M, Curry S, Terreno E, Galliano M, Fanali G, Narciso P, Notari S, Ascenzi P. The extraordinary ligand binding properties of human serum albumin. *IUBMB Life* 2005; 57: 787–96.

Flanagan RJ, Taylor A, Watson ID, Whelpton R. *Fundamentals of Analytical Toxicology.* Chichester: Wiley, 2008.

Goodman LS, Gilman AG. *The Pharmacological Basis of Therapeutics*, 4th edn. New York: Macmillan, 1970.

Hardee GH, Fleitman JS, Otagiri M, Perrin JH. Microcalorimetric investigations of drug–albumin interactions. *Biopharm Drug Dispos* 1984; 5: 307–14.

Israili ZH, Dayton PG. Human alpha-1-glycoprotein and its interactions with drugs. *Drug Metab Rev* 2001; 33: 161–235.

Leonard RF, Knott PJ, Rankin GO, Robinson DS, Melnick DE. Phenytoin–salicylate interaction. *Clin Pharmacol Ther* 1981; 29: 56–60.

McGuire P, Howes OD, Stone J, Fusar-Poli P. Functional neuroimaging in schizophrenia: diagnosis and drug discovery. *Trends Pharmacol Sci* 2008; 29: 91–8.

Mark LC, Burns JJ, Campomanes CI, Ngai SH, Trousof N, Papper EM, Brodie BB. The passage of thiopental into brain. *J Pharmacol Exp Ther* 1957; 119: 35–8.

Mohanakrishnan P, Chignell CF. Copper and nickel binding to canine serum albumin. A circular dichroism study. *Comp Biochem Physiol C* 1984; 79: 321–3.

Ostergaard J, Heegaard NH. Capillary electrophoresis frontal analysis: principles and applications for the study of drug-plasma protein binding. *Electrophoresis* 2003; 24: 2903–13.

Perozzo R, Folkers G, Scapozza L. Thermodynamics of protein–ligand interactions: history, presence, and future aspects. *J Recept Signal Transduct Res* 2004; 24: 1–52.

Rosenthal HE. A graphic method for the determination and presentation of binding parameters in a complex system. *Anal Biochem* 1967; 20: 525–32.

Saidman LJ, Eger EI, 2nd. Uptake and distribution of thiopental after oral, rectal, and intramuscular administration: effect of hepatic metabolism and injection site blood flow. *Clin Pharmacol Ther* 1973; 14: 12–20.

Shand DG, Rangno RE. The disposition of propranolol. I. Elimination during oral absorption in man. *Pharmacology* 1972; 7: 159–68.

Schanker LS, Morrison AS. Physiological disposition of guanethidine in the rat and its uptake by heart slices. *Int J Neuropharmacol* 1965; 4: 27–39.

Solon EG, Balani SK, Lee FW. Whole-body autoradiography in drug discovery. *Curr Drug Metab* 2002; 3: 451–62.

Takano A, Suhara T, Ikma Y, Yasuno F, Maeda J, Ichimiya T, Okubo Y. Estimation of time course of dopamine D2 receptor occupancy from plasma pharmacokinetics of antipsychotics. *International Congress Series* 1264 (Elsevier) 2004, 173.

Upton RN. Cerebral uptake of drugs in humans. *Clin Exp Pharmacol Physiol* 2007; 34: 695–701.

Vernaleken I, Kumakura Y, Cumming P, Buchholz HG, Siessmeier T, Stoeter P, *et al.* Modulation of [18F]fluorodopa (FDOPA) kinetics in the brain of healthy volunteers after acute haloperidol challenge. *Neuroimage* 2006; 30: 1332–9.

Wyngaarden JB, Woods LA, Ridley R, Seevers MH. Anesthetic properties of sodium 5-allyl-5-(1-methylbutyl)-2-thiobarbiturate and certain other thiobarbiturates in dogs. *J Pharmacol Exp Ther* 1949; 95: 322–7.

3

Drug Elimination

3.1 Introduction

The action of a drug may be terminated by redistribution away from its site of action as discussed in Section 2.4.3.1. However, most drugs have a limited duration of action because they are eliminated by metabolism and/or excretion. Some drugs, particularly polar, water-soluble ones, may be excreted in the urine unchanged but lipophilic drugs tend to be highly protein bound, which reduces glomerular filtration and any drug in kidney tubular fluid tends to be reabsorbed back into the body. Consequently, lipophilic drugs would have very long elimination half-lives if they were not metabolized into polar molecules that are more readily excreted.

3.2 Metabolism

The liver is undoubtedly the major site of metabolism for the majority of drugs but most tissues have some metabolizing capacity. Other notable sites of metabolism include:

- Plasma and other body fluids – hydrolysis, mainly of esters
- Nerve terminals – metabolism of endogenous transmitters and drugs with similar structures
- Mucosal cells – responsible for presystemic metabolism of many drugs
- Kidney, lung and muscle – metabolize some drugs
- Brain – capable of glucuronidation of morphine
- Intestinal flora – metabolism of several drugs and hydrolysis of glucuronides.

Drug metabolism can be divided into two, sometimes three, distinct phases. Phase 1 reactions either insert, usually via oxidation, or reveal, by hydrolysis or desalkylation, reactive or functional groups that can undergo phase 2 reactions. For this reason, phase 1 reactions may be referred to as 'functionalization'. In phase 2 reactions, molecules with suitable functional groups are conjugated with endogenous compounds such as glucuronic acid, sulfate, acetate and amino acids, for example. Conjugation may be with products of phase 1 metabolism or if the drug has suitably reactive groups, conjugation may occur directly. Phenacetin is desalkylated to paracetamol which is then conjugated with either

Drug Disposition and Pharmacokinetics, By Stephen H. Curry and Robin Whelpton
© 2011 John Wiley & Sons, Ltd

glucuronate or sulfate:

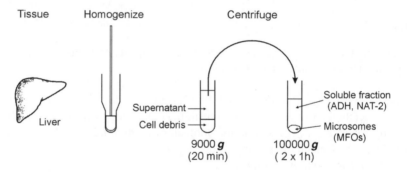

Phenacetin — Phase 1 / Desalkylation → Paracetamol — Phase 2 / Glucuronidation → Paracetamol glucuronide

Clearly, paracetamol can be conjugated directly without introduction of a functional (reactive) group via a phase 1 reaction. The term phase 3 metabolism is sometimes used to describe the further metabolism of phase 2 conjugates, such as occurs with the glutathione conjugate of paracetamol.

3.2.1 Phase 1 metabolism

These reactions include reductions and hydrolyses, but the most common reactions are the oxidations catalysed by the mixed-function oxidase (MFO) system, so called because of the apparent lack of specificity for individual substrates. These enzymes are isolated by subjecting tissue homogenates to high centrifugal forces ($100,000\,g$) after the cell debris has been removed by centrifugation at $9,000\,g$ (Figure 3.1). The pellet

Tissue   Homogenize   Centrifuge

Liver

Supernatant
Cell debris
$9000\,g$
(20 min)

Soluble fraction
(ADH, NAT-2)
Microsomes
(MFOs)
$100000\,g$
(2 x 1h)

Figure 3.1 Isolation of mixed-function oxidases from liver.

comprises microsomes, small particles of smooth endoplasmic reticulum (SER) whilst the supernatant layer (soluble fraction) contains enzymes from the cytosol, such as alcohol dehydrogenase (ADH) and *N*-acetyltransferase type 2 (NAT2).

Microsomal oxidations involve a relatively complex chain of redox reactions and require (i) reduced nicotinamide adenine dinucleotide phosphate (NADPH), (ii) a flavoprotein (NADPH-cytochrome P450 reductase) or cytochrome b_5 (iii) molecular oxygen and (iv) a haem-containing protein, cytochrome P450, so called because in its reduced form it binds with carbon monoxide to produce a characteristic spectrum with a peak at 450 nm. Cytochromes, which comprise a superfamily (Section 3.2.1.1) with nearly 60 genes having been identified in man, catalyse the final step (Figure 3.2). Drug (DH) combines with the CYP, and the iron(III) in the complex is reduced to iron(II) by acquiring an electron from NADPH-P450 reductase. The reduced complex combines with oxygen [(DH)Fe^{2+}O$_2$)] and combination with a proton and a further electron (from NADPH-cytochrome P450 reductase or cytochrome b_5) produces a DH–Fe^{2+}OOH complex. The addition of a proton liberates water and a Fe(III) oxene complex [DH(FeO)$^{3+}$], which extracts hydrogen from the drug with the formation of a pair of free radicals. Finally, the oxidized drug is

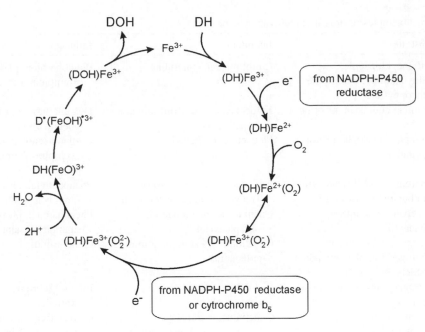

Figure 3.2   Diagrammatic representation of the final stage of incorporation of molecular oxygen into drug (DH) showing the various oxidation states of iron in the haem containing cytochrome.

released from the complex with the regeneration of the P450 enzyme. The last stage of the reaction involves free radicals and the same products can sometimes be produced by treating the drug with Fenton's reagent (a source of hydroxyl radicals). However, treating chlorpromazine, for example, with Fenton's reagent gives equal amounts of 3- and 7-hydroxychlorpromazine, whereas the major hydroxylated metabolite in humans is 7-hydroxychlorpromazine.

3.2.1.1   *Cytochrome P450 superfamily*

Although the Human Genome Project has identified 57 human genes that encode various forms of cytochrome P450, only three families, CYP1, CYP2, and CYP3, appear to be important in drug metabolism. The enzymes are classified on the basis of cDNA cloning according to similarities in amino acid sequence. A family contains genes that have at least a 40% sequence homology. Members of a subfamily (denoted by a letter) must have at least 55% identity. The individual gene products are identified by the final number.

The substrate specificity of CYP isoforms varies, as does that of their inhibitors and the agents that can induce their synthesis (Table 3.1). Note how several inhibitors and inducers appear several times. Ketoconazole and related drugs work by inhibiting fungal cytochromes whilst tranylcypromine binds covalently to monoamine oxidase and probably to other enzymes as well. Therefore, it might be predicted that these drugs would have the effects that they do. The H_2-receptor antagonist, cimetidine, inhibits several CYP isoforms and it is likely that a new drug will be required to show no or minimally significant interaction with cimetidine before approval from the licensing authorities is obtained, unless it can be shown categorically that the new drug will never be administered in a way in which an interaction could occur. Some drugs are both substrates and (competitive) inhibitors. Well-known enzyme inducing agents, phenobarbital, rifampicin (rifampin) and phenytoin feature large in Table 3.1. Note also that components of foods, particularly grapefruit, and herbal remedies, Saint John's wort, can inhibit or induce enzymes. The effects of enzyme inhibition and induction are discussed in Chapter 17.

Table 3.1   Some examples of drug-metabolizing cytochromes

Isoform	Substrates	Inhibitors	Inducers
CYP1A2	Caffeine, clozapine, fluvoxamine, olanzapine theophylline	Ciprofloxacin, cimetidine	Phenobarbital, phenytoin, rifampicin, tobacco smoke
CYP2A6	Cotinine, coumarin, nicotine	Ketoconazole, star fruit juice, grapefruit juice	Dexamethasone, rifampicin, phenobarbital
CYP2B6	Bupropion, cyclophosphamide[a], pethidine	Fluoxetine, orphenadrine	Carbamazepine, cyclophosphamide, phenobarbital
CYP2C9	Losartan[a], NSAIDs phenytoin, tolbutamide, S-warfarin	Amiodarone, ketoconazole, Tranylcypromine, valproic aicd	Phenobarbital, phenytoin, rifampicin, Saint John's wort
CYP2C19	Diazepam, omeprazole, phenytoin	Omeprazole, moclobemide, sulfaphenazole	Phenobarbital, phenytoin, rifampicin, Saint John's wort
CYP2D6	Codeine[a], debrisoquine, haloperidol, phenothiazines, SSRIs	Cimetidine, fluoxetine, quinidine, terbenafine	Glutethimide
CYP2E1	Enflurane, ethanol, halothane, paracetamol	Disulfiram, ethanol (acute), miconazole	Ethanol (chronic), isoniazid, rifampicin
CYP3A4	Clozapine, nifedipine, simvastatin[a]	Cimetidine, erythromycin, ketoconazole, SSRIs, grapefruit juice	Carbamazepine, phenobarbital, rifampicin, Saint John's wort
CYP3A5	Midazepam, caffeine		Dexamethasone

[a] Prodrug.

The distribution of the various CYPs also varies. CYP3A4, which metabolizes a large number of drugs, constitutes approximately 30% of the total CYP content of the liver and 70% of the total CYP in gastrointestinal epithelial cells. Many of its substrates are also transported by P-glycoprotein (P-gp), with which it appears to work in concert, so if a molecule escapes metabolism it is likely to be expelled into the gastrointestinal (GI) lumen by P-gp. CYP3A5 is the most abundant CYP3A form in the kidney. CYP3A7 is a foetal form of the enzyme that is believed to be important in the metabolism of endogenous steroids and *trans*-retinoic acid. It rapidly declines during the first week of life and is rarely expressed in adults. Although CYP2D6 represents only 2% of hepatic CYP, it metabolizes a large number of substrates, including many antipsychotic and antidepressant drugs, β-blockers, and the antihypertensive, debrisoquine. It is also responsible for activating the prodrug, tramadol, and for the *O*-desmethylation of codeine to morphine. It has been suggested that those with a deficit of CYP2D6 fail to obtain pain-relief when taking these drugs (Section 10.5.1).

3.2.1.2   *Other oxidases*

Not all oxidases are cytochromes. Monoamine oxidase (MAO) is a mitochondrial enzyme located in aminergic nerve terminals, the liver and intestinal mucosa. Not only does it deaminate endogenous neurotransmitters, noradrenaline (norepinephrine), dopamine, and serotonin, it is normally responsible for the first-pass metabolism of indirectly acting sympathomimetic amines (IASA) such as tyramine and ephedrine. When MAO is inhibited, for example by antidepressant drugs like tranylcypromine, the absorption of tyramine and subsequent displacement of elevated concentrations of noradrenaline can result in a dangerous hypertensive crisis (Table 17.3). Amphetamine is an IASA but it is a poor substrate for MAO because it has an α-methyl substituent; it tends to inhibit the enzyme.

Flavin-containing mono-oxygenases (FMOs) are microsomal enzymes that catalyse the NADPH-dependent oxidation of a large number of sulfur-, selenium-, and nitrogen-containing compounds, such as the *N*-oxidation of tertiary amines and stereospecific oxidation of sulfides. A genetic failure to express one isoform, FMO3, results in an inability to metabolize trimethylamine which leads to large quantities being excreted in the urine and sweat, a very distressing condition for the sufferers, who smell of putrefying fish.

Other phase 1 enzymes include the soluble enzyme, alcohol dehydrogenase (ADH), the mammalian form of which oxidizes several alcohols as well as ethanol, including methanol, ethylene glycol, and 2,2,2-trichloroethanol, the active metabolite of chloral hydrate. Xanthine oxidase metabolizes 6-mercaptopurine, a metabolite of azathioprine. The conversion of xanthine to uric acid may be inhibited by allopurinol, a drug sometimes used to treat gout.

3.2.1.3   Hydrolysis

Blood contains several esterases, including butyrylcholinesterase (BChE), a soluble enzyme (also known as pseudo- or plasma cholinesterase) and acetylcholinesterase (AChE) which is bound to the red cell membranes. Amides may be hydrolysed by plasma but the liver contains higher amidase activity as well as esterases. Esters are usually rapidly hydrolysed; examples include acetyl and carbamoyl esters (Figure 3.3).

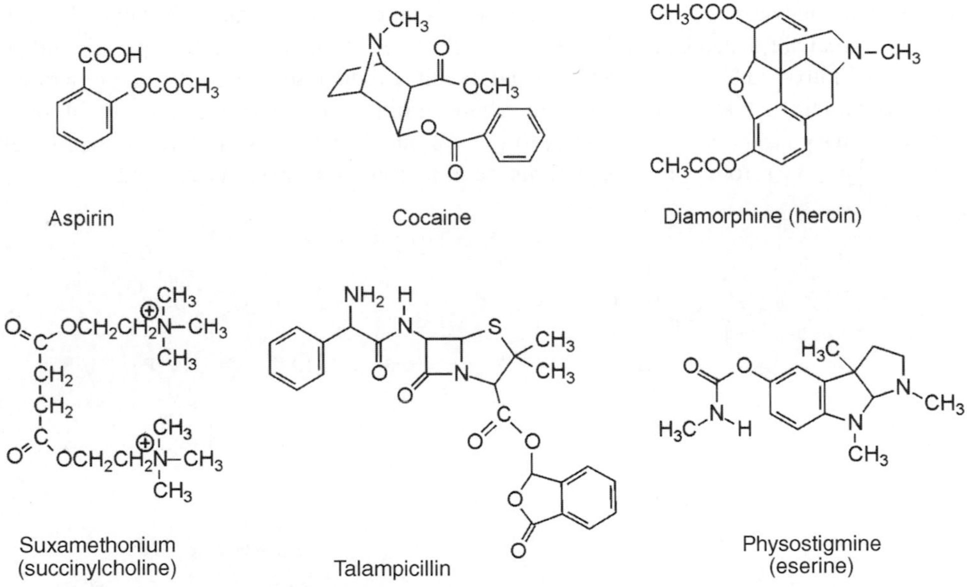

Figure 3.3   Examples of esters that are hydrolysed.

Several of the examples in Figure 3.3 are diesters. Hydrolysis of cocaine gives benzoylecgonine or methylecgonine which are further hydrolysed to ecgonine. Diamorphine is hydrolysed to 6-monoacetyl-morphine (6-MAM) and then morphine. Suxamethonium is hydrolysed to succinylmonocholine and succinic acid. Carbamate esters such as physostigmine are hydrolysed by cholinesterases, but much more slowly than acetylcholine, and so are substrate inhibitors of the enzymes. Talampicillin is an example of a prodrug of ampicillin that was produced to increase the oral bioavailability of the parent drug.

Amides and hydrazides are hydrolysed (Figure 3.4). Procainamide is hydrolysed to *p*-aminobenzoic acid. Amide local anaesthetics, lidocaine (lignocaine) and prilocaine are hydrolysed more slowly than the esters such as procaine and benzocaine. The *o*-toluidine (2-methylaniline) produced from prilocaine is thought to be responsible for the methaemoglobinaemia seen with this anaesthetic. Hydrolysis of hydrazides may lead to reactive hydrazines.

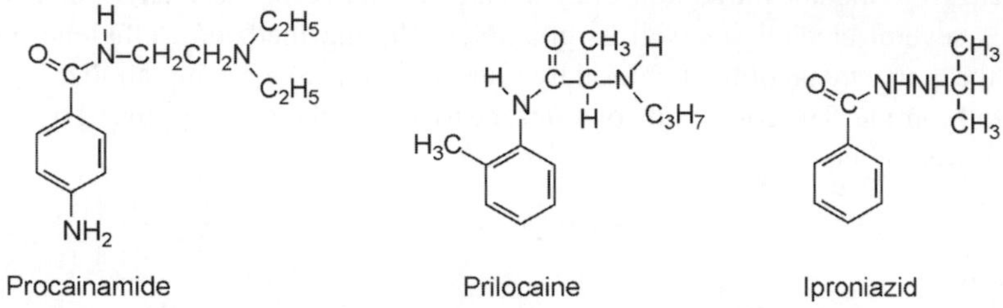

Procainamide   Prilocaine   Iproniazid

Figure 3.4 Examples of amides and hydrazides that are hydrolysed.

3.2.1.4 Reductions

Azo, nitro and dehalogenation reduction are catalysed by microsomal reductases. Several pathways are probably involved that are dependant on NADPH, cytochrome *c* reductase and flavoprotein enzymes. It is thought at azo- and nitro-reductases are different and the nitro-reductases can be divided into those that are oxygen-sensitive and those that are not. The nitro groups in nitrazepam, clonazepam and chloramphenicol are reduced to primary aromatic amines. The azo dye, prontosil, which is inactive *in vivo*, is reduced to sulfanilamide (Figure 3.5) from which the antimicrobial sulfonamides were developed.

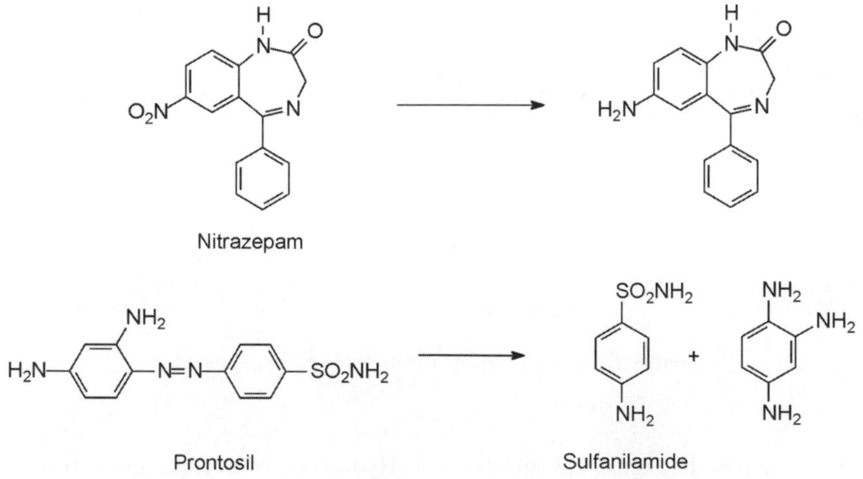

Nitrazepam

Prontosil   Sulfanilamide

Figure 3.5 Examples of nitro and azo reduction.

Choral is reduced to trichloroethanol and the ketone group in warfarin is reduced to a secondary alcohol. These reactions are probably not microsomal and may be reversed oxidations.

3.2.2 *Examples of phase 1 oxidation*

3.2.2.1 *Oxidation of alcohols*

Alcohols may be oxidized via various routes including microsomal CYP2E1 and catalase but in mammals the major route is catalysed by cytosolistic ADH to the corresponding aldehyde or ketone. Ethanol and methanol are oxidized to acetaldehyde and formaldehyde, respectively. Further oxidation by aldehyde dehydrogenase (ALDH) gives either acetic acid or formic acid (Figure 3.6). Formaldehyde and formic acid are much more

$$CH_3CH_2OH \xrightarrow{\text{ADH}} CH_3CHO \xrightarrow{\text{ALDH}} CH_3COOH$$

Ethanol Acetaldehyde Acetic acid

$$CH_3OH \xrightarrow{\text{ADH}} HCHO \xrightarrow{\text{ALDH}} HCOOH$$

Methanol Formaldehyde Formic acid

Figure 3.6 Oxidation of ethanol and methanol by alcohol dehydrogenase (ADH) and aldehyde dehydrogenase (ALDH).

toxic than acetaldehyde and acetic acid, so in methanol poisoning, ethanol may be given to compete with methanol for ADH and so reduce the rate of production of the toxic metabolites. Disulfiram inhibits ALDH so that after consumption of ethanol, acetaldehyde concentrations increase causing flushing, nausea and vomiting. The drug may be used in aversion therapy for alcoholics.

3.2.2.2 *Aliphatic and aromatic hydroxylations*

Microsomal hydroxylations are commonly catalysed by MFOs and occur with both aromatic and aliphatic moieties. Aromatic hydroxylations are exemplified by oxidation of phenobarbital to *p*-hydroxyphenobarbital, which is devoid of pharmacological activity, and by hydroxylation of phenothiazines and tricyclic antidepressants.

Examples of aliphatic hydroxylation include 4-hydroxlyation of debrisoquine (Section 10.5.1) and 3-hydroxylation of 1,4-benzodiazepines (Figure 3.7).

3.2.2.3 *Oxidative desalkylation*

Although cytochrome P450 catalysed oxidations insert an atom of oxygen into the drug, the initial product may be unstable and rearrange. This is the case with oxidative *N*- and *O*-desalkylation reactions. In the case of desmethylations, the methyl group is lost as formaldehyde, which contains the oxygen atom from the initial oxidation. Loss of larger alkyl groups results in the production of the corresponding aldehyde. With *in vitro* metabolism studies, the aldehyde may be trapped and measured as an indictor of the degree of desalkylation. *N*-Desalkylation is common with tertiary and secondary amines (Figure 3.7). Examples of drugs undergoing *O*-desalkylation include phenacetin (described earlier) and codeine (to give morphine).

3.2.2.4 *N- and S-oxidation*

Tertiary amines and sulfides are oxidized to *N*-oxides and sulfoxides, respectively. Sulfoxides may be further oxidized to sulfones. Thioridazine (Figure 3.8) is an interesting example of a drug that forms an *N*-oxide,

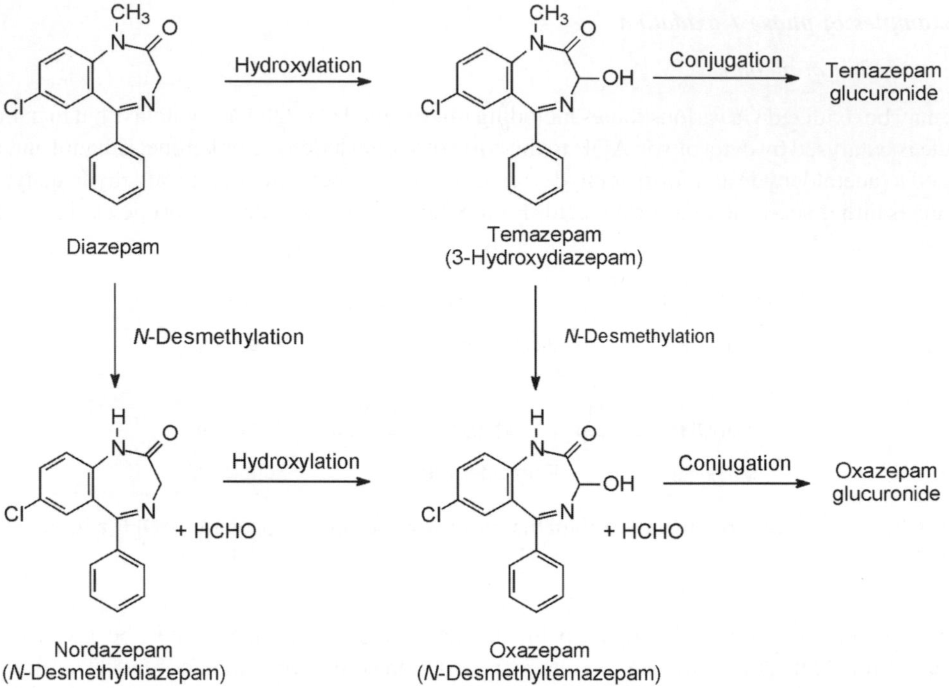

Figure 3.7   Metabolism of diazepam. Aliphatic hydroxylation gives temazepam, whilst *N*-desmethylation produces nordazepam and formaldehyde. Desmethylation of temazepam, or hydroxylation of nordazepam, gives oxazepam. The 3-hydroxy metabolites are conjugated with glucuronic acid and excreted in the urine.

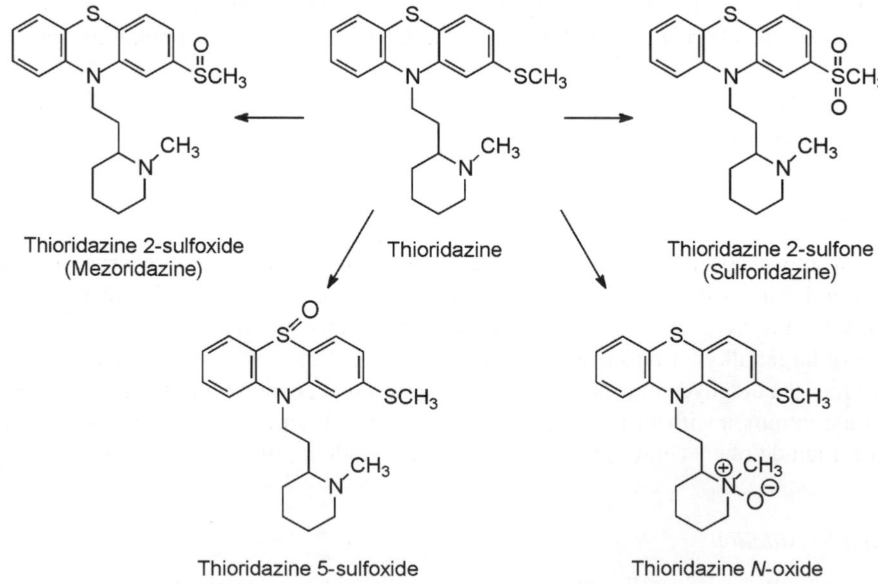

Figure 3.8   Some of the metabolic pathways of thioridazine.

a sulfone and two sulfoxides (more if the stereoisomers are taken into consideration). All phenothiazines form 5-sulfoxides, which are considered to be pharmacologically inactive. However, oxidation of the 2-thiomethyl group in thioridazine gives the 2-sulfoxide (mesoridazine) and 2-sulfone (sulforidazine) which are pharmacologically active and marketed as drugs in some countries.

N-Oxides are labile and are usually easily reduced to the parent amine or may spontaneously desalkylate to the secondary amine. A few drugs are marketed as *N*-oxides, chlordiazepoxide and minoxidil (see Figure 3.13), for example. Because of their labile nature, it is difficult to demonstrate whether the activity resides in the parent compound or one or more of its metabolites.

3.2.2.5 Oxidative deamination

Monoamine oxidase was mentioned earlier, and is responsible for deamination of endogenous amine transmitters, noradrenaline, dopamine and serotonin. Compounds with similar structures, such as ephedrine and tyramine, are deaminated to the corresponding aldehydes, which may be oxidized to carboxylic acids or reduced to alcohols. Other enzymes capable of deamination, include diamine oxidase and cytochromes; CYP2C3 for example, which has been shown to deaminate amphetamine to phenylacetone (Figure 3.9).

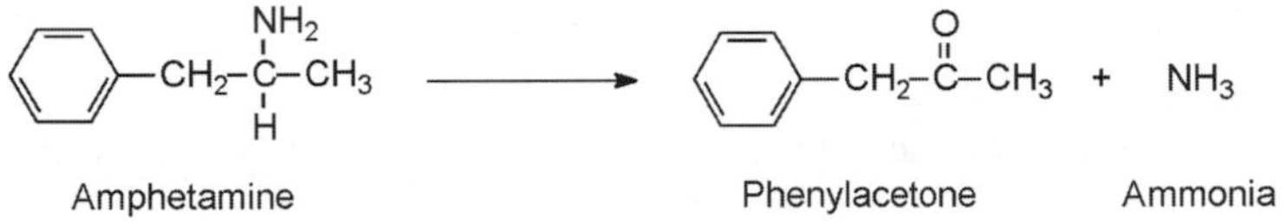

Figure 3.9 Deamination of amphetamine.

3.2.2.6 Desulfuration

Sulfur can be exchanged for oxygen. This exchange in the general anaesthetic, thiopental, produces the anxiolytic barbiturate, pentobarbital. In insects, malathion is converted to the much more toxic malaoxon (an example of lethal synthesis) whereas in mammals the major pathway is hydrolysis to the dicarboxylic acid (Figure 3.10).

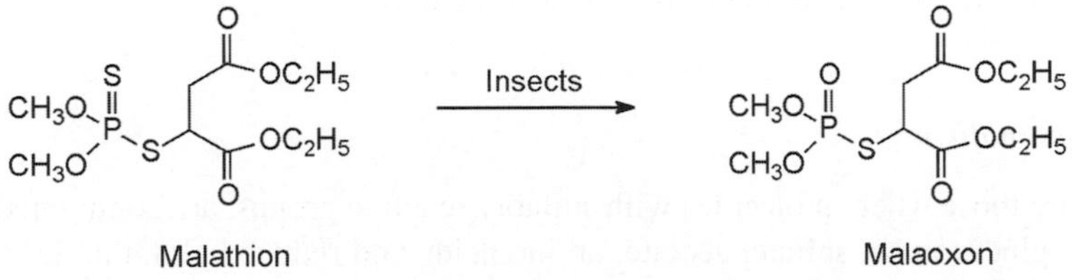

Figure 3.10 Insects convert malathion to malaoxon.

3.2.2.7 Combination of reactions

For those compounds with suitable groups, several of the reactions described above may occur. Figure 3.11 shows some of the metabolic pathways of chlorpromazine. Additionally, hydroxylation may occur at other

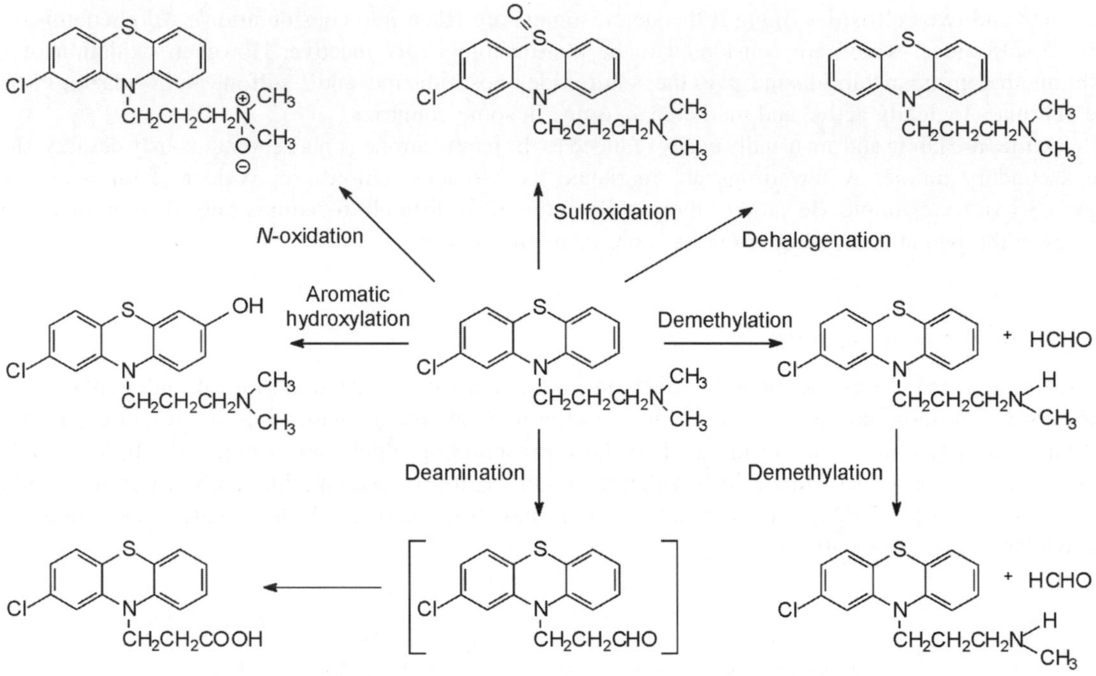

Figure 3.11 Examples of some of the pathways by which chlorpromazine is metabolized.

sites and that, plus conjugation reactions (Section 4.2.4), results in there being over 160 potential metabolites – and chlorpromazine is not the most chemically complex of the phenothiazine drugs.

3.2.3   *Miscellaneous reactions*

In this class are examples such as ring-opening, for example phenytoin to diphenylureidoacetic acid, and ring-formation, for example as occurs with proguanil. Transesterification can occur when ethanol is consumed. This has been shown to occur with methylphenidate (to ethylphenidate) and cocaine, when cocaethylene is formed. The ethyl homologue is active but has a longer half-life than cocaine and it has been suggested that this is why users take their cocaine with alcohol.

3.2.4   *Phase 2 metabolism*

Phase 2 reactions are those where molecules with suitably reactive groups, are conjugated with endogenous substances such as glucuronate, sulfate, acetate, amino acids and reduced glutathione (GSH). Usually the products are considerably less pharmacologically active, more water-soluble and more amenable to excretion, than the original substance. However there are some notable exceptions.

3.2.4.1   *D-glucuronidation*

This conjugation is a two stage process; the first stage being the synthesis of the donor molecule, uridine diphosphate glucuronic acid (UDPGA) in the cytosol. First glucose-1-phosphate is combined with uridine triphosphate and then oxidized to the carboxylic acid:

$$
\begin{array}{c}
\text{glucose-1-phosphate} \\
+ \\
\text{uridine triphosphate}
\end{array}
\xrightarrow[\text{transferase}]{\text{uridyl}}
\begin{array}{c}
\text{uridine diphosphate } \alpha\text{-D-glucose} \\
\text{(UDPG)}
\end{array}
$$

$$
\text{UGPG dehydrogenase} \bigg\downarrow + 2\,\text{NAD}^+
$$

$$
\begin{array}{c}
\text{uridine diphosphate glucuronic acid} \\
\text{(UDPGA)} \\
+ \\
2\text{NADH}_2
\end{array}
$$

The substrate–donor interaction is catalysed by UDP-glucuronyl transferases, which are membrane-bound enzymes. The reaction always produces the β-glucuronide (Figure 3.12).

COOH   HO   COOH

OH   O-(UDP)   HOOC   OH   COOH

UDPGA   Salicylic acid   Salicylic phenolic (ether) glucuronide

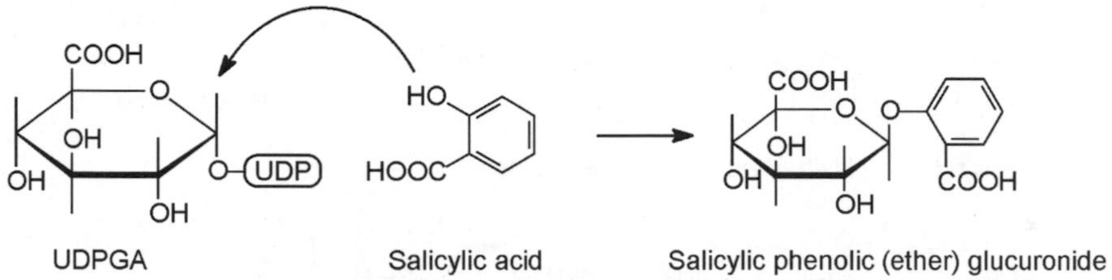

Figure 3.12 Glucuronidation of salicylic acid produces the β-glucuronide.

A large number of different glucuronides are possible, including *ether* glucuronides, formed with phenolic or alcoholic hydroxyl groups, *ester* glucuronides, formed with carboxylic acids, *N-glucuronides* formed with aromatic amines, alicyclic amines and sulfonamides and *S-glucuronides*, formed with various thiol compounds.

Generally, glucuronides are very water soluble, amenable to excretion via the kidney and bile (Sections 3.3.1 and 3.3.2), and pharmacologically inactive. However, morphine glucuronides are remarkable exceptions. There are three morphine glucuronides, 3-, 6-glucuronide and 3,6-diglucuronide. The 3-glucuronide accumulates in plasma to give higher concentrations than the parent compound and the 6-glucuronide is considered to be at least as active as morphine at μ-opioid receptors.

3.2.4.2 Sulfation

Ethereal sulfation is catalysed by several cytoplasmic sulfotransferases, depending on the substrate. The sulfate donor, 3′-phosphoadenosine-5′-phosphosulfate (PAPS), is formed from sulfate and ATP. Sulfate conjugates are usually very water soluble and readily excreted, and probably less pharmacologically active than the parent drug. The antihypertensive, minoxidil is interesting. First, it is an *N*-oxide and, second, sulfation occurs via this group (Figure 3.13). Minoxidil is also used to reduce or to prevent hair loss, and the sulfate conjugate has been shown to be 14 times more potent than minoxidil in stimulating cysteine incorporation in hair follicles.

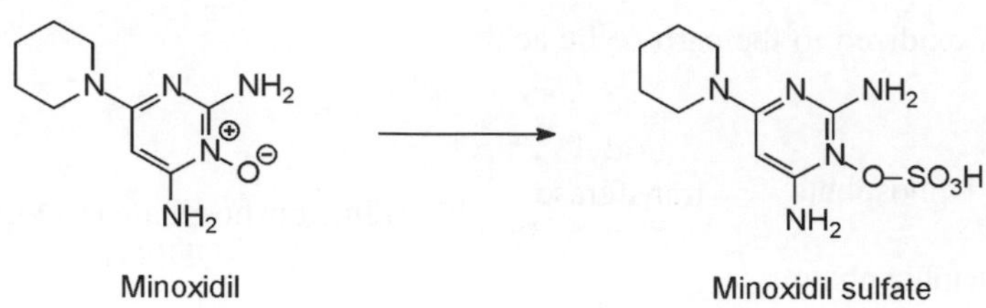

Figure 3.13   Sulfoxidation of minoxidil.

3.2.4.3   N-Acetylation

Aromatic primary amines and hydrazines are *N*-acetylated by *N*-acetyltransferase type 2 (NAT2) which is chiefly expressed in the liver, colon and intestinal epithelium. The donor molecule is acetylCo-A. NAT2 acetylates isoniazid, hydralazine, phenelzine, procainamide, dapsone and several antibacterial sulfonamides, including sulfadimidine (sulfamethazine) (Figure 3.14).

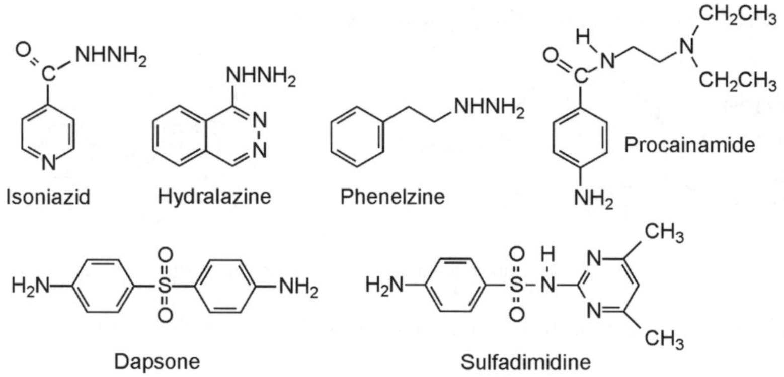

Figure 3.14   Examples of arylhydrazines and arylamines acetylated by human NAT2.

Most acetylated metabolites have reduced pharmacological activity and are considered inactive, this is particularly true of the sulfonamide antimicrobials. An exception is *N*-acetylprocainamide (acecainide) an antiarrhythmic agent that can be given either intravenously or orally.

It would be expected that acetylation would result in reduced aqueous solubility. This is only partially true; the acetyl metabolites of the older sulphonamides were less soluble in urine and caused crystalluria, but not the newer, more acidic, ones. Acetylation reduces the pK_a sufficiently to increase the degree of ionization at physiological pH values which more than compensates for the increase in lipophilicity.

3.2.4.4   Methylation

N-, *O*- and *S*-methylations are known and are common biochemical reactions, occurring with amines, phenols and thiols. The methyl donor is *S*-adenosylmethionine (SAM). Noradrenaline (norepinephrine) and related phenylethanolamines are *N*-methylated by phenylethanolamine-*N*-methyltransferase, which is found in the soluble fraction of adrenal homogenates. Histamine is *N*-methylated by imidazole *N*-methyltransferase.

Several enzymes are involved in *O*-methylation. Catecholol-*O*-methyltransferase (COMT) methylates both endogenous and foreign catecholamines, including isoprenaline (isoproteronol). Phenolic metabolites from phase 1 oxidations may be methylated, for example 7-hydroxychlorpromazine.

A number of exogenous thiols are methylated by *S*-methyltransferase. Endogenous substrates include the thiol containing amino acids, cysteine and homocysteine, and GSH. Inorganic sulfur, selenium and tellurium compounds are methylated to volatile derivatives.

3.2.4.5   Glycine conjugates

Aromatic amino acids, such as benzoic and salicylic acids, are conjugated with glycine to produce hippuric and salicyluric acids respectively. The acids react with coenzyme A, to form an acyl donor that reacts with glycine, and possibly other amino acids (Figure 3.15).

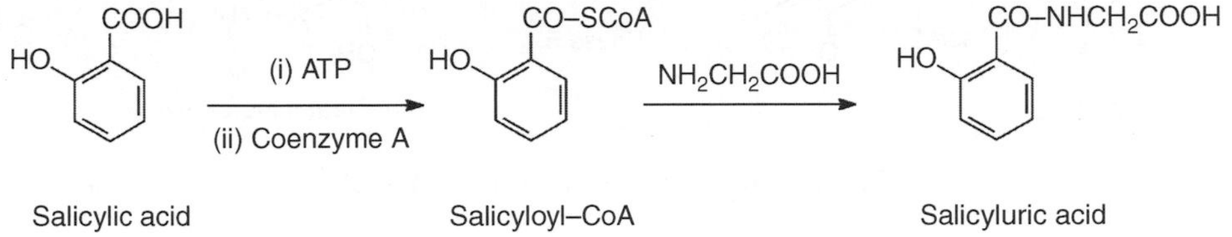

Figure 3.15   Formation of salicyluric acid.

3.2.4.6   Conjugation with glutathione

The reduced form of glutathione is an important reducing agent; one of its roles is to reduce methaemoglobin to haemoglobin. GSH usually acts as a nucleophile and can react chemically or enzymatically via a family of glutathione transferases. Substrates include aromatic nitro and halogen compounds and oxidized products of phase 1 metabolism. Probably the best known reaction of GSH is that with the reactive paracetamol metabolite, NAPQI (Figure 3.16). The GSH conjugate undergoes further (phase 3) metabolism to give the mercapurate derivative, which is excreted in urine.

3.2.5   *Kinetics of metabolism*

The simplest treatment of enzyme kinetics is that developed by Michaelis. The substrate combines with the enzyme and either dissociates or is converted to products:

$$\text{Enzyme} + \text{Substrate} \rightleftharpoons \text{Complex} \rightarrow \text{Products} + \text{Enzyme}$$

This is, in some ways, similar to the binding isotherms that have been derived for drug–receptor and drug–protein interactions but products are released. As the substrate concentration is increased, the enzyme becomes saturated with substrate so that the velocity, v, of the reaction cannot increase anymore [Figure 3.17(a)]. That is, for a given concentration of enzyme there is a maximum velocity, or rate, V_{max}. The rate of reaction v at any substrate concentration, C, is given by the Michaelis–Menten equation:

$$v = \frac{V_{max}.C}{Km + C} \tag{3.1}$$

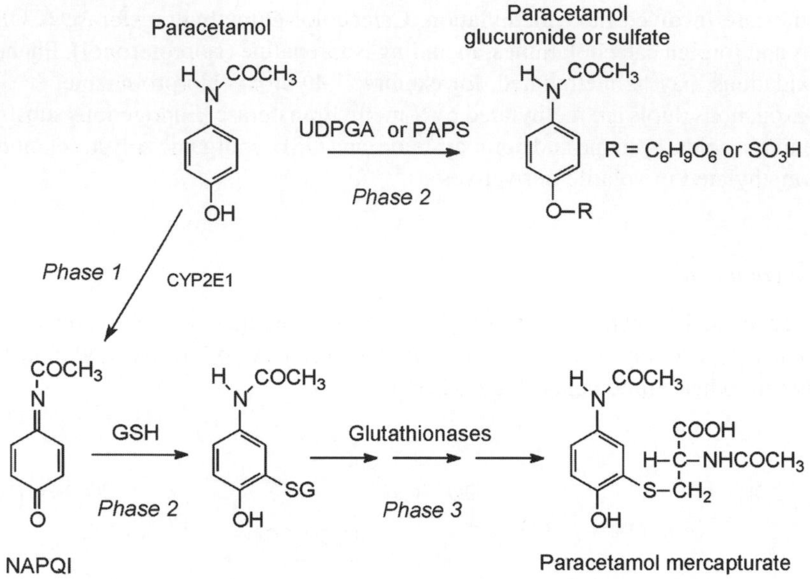

Paracetamol
H COCH3
N
OH
UDPGA or PAPS
Phase 2
Paracetamol
glucuronide or sulfate
R = C6H9O6 or SO3H
O–R
Phase 1
CYP2E1
COCH3
N
O
NAPQI
GSH
Phase 2
SG
Glutathionases
Phase 3
COOH
H–C–NHCOCH3
S–CH2
Paracetamol mercapturate

Figure 3.16 Metabolism of paracetamol.

where *K*m, which is numerically equal to the substrate concentration at one half V_{max}, is known as the Michaelis constant. This is easily demonstrated by substituting $V_{max}/2$ for *v* in Equation 3.1 and rearranging.

Values of *K*m and V_{max} can be derived by iteratively fitting *v* versus *C* values using relatively inexpensive computer software running on personal computers (Section 5.2.2). However, before such facilities were available Equation 3.1 was rearranged so that the constants could be obtained graphically. Lineweaver and Burk used a double reciprocal plot [Figure 17.1(a)]:

$$\frac{1}{v} = \frac{Km}{V_{max}C} + \frac{1}{V_{max}} \tag{3.2}$$

so that a plot of $1/v$ against $1/C$ is a straight line of slope Km/V_{max} [Figure 3.17(b)]. Such plots can be used to investigate whether an enzyme inhibitor is competitive or non-competitive. An alternative 'linearization' of

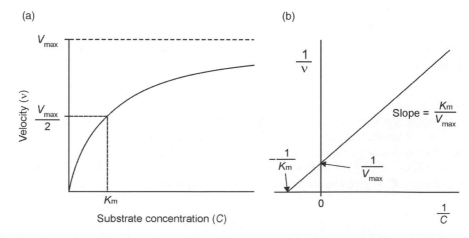

(a)
Vmax
Velocity (v)
Vmax/2
Km
Substrate concentration (C)
(b)
1/v
Slope = Km/Vmax
−1/Km
1/Vmax
0
1/C

Figure 3.17 (a) Relationship between velocity and substrate concentration. (b) Lineweaver–Burk plot.

Equation 3.1 is the Eadie–Hofstee plot where v is plotted against v/C:

$$v = V_{max} - \frac{v}{C} Km \qquad (3.3)$$

and Km is obtained from the slope and V_{max} from the y-intercept.

The velocities are usually obtained by conducting experiments at several initial substrate concentrations and measuring the formation of product or disappearance of substrate over a very short time period (e.g. up to 1 minute). The values can then be substituted into the Lineweaver–Burk or Eadie–Hofstee plot to obtain values of Km and V_{max}. This approach avoids having to increase the substrate concentrations to the point where an accurate estimate of V_{max} can be obtained.

The majority of drugs exhibit first-order elimination kinetics (Section 1.5.3.1), even those that are extensively metabolized. The explanation is that the effective drug concentration is very low compared to that of the enzyme. Therefore at any time little enzyme is bound by drug and so that in effect the non-bound enzyme concentration is constant. At low substrate concentrations $Km \gg C$, and so C makes a negligible contribution to the denominator of the Michaelis–Menten equation, i.e. $Km + C \rightarrow Km$ and Equation 3.1 becomes:

$$v \approx \frac{V_{max}}{Km} C \qquad (3.4)$$

This is the equation of a first-order reaction because the rate is directly proportional to the substrate concentration. Applying a similar argument to the situation when $C \gg Km$, the denominator approximates to C, as it is now Km that makes a negligible contribution to $Km + C$ and:

$$v = \frac{V_{max} C}{C} = V_{max} \qquad (3.5)$$

which is the situation described earlier when the enzyme is saturated. The reaction rate is constant and so the kinetics are zero-order. Ethanol kinetics approximate to zero-order at high concentrations and first-order at very low concentrations, but the kinetics can only be adequately described by the Michaelis–Menten equation at intermediate concentrations. Although most drugs exhibit first-order elimination kinetics at therapeutic concentrations, the anticonvulsant phenytoin is a notable exception and its time course can only be adequately described using Equation 3.1. Km and V_{max} apply when substrate is incubated with enzyme. However, *in vivo* the concentration is usually the plasma concentration and so the constants are not the same and should be referred to as 'apparent Km' and 'apparent V_{max}'.

3.3 Excretion

The organs most involved in excretion are the kidney and liver. Although the liver is an important site of drug metabolism it is also an important site of excretion, drugs and their metabolites being excreted via the bile. Other sites might not be so important with regards to the amounts excreted but may be of toxicological significance. Excretion of drugs in milk may have consequences for a feeding infant. Expired air is used for medico-legal estimates of blood alcohol whilst measurement of drugs in saliva, sweat, hair and nails are often of forensic interest.

3.3.1 Urine

The functioning unit of the kidney is the nephron, there being over 1.2 million in a human kidney (Figure 3.18).

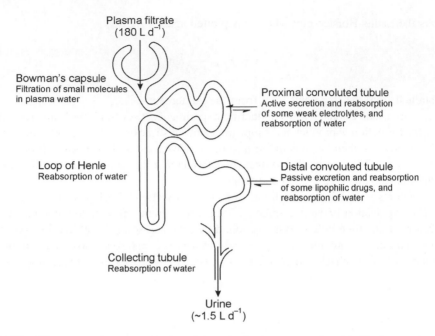

Figure 3.18   Diagrammatic representation of the principal processes involved in the renal excretion of drugs.

3.3.1.1   *Glomerular filtration*

The Bowman's capsule filters small molecules up to the size $M_r = {\sim}66,000$. Peptides may be filtered. Appoximately 90% of the insulin ($M_r = {\sim}6,000$) in plasma water appears in the ultrafiltrate whereas the figure for serum albumin ($M_r = {\sim}69,000$) is normally $<0.1\%$. Approximately 180 L of plasma is filtered per day, equivalent to 13–15 times the volume of extracellular fluid. The plasma filtrate has the composition of plasma water, so small, unbound drugs are filtered, but bound drug is retained.

3.3.1.2   *Active tubular secretion*

The proximal convoluted tubular cells have at least three types of transport proteins. Organic anion transport proteins (OATP) and organic cation transporters (OCT) are thought to be in the basolateral membrane while P-gp is probably in the apical membrane. Many acidic drugs and their glucuronide and glycine conjugates are actively transported. Examples include penicillins, some sulfonamides and thiazide diuretics. Examples of basic drugs actively secreted include quinine, pethidine (meperidine), amiloride and triamterene.

Penicillins have short elimination half-lives ($\sim$0.5–1 h), which can be increased by co-administration of probenecid which competes for the OATP so that the renal excretion of the penicillin is reduced. Probenecid is no longer used for this purpose but it is sometimes used as a uricosuric drug, because it competes for the carrier that actively *reabsorbs* uric acid in the PCT, and so increases the excretion of uric acid.

3.3.1.3   *Passive diffusion*

To some extent the kidney tubule behaves like a typical lipid barrier. Lipophilic molecules may diffuse in either direction and, if it were a static system, the concentrations in tubular fluid and plasma water would equilibrate. However, as a very large proportion of the filtered water is reabsorbed (Figure 3.18), the concentration of the tubular fluid markedly increases so the net movement of drug or metabolite is from tubular fluid to plasma, that is lipophilic molecules are reabsorbed. The reabsorption of weak electrolytes will be affected by the degree of ionization and the pH-partition hypothesis applies (Section 3.3.1.5).

Lipophilic drugs tend to be highly protein bound; they are hydrophobic and tend whenever possible to move from aqueous to non-aqueous environments. Protein binding reduces glomerular filtration and that, coupled with tubular reabsorption, means that little parent drug appears in the urine. This nicely illustrates the need for prior metabolism of such drugs as an aid to excretion – albeit as metabolites.

3.3.1.4 Renal clearance

The functioning of the kidney and how it handles substances has been investigated and described in terms of the physiological concept, renal clearance:

$$CL_R = \frac{U}{P} \times \text{urine flow rate} \tag{3.6}$$

where U is the concentration of substance in urine and P the concentration of substance in plasma. The ratio U/P is dimensionless so the units of clearance are those of urine flow – usually mL min^{-1}. Renal clearance can be defined as the volume of plasma flowing through the kidneys from which substance is removed per unit time, usually per minute. The two kidneys combined receive $\sim$20–25% of the cardiac output $\sim$1200 mL min^{-1} of blood, or $\sim$650 mL min^{-1} plasma. If 10% of a substance was removed from the plasma as it passed through the kidneys, then the clearance would be 65 mL min^{-1}, or if it were 20% it would be 130 mL min^{-1}. The proportion removed is known as the extraction ratio, E, and so

$$CL_R = EQ \tag{3.7}$$

where Q is the plasma flow rate. It follows from Equation 3.7 that the maximum value of renal clearance cannot exceed the plasma flow, when $E = 1$.

The renal clearance of a substance that is freely filtered, not actively secreted by renal tubular cells and not reabsorbed is a measure of the glomerular filtration rate (GFR). Inulin is a highly water-soluble polysaccharide that is not bound to plasma proteins; it is not secreted, nor reabsorbed, and is used to assess GFR. Its value is maximal in males of about 20 years (125 mL min^{-1}, for both kidneys combined). It is less in women and declines with age. Creatinine clearance is sometimes used to assess GFR or serum creatinine is taken as an indicator of renal function, the premise being that reduced filtration is reflected in elevated serum creatinine. However, creatinine is a metabolite of creatine, chiefly found in skeletal muscle and can vary, so exogenously administered inulin is a more accurate method of measuring GFR. The interest in GFR is that it is an indicator of kidney function and doses of some drugs are adjusted on the basis of GFR.

p-Aminohippuric acid (PAH) is actively secreted and totally cleared as it passes through the kidney, that is, its extraction ratio is 1. This means that renal clearance of PAH equals the plasma flow through the kidneys, and PAH has been used to measure this. Kidney *blood* flow will be plasma flow divided by $(1 - H)$, where H is the haematocrit.

Because of tubular reabsorption, it is not always obvious from the renal clearance whether a drug undergoes active tubular secretion. Penicillin G has a renal clearance of approximately 350 mL min^{-1} and as this far exceeds GFR must be actively secreted into tubular fluid. However, the converse is not true. A drug may be actively secreted but reabsorption may be such that $CL_R <$ GFR. Some indication may be obtained by comparing CL_R with the filtration rate of the unbound fraction. Competition between the protein carriers and plasma proteins probably means that plasma protein binding has little, if any, effect on active tubular secretion of drugs.

3.3.1.5   *Effect of urine pH and flow rate*

The pH of urine may vary, depending on diet and other factors, but pH 6.3–6.6 is a reasonable range under normal physiological conditions. Changing urine pH, say by administering sodium bicarbonate (sodium hydrogen carbonate, $NaHCO_3$) to make it alkaline, can have a major effect on the excretion of some weak acids and bases. The ionization of salicylic acid ($pK_a = 3.0$) at pH 6.3 is 99.95% whereas at pH 7.3 it is 99.995% ionized. This extremely small difference in the degree of ionization, represents a 10-fold difference in the proportion non-ionized (0.05–0.005%). The difference at pH 8.3, a value that can be easily obtained by giving sodium bicarbonate, is 100-fold.

Similar calculations can be performed for a base like amphetamine ($pK_a = 9.8$). At pH 7.8 the percentage non-ionized is 1% but at pH 4.8 the percentage falls to 0.001%, which makes a major change to the amount not reabsorbed and hence renal clearance. A similar situation exists for methamphetamine ($pK_a = 9.9$) and this is reflected in the amounts excreted at different pH values (Figure 3.19).

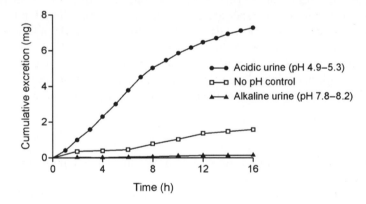

Figure 3.19   Effect of urine pH on amount of methamphetamine excreted in a single subject. (Redrawn from Beckett and Rowland, 1965).

3.3.2   *Biliary excretion*

Drugs excreted in the bile enter the duodenum via the hepatic duct. The rate of release of bile is not uniform but is increased in response to food. Those drugs and metabolites that remain in the GI tract will be removed from the body in the faeces; however there may be some reabsorption back into the body via the hepatic portal system (Section 2.3.1). When temoporfin, an anticancer drug, was injected intravenously, over 99.9% of the dose was recovered in the faeces, demonstrating how important biliary excretion can be. It also illustrates the important fact that finding an orally administered drug in faeces does not necessarily indicate that it has not been absorbed.

Bile canaliculi run between adjacent rows of hepatocytes to join the branching bile duct. Drugs may diffuse passively into the bile but this will result in relatively low concentrations similar to those of the unbound drug. Thus one would expect most drugs to be present in bile to some extent. Because the bile flow is only $\sim$0.5–0.8 mL min^{-1} the biliary clearance, CL_{Bile}, of such drugs will be low:

$$CL_{Bile} = \frac{\text{Concentration in bile}}{\text{Concentration in plasma}} \times \text{bile flow rate} \tag{3.8}$$

However, some drugs and, particularly, water-soluble metabolites such as glucuronides, are actively secreted into bile and can attain bile/plasma ratios in excess of 500 and the biliary clearance of these molecules may be several hundred mL min^{-1}. It is thought that there are at least three types of transport proteins, cationic and anionic ones and P-gp. There appears to be a molecular weight cut-off for active transport with substrates having to have M_r values > 325. This figure is for the rat and there are species variations. Above the minimum, the degree of biliary excretion increases with increasing M_r. Another feature is that polar, acidic molecules make good substrates. This is supported by the fact high concentrations of glucuronide metabolites are often found in bile. Glucuronides are polar and acidic and glucuronidation increases the molecular mass of the parent drug by 193 g mol^{-1}. It is likely that active transport has evolved to transport bile acids and compounds such as bilirubin glucuronide.

Suitable molecules may be actively transported without prior metabolism. The water-soluble cardiac glycoside, ouabain ($M_r = 728.8$) can have a bile:plasma ratio as high as 500. It is likely that temoporfin ($M_r = 680.7$) is actively transported into bile. Sulfobromophthalein ($M_r = 705.6$) is used to test liver function.

3.3.3 Expired air

The lungs are an important site for the excretion of volatile drugs and metabolites. The majority of gaseous and volatile general anaesthetics are excreted in expired air and because they are administered by the same route, the dose and hence the level of anaesthesia can be controlled by adjusting the partial pressures of these agents in the breathing apparatus. Deep alveolar air concentrations tend to equilibrate with those in blood, as would be predicted by Henry's law. This is the basis of breath alcohol measurements. At 34 °C the distribution between blood and alveolar air is 2100 : 1, that is 2100 L of air contain the same amount of ethanol as 1 litre of blood. Other volatile liquids are excreted via the lungs. For example, it has been shown that $\sim$40% of an oral dose of benzene is excreted in expired air. Inorganic selenium salts are metabolized to dimethyl selenide (Section 3.2.1.2), which is exhaled, giving a garlic smell to the breath. Extensive metabolism of an organic compound will eventually produce carbon dioxide which may be excreted in expired air. Study protocols for metabolism studies with ^{14}C-labelled drug may include trapping any exhaled $^{14}CO_2$.

3.3.4 Saliva

Excretion of drugs in saliva appears to occur largely by passive diffusion. The pH of saliva can vary from 5.5, the lower limit for parotid saliva, to 8.4 in certain ruminants. The average pH of 'mixed' saliva is $\sim$6.5, about 1 unit lower than plasma pH. Thus the unbound concentrations of basic drugs are usually higher in saliva than in plasma water whist the converse is true of weak acids. The saliva (S) to plasma (P) equilibrium ratio for an acid will be:

$$\frac{S}{P} = \frac{\left(1 + 10^{(pH(s)-pK_a)}\right)f_{u(p)}}{\left(1 + 10^{(pH(p)-pK_a)}\right)f_{u(s)}} \tag{3.9}$$

where pH(s) is the saliva pH, pH(p) is the plasma pH and $f_{u(p)}$ and $f_{u(s)}$ are the free fractions in plasma and saliva, respectively. There is less protein in saliva than in plasma – what there is, being chiefly antimicrobial and digestive enzymes secreted by the salivary glands. Consequently, the amount of drug-binding protein in saliva is usually considered to be insignificant, so $f_{u(s)}$ is taken to be unity. For salicylic acid, $pK_a = 3.0$, which is 50–90% protein bound in plasma, depending on the dose, the S/P ratio would be expected to be 0.06–0.01 for a saliva pH of 6.5. The equivalent equation for bases is:

$$\frac{S}{P} = \frac{(1 + 10^{(pK_a - pH(s))})f_{u(p)}}{(1 + 10^{(pK_a - pH(p))})f_{u(s)}} \tag{3.10}$$

Pethidine, a weak base, $pK_a = 8.7$, is 40–50% bound to plasma proteins so, for a saliva pH of 6.5, the S/P ratio calculates to be 4.6–3.8. This is consistent with reports that pethidine concentrations in saliva are higher than those in plasma.

There is little evidence that active transport is important for secretion of drugs into saliva. Penicillin may be secreted as a study showed inhibition by probenecid, and metoprolol concentrations have been reported as being higher than predicted by Equation 3.10. Active transport of ions results in saliva being hypotonic, having less sodium but higher concentrations of potassium than plasma. S/P ratios for lithium are 2–3 and vary with plasma concentration, indicating that there is active secretion. The characteristics of this transport have been reported as being very similar to those for potassium.

The interest in salivary excretion has arisen (i) because it was seen as a non-invasive method of sampling for pharmacokinetic studies, for example, in calculating antipyrine half-lives, (ii) as a way of monitoring drugs, both therapeutic and drugs of abuse, and (iii) a method of estimating the plasma concentrations of the unbound drug. However, there are several caveats and considerations. Three pairs of glands are responsible for most of the saliva that comprises oral fluid. In the resting state the relative percentage flows are parotid (21), submandibular (70), sublingual (2) with minor glands making up the balance. On mechanical stimulation the percentage produced by the parotid increases (58) whilst that for the submandibular decreases (33). However, when flow is stimulated by citric acid, the parotid and submandibular flows are similar at approximately 45% of total. Salivary flow is under autonomic control. Stimulation of β-receptors increases the amount of protein and mucin and results in a more viscous fluid whereas α-adrenoceptor excitation increases protein. Muscarinic stimulation increases saliva flow which is accompanied by increases in sodium and bicarbonate concentrations leading to increased pH. Clearly drugs acting at these receptors will influence the composition of the oral fluid. Saliva flow shows diurnal variation and its composition may be affected by disease. Other issues include:

- Saliva concentrations may be low, particularly for highly protein bound drugs.
- Flow rate ~ 1 mL min^{-1} – may result in insufficient sample, and/or long collection periods.
- Stimulating flow changes the composition – move to more constant pH 7.4. Drugs may be lost on materials chewed to stimulate flow.
- Effect of drugs on salivary glands.
- Some individuals, subjects and experimenters, find saliva unpleasant and foaming and frothing may be problematic. Special collecting aids overcome some of the problems.

Despite the problems, it is possible to use saliva for kinetic investigations when it is usual to validate against blood samples (Figure 3.20). However, it is drug monitoring that is driving the increasing interest in oral fluid analysis. For drugs of abuse, many of which are basic, sampling oral fluid is non-invasive and is done under supervision to reduce the risk of adulteration of the sample. Oral fluid is useful for situations where venepuncture may be difficult, such as the nervous patient or more importantly, for therapeutic drug monitoring in children, who are often less embarrassed than adults about spitting into a container. It has been suggested that phenytoin concentrations in saliva are a good approximation to the unbound concentration in serum and that this is a better indicator for dose adjustment than total serum concentrations.

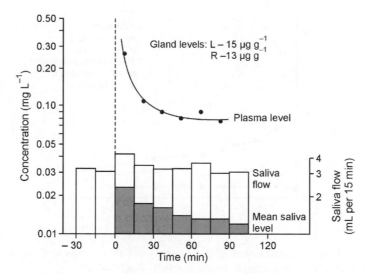

Figure 3.20   Concentrations of clonidine in plasma, submaxillary gland tissue (L = left, R = right) and submaxillary gland saliva, in a cat after 250 mg kg^{-1}, intravenously (Cho and Curry, 1969).

3.3.5   *Stomach and intestine*

A number of basic drugs, following i.v. doses are excreted into gastric juice in the stomach. This is predictable from the pH-partition hypothesis. Amongst the drugs studied this has been demonstrated for quinine and nicotine. Water-soluble drugs and ionized forms of weak electrolytes are likely to be excreted into the intestine in a way analogous the excretion to drugs into the stomach. With both the stomach and the intestine, diffusion from blood into the GI lumen is an essential part of the reversible diffusion reaction involved in the absorption of drugs. When the concentration gradient is favourable, for example after parenteral administration, the net movement will be into the GI lumen.

3.3.6   *Breast milk*

Excretion in milk is a minor pathway in terms of the amount of drug eliminated from the mother. It is of interest because of the potential effects of the drug on the suckling infant and the environmental exposure of populations to veterinary drugs in cows' milk.

   A number of drugs appear in milk, usually in small concentrations suggesting that the mechanism of transfer is passive diffusion i.e. the breast behaves as a lipid membrane. Breast milk is slightly more acidic on average (pH 7.2) than plasma and contains some protein, chiefly whey proteins, lactoferrin and α-lactalbumin. Albumin is present at ~0.4 g L^{-1}, approximately 1/10th of the concentration in plasma. Assuming drugs enter milk by passive diffusion, a method for estimating the amount of a drug that will pass into milk has been proposed (Atkinson and Begg, 1990). Factors taken into consideration are the pK_a, the degree of protein binding in plasma and the lipophilicity, assessed as log D$_{7.2}$, (octanol partition coefficient at pH 7.2). The method predicted a milk/plasma ratio of 2.33 for venlafaxine, which was in good agreement with an average observed value of 2.5. Despite the high concentration of venlafaxine in milk (Figure 3.21), the proportion of the mothers' dose consumed by the infants ($n = 6$) was calculated to be <6% (mean = 3.2%).

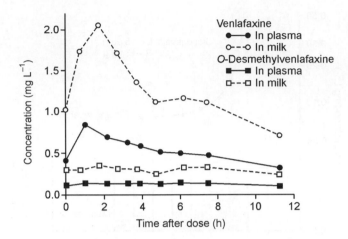

Figure 3.21   Steady-state venlafaxine and *O*-desmethylvenlafaxine concentrations in plasma and milk in a single subject. (Redrawn from the data of Ilett *et al.*, 2002.)

Interestingly, as the recommendations to breast feed increase, so the number of studies showing that the risk from maternal drug use is small, is increasing. However, it is wise to treat each case individually and to monitor the wellbeing of the infant. The death of a 13-day-old breast fed baby has been ascribed to the codeine his mother was taking for post-partum pain (Madadi *et al.*, 2007). This was an unusual case because the mother was an ultrafast metabolizer of codeine to morphine (Section 10.5.1) and also rapidly converted morphine to morphine 6-glucuronide (Section 10.8.1).

3.3.7   Other routes of excretion

Amongst the remaining minor routes of excretion are sweat, keratinous tissues and genital secretions. The increasing interest in these routes of excretion is because the samples provide a way of demonstrating exposure to drugs, either by detecting the drug directly or one or more of its metabolites.

3.3.7.1   Sweat

There are two kinds of sweat gland, eccrine and apocrine. The latter, which form during puberty are associated with hair in the pubic region and under the armpits, and secrete a thick liquid into the hair follicle. Eccrine glands produce a watery sweat that is $\sim$99% H_2O. It as long been known that drugs are excreted in sweat. This was demonstrated for quinine as early as 1844 and for morphine in 1942. There appear to have been few systematic studies as to the mechanisms involved, but a simple diffusion model probably applies as some basic drugs have sweat/plasma ratios >1 whilst weak acids have ratios <1 as would predicted from the pH-partition hypothesis for distribution drugs between plasma water and sweat, pH 5.8 (Johnson and Maibach, 1971).

3.3.7.2   Hair and nails

Drugs and their metabolites enter hair cells, probably by passive diffusion, as the hair is growing. Because the cells subsequently die, the drugs not only remain in the hair they are not metabolized and may be measured, months, years and even centuries later. This explains the forensic interest in measuring drugs in hair as a means of assessing previous exposure. For example, by examining segments of hair it is possible to determine periodic exposure to drugs and poisons.

3.3.7.3   *Genital secretions*

Obviously, excretion of drugs in vaginal secretions or seminal fluid is not going to play a major role in the removal of a drug from the body. Interest in these fluids arises because of the potential for drugs to interfere with sperm motility, have teratogenic effects or provide forensic evidence in a case of sexual assault. The excretion of drugs in semen has been reviewed by Pichini *et al.* (1994). In man approximately 30% of seminal fluid originates from the prostate and $\sim$60% from seminal vessels. The mechanism by which drugs enter seminal fluid is probably by passive diffusion and the equilibrium plasma:seminal fluid ratio of weak electrolytes will be given by modified forms of the Henderson–Hasselbalch equations, analogous to Equations 3.9 and 3.10. However, the situation is complicated by the fact that prostatic fluid is acidic (pH $\sim$ 6.6) while that of vesicular fluid is alkaline (pH $\sim$ 7.8). Some antibiotics have been shown to have fluid/plasma ratios >1, for example ciprofloxin (5.8) and norfloxacin (5.4) but for many other drugs the ratio is less than one, for example, carbamazepine (0.4–0.7), phenytoin (0.17) and valproic acid (0.07), possibly reflecting the effect of plasma protein binding.

3.3.8   *Cycling processes*

Drugs that are excreted into saliva or the stomach are likely to be reabsorbed from the GI tract, particularly those that are readily absorbed when given orally. Therefore, excretion via these routes will make little contribution to the overall removal of such drugs from the body. Instead these drugs will 'trapped' in a cycling process: blood $\rightarrow$ saliva $\rightarrow$ intestine $\rightarrow$ blood, or blood $\rightarrow$ stomach $\rightarrow$ intestine $\rightarrow$ blood.

3.3.8.1   *Enterohepatic cycling*

This is a further, and possibly more significant, form of cycling. Drugs that are excreted in bile may be reabsorbed from the intestine, in much the same way as described above. However, as explained previously (Section 3.3.2), a number of drugs are excreted in bile as glucuronide metabolites. Water-soluble glucuronides are not normally absorbed from the GI tract. However, if the conjugate is hydrolysed back to the original drug by β-glucuronidases in intestinal micro-organisms then drug can be reabsorbed, conjugated and secreted again leading to cycling of drug and drug–conjugate (Figure 3.22). Obviously

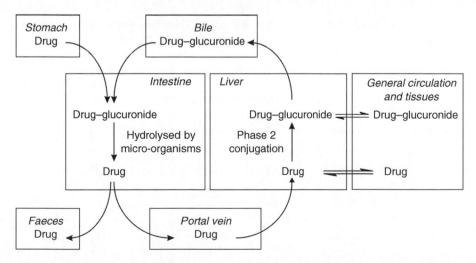

Figure 3.22 Enterohepatic recirculation. Orally administered drug enters the intestine from the stomach and is absorbed into the hepatic portal vein, carried to the liver and converted to the glucuronide which is secreted in the bile. Intestinal micro-organisms hydrolyse the glucuronide liberating parent drug, which can be reabsorbed.

only drugs that undergo phase 2 glucuronidation directly without a prior phase 1 reaction can do this. Examples of such drugs include stilboestrol, chloramphenicol, phenolphthalein and morphine.

3.3.8.2   Significance of cyclic processes

A drug trapped in a cycle persists in the body for longer than would otherwise occur and this reduces the rate of decline in plasma concentrations to some extent. Because the excretion of drugs via saliva and the stomach usually represent a steady release of small proportions of the dose it is most unlikely that the cycling will lead to any discernable fluctuations in plasma concentrations. On the other hand, the release of bile in man is pulsile and greater quantities are released in response to food, particularly fatty meals. Some studies have reported increases in plasma concentrations after meals, which have been ascribed to enterohepatic cycling. Rises in plasma concentrations of β-blocking drugs, benzodiazepines and phenothiazines have been observed at late times following intravenous injection.

Enteroheptic cycling will increase the half-life of a drug but the effect on the activity may be small if only a small proportion of the dose is being cycled. Furthermore, for the majority of drugs, drug molecules in the cycle are not available to the receptors, apart possibly from the laxative, phenolphthalein and the anti-diarrhoeal, morphine.

References and further reading

Aps JK, Martens LC. Review: The physiology of saliva and transfer of drugs into saliva. *Forensic Sci Int* 2005; 150: 119–31.

Atkinson HC, Begg EJ. Prediction of drug distribution into human milk from physicochemical characteristics. *Clin Pharmacokinet* 1990; 18: 151–67.

Beckett AH, Rowland M. Urinary excretion of methylamphetamine in man. *Nature* 1965; 206: 1260–1.

Cho AK, Curry SH. The physiological disposition of 2(2,6-dichloroanilino)-2-imidazoline (St-155). *Biochem Pharmacol* 1969; 18: 511–20.

Gibson GG, Skett P. *Introduction to Drug Metabolism*, 3rd edn. Cheltenham: Nelson Thornes, 2001.

Human Cytochrome P450 (CYP) Allele Nomenclature Committee: http://www.cypalleles.ki.se/ (accessed 16 November 2009).

Ilett KF, Kristensen JH, Hackett LP, Paech M, Kohan R, Rampono J. Distribution of venlafaxine and its *O*-desmethyl metabolite in human milk and their effects in breastfed infants. *Br J Clin Pharmacol* 2002; 53: 17–22.

Johnson HL, Maibach HI. Drug excretion in human eccrine sweat. *J Invest Dermatol* 1971; 56: 182–8.

Madadi P, Koren G, Cairns J, Chitayat D, Gaedigk A, Leeder JS, *et al*. Safety of codeine during breastfeeding: fatal morphine poisoning in the breastfed neonate of a mother prescribed codeine. *Can Fam Physician* 2007; 53: 33–5.

Nelson DR. Cytochrome P450 Home page: http://drnelson.utmem.edu/CytochromeP450.html (accessed 16 November 2009).

Pichini S, Zuccaro P, Pacifici R. Drugs in semen. *Clin Pharmacokinet* 1994; 26: 356–73.

4

Elementary Pharmacokinetics

4.1 Introduction

Pharmacokinetics is the science of using models (especially mathematical equations) to describe the time course of a drug and/or its metabolites in the body, usually in terms of the concentrations in plasma, although other fluids, such as whole blood or urine may be used. Further information may be gained from analysis of tissues or faeces. Probably the most important applications of the subject are in new drug development and in understanding and controlling optimum drug use. It is generally assumed that there is a relationship between the concentration of drug in plasma and the effects that are elicited. This is illustrated in Figure 4.1, which represents the time course of a drug in plasma following a single dose.

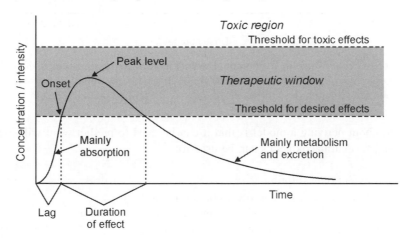

Figure 4.1   Model curve showing a theoretical perfect relationship between drug effect (intensity) and drug concentration following a single dose by any route other than intravenous injection.

If there is a minimum plasma concentration below which there is no therapeutic effect and a higher one, associated with the onset of unwanted, toxic, effects the intermediate concentrations must be those associated with wanted, therapeutic effects. This region is referred to as the 'therapeutic range' or 'therapeutic window'. If a suitably sized single oral dose, for example of aspirin for suppression of recurring inflammatory pain, is taken, there will be a lag-time while the tablet or capsule disintegrates and until sufficient drug has been absorbed for the plasma concentration to reach the threshold for effect to begin. The effect will cease when elimination has reduced the concentration to below the minimum concentration for the desired effect. The duration of effect will be the time between these two points. A larger dose may be taken to increase

Drug Disposition and Pharmacokinetics,   By Stephen H. Curry and Robin Whelpton
© 2011 John Wiley & Sons, Ltd

the intensity, or even the duration of effect, but it is clear from Figure 4.1 that there is a limit to the size of a single dose if toxic concentrations are to be avoided. Doubling the dose generally doubles the peak concentration, and may increase the intensity of the effect, but it will not necessarily double the duration of action. To obtain a longer duration of action, it is necessary to give smaller doses at regular intervals and if the dose and dosage intervals are chosen appropriately the concentrations can be maintained within the therapeutic window (Figure 4.2). The curves of this figure were generated using a simple pharmacokinetic model which will be explained later in this chapter.

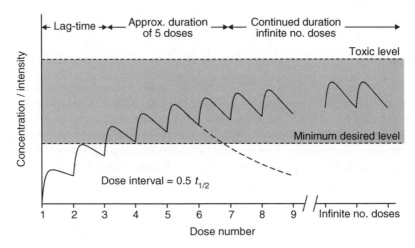

Figure 4.2 Model curves showing a theoretical perfect relationship between drug effect (intensity) and drug concentration during a series of eight doses, with a dose interval $0.5 \times t_{1/2}$ of the drug. The broken line following the fifth dose represents the decay if that had been the last dose.

The crucial point about deriving a model is that it can be used to predict what will happen when the parameters are changed. Pharmacokinetics can be used to:

- Assess the rate and extent of absorption
- Predict the effects of changing doses and routes of administration
- Model the effect of enzyme induction/inhibition and changes in excretion
- Model relationships between concentrations and effect (PK/PD).

Various types of model may be used, but probably the best known, and most widely used, are *compartmental* models. These assume that the drug in the body can be treated as if it were in (a) homogeneous solution(s) in one or more pools or 'compartments'. The concept of compartments is discussed in more detail in Section 5.1.5.

4.2 Single-compartment models

The simplest of compartment models is one where the drug is considered to be present as a well-stirred, homogeneous solution, rather like the solution of a dye in the flask depicted in Figure 4.3(a). The volume of the flask is analogous to the apparent volume of distribution of the drug, as discussed in Section 2.4.1.1. A drug is eliminated by metabolism and/or excretion as discussed earlier in Chapter 3 and, in the figure these processes are represented by the water flowing through the flask. In pharmacokinetics, the model is usually illustrated as a box or circle with an input and an output [Figure 4.3(b)].

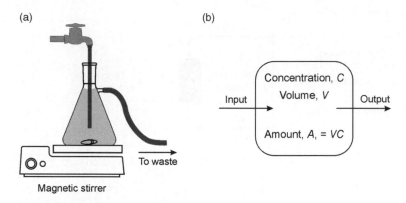

Figure 4.3 (a) Single compartment models are based on the concept of a well-stirred solution of drug. (b) Typical representation of a single compartment model.

Without any recourse to mathematics, it should already be apparent from Figure 4.3(a) that if dye (representing the drug) is rapidly introduced into the flask (representing a bolus i.v. injection) the flow of water will gradually wash out the dye and, furthermore:

- The faster the flow of water, the faster the dye will be removed, and
- The larger the flask, the longer it will take to remove the dye; although of course the initial concentration will be lower.

Intuitively, dye is being removed according to first-order elimination kinetics, observable in the flask as a gradual reduction in the colour of the solution, and this can be confirmed by writing a differential equation of the form of Equation 1.7 described in Section 1.5.3.1. The concentration, C is declining exponentially, and so is the amount, A. Because the volume is constant, $A = V \times C$, and the rate of elimination can be written:

$$-\frac{\mathrm{d}A}{\mathrm{d}t} = \lambda A = \lambda VC \tag{4.1}$$

where λ is a first-order elimination rate constant.

4.2.1 Systemic clearance

Systemic clearance, CL, also known as plasma clearance or whole body clearance, can be defined as the volume of plasma from which drug is removed per unit time, usually expressed as $\mathrm{mL\,min^{-1}}$ or $\mathrm{L\,h^{-1}}$. Clearance is the sum of all the *organ* clearances, so if a drug is eliminated by the kidney and the liver and no other routes:

$$CL = CL_R + CL_H \tag{4.2}$$

where CL_R and CL_H are the renal and hepatic clearances, respectively. Organ clearance is the volume of plasma flowing through the organ from which the drug is removed per unit time. The organ clearance cannot exceed the plasma flow rate through the organ (Equation 3.7, Section 3.4.1.1.). This is represented in Figure 4.4.

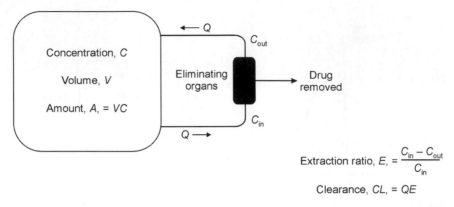

Extraction ratio, $E_r = \dfrac{C_{in} - C_{out}}{C_{in}}$

Clearance, $CL_r = QE$

Figure 4.4   Schematic diagram to illustrate the concept of systemic clearance. The extraction ratio is the proportion of drug removed as it flows through the eliminating organs.

The clearance can be used to calculate the rate of elimination of a drug from its plasma concentration because:

$$\text{rate of elimination} = C \times CL \qquad (4.3)$$

Provided CL is constant, this is a first-order equation because the concentration is declining exponentially and therefore Equations 4.1 and 4.3 are equal:

$$C \times CL = \lambda VC \qquad (4.4)$$

and cancelling C from either side gives:

$$CL = \lambda V \qquad (4.5)$$

Because the elimination kinetics are first-order, the half-life is given by Equation 1.10, which can be rearranged to:

$$\lambda = \frac{0.693}{t_{1/2}} \qquad (4.6)$$

This, when substituted into Equation 4.5 and again rearranged, produces:

$$t_{1/2} = \frac{0.693\,V}{CL} \qquad (4.7)$$

In Section 1.5.3 it was shown that Equation 4.6 arises from simple equations for first-order decay of concentrations in plasma incorporating the measured concentration at any time t, the theoretical concentration when $t = 0$, and using natural or common logarithms (Equation 1.10).

According to Equation 4.7, $t_{1/2}$, is a *dependent* variable that increases when the apparent volume of distribution increases and decreases when clearance increases. This is in agreement with the model of Figure 4.3(a), where the volume of the flask is analogous to V and the flow rate of water represents CL. Because no dye is returned to the flask, the extraction ratio, $E_r = 1$ and so the clearance equals the flow, Q. A more general method for obtaining CL using the area under the plasma concentration–time curve (AUC) is described later (Equation 4.14, Section 4.2.3.2. and Section 5.1.1.4).

4.2.2   *Why drugs have different elimination half-lives*

The Büchner flask of Figure 4.3(a) may seem far removed from the complex anatomy and physiology of the body, but the single-compartment model, although it is not always applicable, is a useful introduction to PK modelling. First, Equation 4.7 explains why drugs have different half-lives; it is because they have different volumes of distribution and/or systemic clearance. The apparent volume of distribution will depend upon the nature of the drug (Table 2.1) and the individual. Lipophilic drugs tend to have large apparent volumes of distribution and, if they enter adipose tissue, will be more widely distributed in someone who is obese. Also, changes in binding may result in changes in the distribution of drugs that are bound to plasma protein (Section 2.4.3.1). Clearance is a measure of how well the eliminating organs remove the drug. Hepatic clearance will increase when enzymes are induced and decrease in the presence of an enzyme inhibitor. Similarly, alkalinization of urine will increase the renal clearance of salicylate (Section 3.3.1.5). The clearance of chlorpromazine is higher than that of penicillin G, but being much more lipophilic, the volume of distribution of chlorpromazine is much higher and this results in a considerably longer half-life (Box 4.1).

Box 4.1   Effect of V and CL on elimination half-life

Penicillin G	*Chlorpromazine*
$V = 0.2 \, \mathrm{L \, kg^{-1}}$	$V = 20 \, \mathrm{L \, kg^{-1}}$
$CL_R = 320 \, \mathrm{mL \, min^{-1}}$	$CL_H = 600 \, \mathrm{mL \, min^{-1}}$
$t_{1/2} = \frac{0.693 \times 0.2}{320} \times \frac{1000}{60} \times 70^a$	$t_{1/2} = \frac{0.693 \times 20}{600} \times \frac{1000}{60} \times 70^a$
$t_{1/2} = 0.5 \, \mathrm{h}$	$t_{1/2} = 27 \, \mathrm{h}$

^{a}70 kg subject

4.2.3   Intravenous administration

When a drug is injected as a rapid intravenous dose (often referred to as a bolus dose), mixing within its volume of distribution is considered to be instantaneous – compartments are assumed to be well-stirred homogeneous solutions of drug. Clearly this must be an approximation because it takes a finite time to give the injection and a finite time for the drug to be distributed, even within blood. However, this may take less than a minute [Figure 4.5(a)] which is a negligible amount of time compared with the half-lives of the majority of drugs. There is no absorption involved and, after the initial mixing, the plasma concentrations show a monophasic exponential decline [Figure 4.5(b)]. If serial blood samples are taken and the plasma concentrations, C, of the drug are measured, then the rate constant, λ, can be obtained from the slope of the plot of ln C versus time. Back-extrapolation of the line to $t = 0$, gives a value for ln C_0 and hence C_0. This concentration is the theoretical plasma concentration of the drug, at $t = 0$, assuming instantaneous equilibration within its volume of distribution. At this time, *all* of the dose is in the body, as there has been no elimination, and the apparent volume of distribution, V, can be calculated as described in Section 2.4.1.1.

4.2.3.1   Half-life

The half-life, as defined earlier, is the time taken for the initial concentration to fall by one half. It would be better to refer to it as 'half-time', as there is no 'life' involved. However it is probably futile to rail against the use of half-life, as the term is used almost universally. There is, of course, a direct analogy with the half-life concept used with the decay of radioactive nuclei. Sometimes $t_{1/2}$ is qualified, for example, 'biological' or 'metabolic' half-life, adjectives that can be misleading, or worse, incorrect. Because it is measured in plasma, 'plasma half-life' has been suggested, but it is possible that $t_{1/2}$ refers to rising plasma concentrations, say after an oral dose. The important thing is to ensure that if a description is used, it is unambiguous.

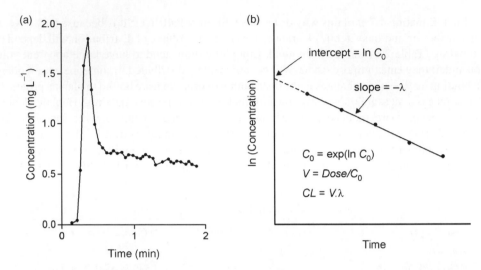

Figure 4.5   (a) Concentrations of phenylbutazone in one jugular vein of an anaesthetized sheep given a rapid intravenous dose in the other jugular vein (redrawn from McQueen and Wardell, 1971). (b) Back-extrapolation of experimental data, ln C versus t, to obtain a value of ln C_0 and hence C_0.

Half-life is a useful concept when applied to first-order reactions as it is constant. Thus, if 50% is lost in $1 \times t_{1/2}$ a further 50% of what is remaining will be lost in the next $t_{1/2}$, i.e. after $2 \times t_{1/2}$, 75% will have been lost and 25% of the original amount will remain. Note that the amount remaining, A, can be readily calculated from the number of half-lives that have elapsed, n,

$$A = \frac{A_0}{2^n} \tag{4.8}$$

where A_0 is the original amount. Note that n need not be an integer (Table 4.1). After five half-lives nearly 97% of the dose has been eliminated and after seven half-lives, >99% has been eliminated. Because the rate of elimination is higher when the concentrations are high shortly after the injection, almost 30% of the dose is lost in a time equivalent to $t_{1/2}$ divided by two.

Table 4.1   Proportion of dose eliminated or remaining as a function of the number of half-lives elapsed following an i.v. bolus dose of drug

Time elapsed (Number half-lives, n)	Amount eliminated (%)	Amount remaining (%)
0.5	29.3	70.7
1	50	50
1.5	64.6	35.4
2	75	25
3	87.5	12.5
5	96.87	3.13
7	99.22	0.78

Sometimes it is more convenient to obtain an estimate of the elimination half-life and to use Equation 4.6 to calculate the elimination rate constant, particularly when the concentration data are plotted on a logarithmic scale (Figure 4.6). Note how the half-life is constant and can be obtained by choosing any convenient concentration that is easily divisible by 2. It is worth noting that the 'slope' of the line is not, as is frequently claimed equal to $-\lambda/2.303$. Any attempt to derive a slope between two time points will give the average slope of the C versus t curve, between those times.

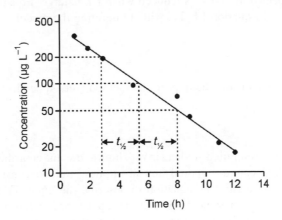

Figure 4.6 Prednisolone concentrations in a kidney transplant patient after an i.v. dose of 20 mg. (Redrawn from Gambertoglio *et al*, 1980).

4.2.3.2   *Area under the curve*

The area under the plasma concentration–time curve (*AUC*) can be computed from the integral of the equation defining the decay (see Appendix):

$$AUC = \int_0^\infty C \mathrm{d}t \qquad (4.9)$$

C is given by Equation 1.8

$$C = C_0 \exp(-\lambda t) \qquad (1.8)$$

so:

$$AUC = \int_0^\infty C_0 \exp(-\lambda t)\,\mathrm{d}t \qquad (4.10)$$

$$AUC = \frac{C_0}{\lambda} \qquad (4.11)$$

But from Equation 2.4, $C_0 = D/V$, so:

$$AUC = \frac{D}{\lambda V} \qquad (4.12)$$

Therefore the *AUC* is directly proportional to the dose injected and inversely proportional to the volume of distribution and the elimination rate constant. This is to be expected as a larger volume will lead to lower plasma concentrations and the rate of elimination will be higher when λ is larger, again reducing the plasma concentrations. Furthermore, $CL = \lambda V$ (Equation 4.5) so:

$$AUC = \frac{D}{CL} \tag{4.13}$$

and again, as might be expected, the AUC is reduced when CL is large. Equation 4.13 provides a way of calculating CL without use of Equation 1.8, i.e. without defining the model,

$$CL = \frac{D}{AUC} \tag{4.14}$$

because the AUC can be derived using the trapezoidal method (see Appendix).

4.2.4   Absorption

During drug absorption, the concentration gradient is very high and so the contribution of diffusion back from the pool of unabsorbed drug can be ignored and absorption can be considered as a non-reversible phenomenon. The model now requires an equation to describe the input (Figure 4.1). Absorption from the GI tract can be very variable for some drugs and the input function can be complex. The kinetics of absorption after i.m. injections are often first-order, and absorption after oral dosing may approximate to first-order kinetics. Thus, the rate of change of the amount of drug in the body is given by the rate at which it is being absorbed and the rate at which it is being eliminated:

$$\frac{dA}{dt} = k_a A_a - \lambda A \tag{4.15}$$

where k_a is the first-order rate constant of absorption and A_a is the amount of drug remaining to be absorbed. Provided all of the drug leaving the site of administration appears in the plasma, the rate of disappearance from the site of administration, which is basically inaccessible, must equal the rate of appearance in plasma. The rates of both processes can be accessed by k_a. Integration of Equation 4.15 produces:

$$A = \frac{A_0 k_a}{k_a - \lambda} [\exp(-\lambda t) - \exp(-k_a t)] \tag{4.16}$$

where A_0 is the dose, D. Dividing by V converts the amounts into plasma concentrations, and as all of the dose may not reach the systemic circulation, Equation 4.16 can be rewritten:

$$C = \frac{FD}{V} \frac{k_a}{k_a - \lambda} [\exp(-\lambda t) - \exp(-k_a t)] \tag{4.17}$$

where F is the fraction of the oral dose which reaches the systemic circulation (Section 8.3). Equation 4.17 can be written as:

$$C = C_0' \exp(-\lambda t) - C_0' \exp(-k_a t) \tag{4.18}$$

where $C_0' = \dfrac{FD}{V} \dfrac{k_a}{k_a - \lambda}$.

Thus it is clear, via Equation 4.18, that Equation 4.17 represents a concentration–time curve that is the *difference* of two exponential terms that have a common intercept on the y-axis [Figure 4.7(a)]. Both exponential terms asymptote to zero as t increases but the one with the largest rate constant decays the fastest, so that at later times Equation 4.18 approximates to a single exponential, as can be seen in Figure 4.7(b).

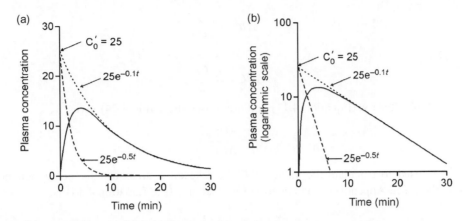

Figure 4.7 Simulated curves for first-order input into a single-compartment model, (a) linear scales, (b) semilogarithmic scales. $C_0' = 25$ arbitrary units, $k_a = 0.5\,\text{min}^{-1}$, $\lambda = 0.1\,\text{min}^{-1}$.

The maximum concentration, C_{max}, occurs when the rates of input and output are equal, and $dC/dt = 0$. The time of the peak, t_{max}, is given by differentiating Equation 4.17 with respect to t and rearranging:

$$t_{max} = \frac{1}{k_a - \lambda} \ln \frac{k_a}{\lambda} \tag{4.19}$$

Substitution into Equation 4.17 gives C_{max}, or it can be computed from:

$$C_{max} = \frac{D}{V}\left(\frac{\lambda}{k_a}\right)^{[\lambda/(k_a-\lambda)]} \tag{4.20}$$

For linear examples, t_{max} is independent of the dose, but C_{max} is directly proportional to the dose. This relationship does not hold for non-linear examples (Section 4.3). Also, the peak is not the time that absorption ceases and elimination starts. According to the model both start at $t = 0$ and continue to $t = \infty$. Absorption continues to make a significant contribution to the concentration–time curve at times beyond t_{max} as can be seen in Figure 4.7. The faction of the dose absorbed, f_{A_a} at t_{max} is a function of the relative sizes of the rate constants of absorption and elimination (Kaltenbach *et al.*, 1990):

$$f_{A_a} = 1 - \left(\frac{k_a}{\lambda}\right)^{-k_a/(k_a-\lambda)} \tag{4.21}$$

When $k_a = 2\lambda$, then 75% of the dose is absorbed by t_{max}, whereas for 90% of the dose to be absorbed when $t = t_{max}$, when k_a must equal 8λ. Therefore, it is important when performing pharmacokinetic experiments to ensure that data are collected for sufficient time for accurate assessment of the elimination rate constant, i.e. when absorption is making a negligible contribution to the concentration–time data. (Section 5.2.3).

4.2.4.1 Area under the curve

The *AUC* for the oral case is obtained in the same way as for the intravenous case, that is integration of the plasma concentration–time curve, Equation 4.17:

$$AUC = \int_0^\infty \frac{FD}{V}\frac{k_a}{k_a - \lambda}[\exp(-\lambda t) - \exp(-k_a t)]dt \tag{4.22}$$

$$AUC = \frac{FD}{\lambda V} \tag{4.23}$$

and, as $C_0 = D/V$:

$$AUC = F\frac{C_0}{\lambda} \tag{4.24}$$

Comparison of *AUC* after i.v. (Equation 4.11) and oral doses (Equation 4.24) gives:

$$F = \frac{AUC_{\text{p.o.}}}{AUC_{\text{i.v.}}} \tag{4.25}$$

Equation 4.25 is used to calculate the value of *F*, the respective areas usually being obtained by the trapezoidal method (see Appendix). The equation for clearance (see Equation 4.14) is:

$$CL = F\frac{D}{AUC} \tag{4.26}$$

This means that *CL* can *only be obtained after intravascular injection*, when *F* is known to be 1, otherwise the ratio of *D/AUC* gives *CL/F*. Similarly, because the amount of drug reaching the systemic circulation (*F*) can only be accessed from an i.v. injection it is not possible to determine the apparent volume of distribution from oral administration. There is, however, a concept known as 'apparent oral clearance', which uses these equations and literature values of *F*, obviously most of them less than unity, to calculate the clearance as if from an intravenous dose.

4.2.4.2 Special cases: $k_a < \lambda$ and $k_a = \lambda$

Drugs are usually formulated to be absorbed rapidly to give a rapid onset of effect. Consequently, one would expect k_a to be greater than λ. However, k_a can be less than λ. Indeed, that is the principle of sustained-release preparations. A growth and decay still occurs because when there is a large pool of drug to be absorbed, the *rate* of absorption still exceeds that of elimination (which initially is zero). In these special cases, because the absorption is rate-limiting, the half-life estimated from the decay phase is that for absorption and the rate constant of elimination is obtained from the steeper exponential curve. When the rate constants are substituted into Equation 4.17 the concentration curve is of the same form as that of Figure 4.7, but the rate constants appear to have been exchanged [Figure 4.8(a)]. This phenomenon has been referred to as 'flip-flop'. It is not possible from the concentration–time data to assign the rate constants to elimination or absorption. The only way to know which is which is from unambiguous data, such as with an independent i.v. dose. In the example of Figure 4.8(a), the rate constant of elimination was assigned by giving an intramuscular injection of a preparation that was not sustained-release.

If the absorption and elimination rate constants are equal then clearly Equation 4.17 is no longer applicable, as it reduces to zero at all times! By writing differential equations in which the rate constants are equal and *then* differentiating, an equation can be obtained for this situation (Gibaldi and Perrier, 1982):

$$C = \frac{FD}{V}kt\exp(-kt) \tag{4.27}$$

where $k = k_a = \lambda$. For this unusual situation:

$$t_{\max} = 1/k \tag{4.28}$$

and

$$C_{\max} = \frac{0.37FD}{V} \tag{4.29}$$

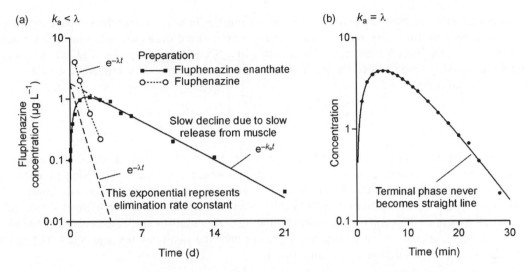

Figure 4.8 Examples of special cases. (a) Fluphenazine enanthate was developed so that the rate of release from muscle is rate limiting. Resolution of the data into its two exponentials makes λ appear to be the rate constant of absorption, but the decline after fluphenazine (open circles) confirms it is the elimination rate constant. (b) Methylene blue simulation of $k_a = \lambda$. The terminal decline never becomes straight and the parameters cannot be resolved using the method of residuals, however they can be resolved by iterative curve fitting to Equation 4.27 (solid line).

A crucial point about Equation 4.27 is that a plot of ln C versus time has no linear portion, so it is impossible to solve for k graphically. Although it is most unlikely that this situation would occur, it is easy to simulate using the dye model by connecting together two flasks of equal volume in series, so that the output of the first becomes the input of the second. This example illustrates the potential problem of solving for k_a and λ, when the values are similar.

4.2.4.3 Special case: lag time

The rise in plasma drug concentrations does not, in fact, commence immediately after an oral dose. Inevitably there is a lag phase because, even after administration of an oral solution, it takes a finite time for drug to reach its site of absorption. The lag time will be longer for tablets and capsules that have to disintegrate, and delayed gastric emptying may delay absorption even further. The lag time after consumption of enteric coated tablets may be several hours. The lag time, t_{lag}, moves the plasma concentration–time curve to the right so that Equation 4.17 becomes:

$$C = \frac{FD}{V}\frac{k_a}{k_a - \lambda}\left\{\exp\left[-\lambda(t - t_{lag})\right] - \exp\left[-k_a(t - t_{lag})\right]\right\} \qquad (4.30)$$

An estimate of t_{lag} can be obtained by iterative curve-fitting of Equation 4.30 or, if the method of residuals is used the construction lines intersect at $t = t_{lag}$, rather than intersecting at the y-axis where $t = t_0$ (Section 5.2.1).

4.2.5 Infusions

Equations exist for the oral single dose situation in which absorption is at a constant rate, i.e. with zero order kinetics, and elimination occuring with first order kinetics. This can be important with controlled-release oral dosage forms. However, the principles are easily demonstrated by considering intravenous infusions.

For drugs that are given intravenously, an infusion over minutes or hours, rather than a bolus dose, may be appropriate, especially for safety reasons. Also, orally-administered drugs are often taken for extended periods of time, sometimes for the rest of a patient's life. Second and later doses of oral drugs given repeatedly are usually given before the previous dose has been completely eliminated, so that the drug accumulates in the body.

4.2.5.1   Zero-order input

Drugs may be infused using a motorized syringe pump and miniaturized versions are available for ambulatory patients. A constant rate infusion represents zero-order input, R_0, into the single-compartment model. At the start of the infusion the plasma concentration will rise as the infusion proceeds, however, for a drug that is eliminated according to first-order kinetics, it will not continue to rise indefinitely but 'level off' to what is referred to as the steady-state concentration, C^{ss}. This must be the case because as the plasma concentration increases the rate of elimination increases until the two rates become equal. The equation defining the plasma concentration during an infusion is:

$$C = C^{ss}[1 - \exp(-\lambda t)] \qquad (4.31)$$

This is the equation of an exponential curve that asymptotes to C^{ss}. It is the same shape as the decay curve at the end of the infusion, but inverted. Because the curves are the same shape, it can be appreciated that during the infusion the concentration will be 50% of C^{ss} at $t = 1 \times t_{1/2}$, 75% of C^{ss} at $t = 2 \times t_{1/2}$, and so on. In fact, the values are the same as those for the percentage eliminated in Table 4.1, therefore after $7 \times t_{1/2}$, the concentration will be $>0.99C^{ss}$, even though according to Equation 4.31, C^{ss} is only achieved at infinite time [Figure 4.9(a)].

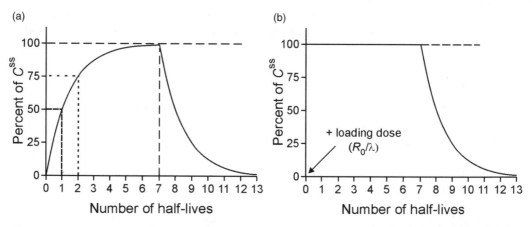

Figure 4.9   Comparison of attainment of steady-state concentrations as a function of elimination half-life (a) following a constant rate infusion for $7 \times t_{1/2}$ and (b) the same infusion with a loading dose injected at the start of the infusion.

Post-infusion the plasma concentration, $C_{postinf}$, decays according to an equation analogous to Equation 1.8, but rather than C_0, the concentration is concentration when the infusion was stopped, and the time is the time elapsed $(t - T)$ since the time at the end of the infusion period, T:

$$C_{postinf} = C^{ss}[1 - \exp(-\lambda T)]\exp[-\lambda(t - T)] \qquad (4.32)$$

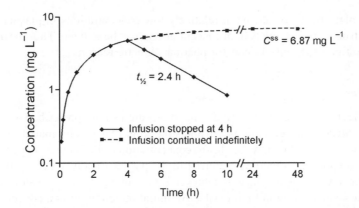

Figure 4.10 Amikacin infusion: simulated curves derived using Equations 4.5, 4.34, 4.31 and 4.32.

Figure 4.10 illustrates the principles of infusion using a simulation for the antibiotic, amikacin. In this type of simulation we ask 'what if' questions, in a sense the opposite of data analysis. The amikacin simulation shows concentrations in plasma to be expected if amikacin is infused at 0.625 mg min^{-1} (or 0.0089 mg min kg^{-1}) into a 70 kg subject in whom the clearance is 1.3 mL min^{-1} kg^{-1}. The apparent volume of distribution is 0.27 L kg^{-1}, the C^{ss} that results is approximately 7 mg L^{-1}. When the infusion is stopped, the drug concentration falls with a half-life of 2.4 hours. This simulation shows two 'what if' questions being asked, in that the effect of infusion to steady-state is compared with infusion for 4 hours after which the concentrations decline.

4.2.5.2 Loading dose

Because drugs with long elimination half-lives will approach steady-state slowly this may result in an unacceptable delay before therapeutic concentrations are achieved. This delay can be overcome by injecting a large bolus, known as a 'loading-dose', at the start of the infusion. The aim is to reach steady-state conditions immediately, and so the loading-dose must be such that the injection gives a plasma concentration of C^{ss}. This amount is, as always, the concentration multiplied by the apparent volume of distribution, so:

$$\text{loading dose} = VC^{ss} \tag{4.33}$$

The rate of elimination is the amount of drug multiplied by the elimination rate constant, and at steady-state this equals the rate of infusion, so:

$$R_0 = \lambda VC^{ss} \tag{4.34}$$

and so doubling the rate of infusion would double C^{ss}. Combining Equations 4.33 and 4.34 gives:

$$\text{loading dose} = R_0/\lambda \tag{4.35}$$

This equation was used to calculate a loading dose for the data of Figure 4.9(b).

Loading doses, both by injection and orally, are widely used with antibacterial drugs, such as sulfona-mides, and in cancer therapy, such as with trastuzumab. They can also be used with non-steroidal anti-inflammatory drugs such as naproxen. However, there are situations in which pharmacokinetic considera-tions imply that a loading dose might make sense, but pharmacological considerations dictate otherwise. This occurs with tricyclic antidepressants, in that with these drugs the patient must initially develop a tolerance to the acute cardiovascular effects, occurring at relatively low concentrations in plasma, so that the antide-pressant effect, which occurs at relatively high concentrations, can be utilized. Thus the desired effect cannot be obtained after the first dose, only in part for pharmacokinetic reasons.

4.2.6   Multiple doses

When a sustained effect is required from oral dosing, drugs are usually prescribed as divided doses, taken over a period of time. Steady-state conditions will apply for drugs that are eliminated according to first-order kinetics, as they do for infusions, but now the plasma concentrations will fluctuate between doses. Because of this fluctuation this situation is sometimes referred to as 'pseudo-steady-state', or the 'pharmacokinetic steady-state'. The rate at which steady-state is achieved is the same as for the infusion, >99% after $7 \times t_{1/2}$. This is demonstrated schematically in Figure 4.11. An initial injection gives a concentration of 100 units, which falls to 50 units in $1 \times t_{1/2}$, when the dose is repeated. This gives 150 units, which falls to 75 in the next half-life, a further injection takes the concentration to 175 units, and so on, until at steady-state the peak and trough concentrations are 200 and 100 units, respectively.

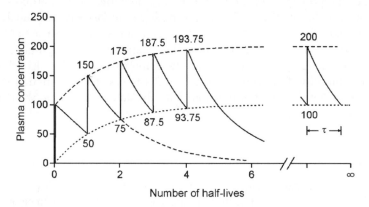

Figure 4.11   Modelled curve showing repeated i.v. injections at intervals equivalent to the half-life. Note the steeper decline in the curves as the concentration increases.

At steady-state, the maximum (peak) and the minimum (trough) concentrations after i.v. bolus doses can be calculated:

$$C_{\max}^{\text{ss}} = \frac{D}{V}\left(\frac{1}{1 - \exp(-\lambda\tau)}\right) \tag{4.36}$$

$$C_{\min}^{\text{ss}} = \frac{D}{V}\left(\frac{1}{1 - \exp(-\lambda\tau)}\right)\exp(-\lambda\tau) \tag{4.37}$$

where τ is the dosing interval. The 'average' steady-state is not the mean of the peak and trough concentrations but is based on the area under the curve as this is considered to give an estimate of the patient's exposure to the drug. It is the area under the curve divided by the dosage interval, τ, so from Equations 4.11 and 4.12:

$$C_{\text{av}}^{\text{ss}} = \frac{AUC}{\tau} = F\frac{C_0}{\lambda\tau} = \frac{FD}{V\lambda\tau} \tag{4.38}$$

The equation actually applies to both intravenous doses (when $F = 1$) and oral doses. With oral doses, because the average concentration is proportional to the dose and inversely proportional to τ then, for example, giving half the dose twice as often will give the same average concentration, but the peak to trough fluctuations will be less. It should be noted that the percent increase from the trough level in this situation is greater than the percent decrease from the peak. Thus, there is no validity in a statement such as: 'The concentration fluctuated by (X) percent.' Some authors favour the use of a 'fluctuation index, FI,' using a formula of the type:

$$FI = \frac{C_{max} - C_{min}}{C_{min}} \times 100 \tag{4.39}$$

where C_{max} and C_{min} are the peak and tough concentrations, respectively. There is, however, no unanimity on how such indices should be calculated, and one particularly interesting alternative uses the areas above and below the arithmetic or geometric mean steady-state concentration, so caution should be exercised in using the term fluctuation index.

Equations for multiple-dosing with first-order input (oral, i.m.) have been derived and can be found in standard texts such as Gibaldi and Perrier (1982). The situation is complex if the doses are given during the absorptive phase, however the equations that assume doses are given post-absorption are simpler and relatively easy to use. For example the concentration at any time t in the nth dosage interval is given by:

$$C_n = \frac{FDk_a}{V(k_a - \lambda)} \left[\left(\frac{1 - \exp(-n\lambda\tau)}{1 - \exp(-\lambda\tau)} \right) \exp(-\lambda t) - \left(\frac{1 - \exp(-nk_a\tau)}{1 - \exp(-k_a\tau)} \right) \exp(-k_a t) \right] \tag{4.40}$$

where τ is the dosing interval. A loading dose can be calculated:

$$\text{loading dose} = D \left(\frac{1}{1 - \exp(-\lambda\tau)} \right) \tag{4.41}$$

In this situation, D, the amount of drug given at the second and subsequent dose intervals, is often referred to as the 'maintenance' dose to distinguish it from the loading dose. The concentration−time curves of Figures 4.2 and 4.11 were simulated using Equations 4.40 and 4.41.

Note how in Figure 4.12 the drug with the shorter half-time approaches steady-state more quickly, as would be expected. However, the peak−trough fluctuations are greater because, having a shorter half-life, a

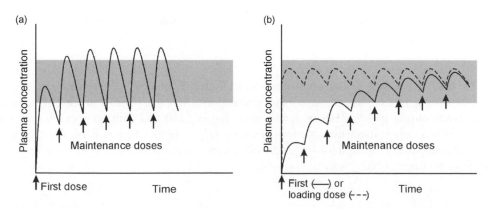

Figure 4.12 Examples of multiple dosing at regular intervals. (a) Drug with a short elimination half-life and (b) drug with a long elimination half-life, also showing the effect of a loading dose (broken line). The shaded area represents the therapeutic window.

greater amount of drug is eliminated between doses. As a consequence it may prove difficult to maintain the concentrations within a narrow therapeutic window. This situation can arise with morphine, when the patient may suffer pain at pre-dose, trough, concentrations and show signs of respiratory depression at peak concentrations. One way of dealing with such a situation is to give smaller doses more frequently as this reduces the difference between peak and trough levels whilst maintaining the same average concentration (Equation 4.38). An alternative is to use sustained-release preparations so that the 'effective' half-life is longer and the time course resembles that of Figure 4.12(b), when it takes longer to reach steady-state but the fluctuations are markedly reduced. A loading dose may be used to overcome any problems due to the delay in onset of effects.

Figure 4.13 shows the results of a simulation of lithium concentrations in a patient given 900 mg daily over a prolonged period of time. The half-life of lithium in this patient was 18 hours, the apparent volume of distribution was $0.66\,\mathrm{L\,kg^{-1}}$, and the clearance was $0.35\,\mathrm{mL\,min^{-1}\,kg^{-1}}$. The bioavailability ($F$) was unity. The figure shows the concentrations after the first dose, then the rise towards the pharmacokinetic steady-state – at this dosage in this patient the concentrations would have overshot the therapeutic range of 0.5–$1.25\,\mathrm{mmol\,L^{-1}}$ and the dose would have needed to be reduced in order to reduce the risk of toxicity.

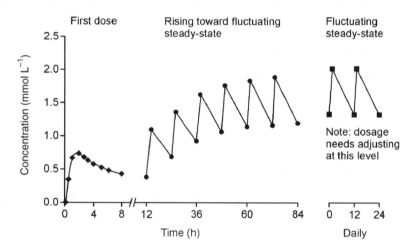

Figure 4.13   Calculation of lithium steady-state serum concentrations from the data of the first oral dose.

Whether or not fluctuation is desired often depends on the clinical objective. For example, a sleep aid drug taken night after night needs to have the maximum fluctuation achievable in order to prevent the occurrence of residual effects during the day. Hence, temazepam ($t_{1/2} \sim 12\,\mathrm{h}$) and with no active metabolites may be considered to be a more suitable hypnotic than diazepam ($t_{1/2} \sim 48\,\mathrm{h}$), which also has an active metabolite that accumulates in plasma (Section 6.2.4). On the other hand, an anticonvulsant is likely to be at its best if there is little or no fluctuation.

4.3   Non-linear kinetics

A feature of first-order kinetics is that C^{ss} is proportional to the dose administered; however there are situations when this is not the case and the kinetics are referred to as non-linear. The plasma concentrations may increase disproportionately or may not increase as much as would be predicted from linear models (Figure 4.14). Antiepileptic drugs are examples of drugs exhibiting non-linear kinetics. Phenytoin metabolism tends to saturation at doses in the therapeutic range, whist gabapentin shows saturable absorption. Valproic acid concentrations may be less than predicted due to dose-dependent binding to plasma proteins, although the clinical effects may be greater (Section 2.5.3). Carbamazepine concentrations fall on repeated dosing because of auto-induction of enzymes. Phenobarbital shows linear kinetics at therapeutic doses but non-linear kinetics in overdose.

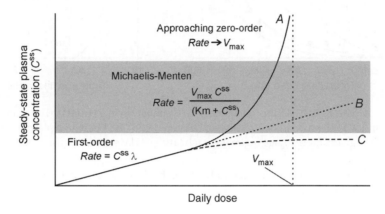

Figure 4.14 Comparison of non-linear kinetics. A: saturation of elimination, B: linear kinetics, C: saturation of absorption. Shaded area represents the therapeutic window.

Because half-life is dependent on apparent volume of distribution and clearance, non-linearity must be a result of non-linearity in one or both of these variables. The rate of elimination is $C \times CL$ (Equation 4.3), but for a drug eliminated according to Michaelis–Menten kinetics the rate is also given by Equation 3.1, so putting the equations equal to each other and dividing each side by C gives:

$$CL = \frac{V_{max}}{Km + C} \tag{4.42}$$

As concentration increases to the point where $C \gg Km$ (zero-order case) then:

$$CL = \frac{V_{max}}{C} \tag{4.43}$$

Thus, for non-linear kinetics where enzyme is becoming saturated with substrate, clearance will decrease with increasing drug concentrations. In the case of valproic acid, concentration dependent changes in protein binding result in concentration dependant changes in the apparent volume of distribution and hence nonlinear kinetics.

4.3.1 *Phenytoin*

Non-linear kinetics, such as those exhibited by phenytoin, can be modelled using the Michaelis–Menten equation. At low doses the kinetics approximate to first-order but the kinetics at higher doses can only be adequately described by Equation 3.1. As the dose is further increased the kinetics approach zero-order. This can result in C^{ss} increasing very rapidly for small increases in dose, making it difficult the maintain concentrations in the therapeutic window (Figure 4.14). Individual variations in apparent Km and V_{max} complicate things further [Figure 4.15(a)]. If it is necessary to individualize the dose a modification of

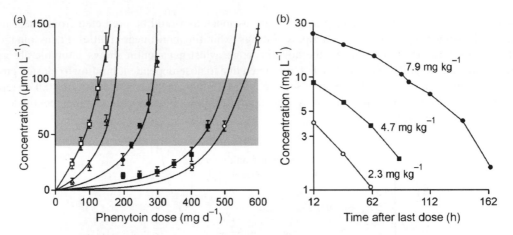

Figure 4.15   (a) Steady-state serum concentrations of phenytoin in 5 patients. The shaded area is the therapeutic window. (Redrawn from Richens and Dunlop, 1975) (b) Phenytoin plasma concentrations in a single subject given three dosage regimens, each lasting 3 days. (After Gerber and Wagner, 1972).

Equation 3.3 (Eadie–Hofstee) can be used. At steady-state, the daily dose, R, can be used as an estimate of the rate of elimination (mg day^{-1}) so:

$$R = V_{max} - \frac{R}{C^{ss}} Km \qquad\qquad (4.44)$$

Because there are two unknowns, C^{ss} values must be obtained for at least two dosage rates, when Equation 4.44 can be solved for V_{max} and Km. Alternatively a plot of R/C^{ss} versus R is a straight line of slope $-Km$ and y-intercept, V_{max}. The values can be used to calculate a daily dose that will give a concentration in the therapeutic range for that particular patient.

Non-linear kinetics are apparent from curvature of the ln C verses t plot [Figure 4.15(b)]. At high concentrations, the rate of elimination may be approaching V_{max} but, because the *proportionate* change is less than that at lower concentrations, the slope is less. As the concentration falls, the rate is less but the proportionate change tends to become constant as the kinetics approximate to first-order, when the slope of lines of a semilogarithmic plot will be the same, irrespective of the initial concentration.

4.3.2   Ethanol

Ethanol is metabolized primarily by liver alcohol dehydrogenase (ADH), although other routes of elimination exist, including metabolism by other enzymes, and excretion of unchanged drug in urine and expired air. Because the Km of human ADH is very low, and the 'dose' of ethanol, by pharmacological standards is high, the enzyme becomes saturated even with moderate drinking. In the 1930s Erik Widmark used a single-compartment model and zero-order kinetics to model blood ethanol concentrations. However, with advances in analytical methods it became clear that at low concentrations the kinetics were more akin to first-order decay. This led to the suggestion that at concentrations above $0.2\,\mathrm{g\,L^{-1}}$ the kinetics were zero-order whilst those below this concentration were first-order. Inspection of Equation 3.1 clearly shows that there cannot be an abrupt switch from first-order to zero-order kinetics and this can be demonstrated by substituting values into the equation (Table 4.2). If first-order kinetics applied, then the values calculated below $0.2\,\mathrm{g\,L^{-1}}$ would be directly proportional to the concentration. Clearly they are not. Similarly, if the kinetics were zero-order above $0.2\,\mathrm{g\,L^{-1}}$, then the rate would be constant. Again this is not the case, although the rate does asymptote to V_{max} at very high concentrations.

Table 4.2 Calculated rates of ethanol metabolism as a function of concentration

C (g L^{-1})	0.01	0.02	0.05	0.1	0.2	0.04	0.8	1.6	4.0	8.0
Rate (g h^{-1})	0.73	1.33	2.67	4.0	5.33	6.4	7.11	7.54	7.8	7.9

$Km = 0.1\,g^{-1}$, $V_{mx} = 8\,g\,h^{-1}$

In a theoretical decay curve for ethanol [Figure 4.16(a)], the initial part of the curve is almost linear, reflecting approximately zero-order elimination, but the line becomes progressively more curved. Plotting the data with a logarithmic y-axis [Figure 4.16(b)] gives the typical shape expected for Michaelis–Menten elimination kinetics. At high concentrations the elimination rate approaches V_{max}, but the *proportionate* change is small, hence the shallow slope of the log-transformed curve. At later times (lower plasma concentrations) the rate of elimination is much lower, but the proportionate change is greater and will become constant when the elimination is first-order.

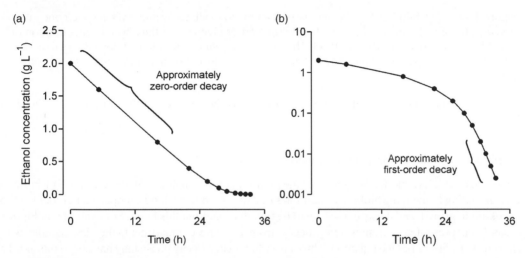

Figure 4.16 Simulated decay curve for ethanol ($C_0 = 2\,g\,L^{-1}$, $V_{max} = 8\,g\,h^{-1}$, $Km = 0.1\,g\,L^{-1}$).

4.4 Relationship between dose, onset and duration of effect

For a drug exhibiting single-compartment kinetics there are some simple theoretical relationships between the dose given and the duration and intensity of effect. Assuming that after an intravenous dose the effect ceases at a concentration C_{eff}, then the duration of effect increases by $1 \times t_{1/2}$ for every doubling of the dose [Figure 4.17(a)]. If the y-axis were dose and a horizontal line was drawn representing the duration of action then, the duration of action of the lowest dose would be less than that at higher doses, which would double with each doubling of the dose. Similarly, if at some concentration or time after the injection the effect wears off and the dose is repeated then the time for the second dose to wear off will be longer, because there is now more drug in the body. However if subsequent doses are given the time to reach the concentration at which the effect ceases is now constant between doses [Figure 4.17(b)]. If a horizontal line were drawn on Figure 4.17(a) then the duration of action of the lowest dose would be less than that at higher doses.

In the discussion above it has been assumed that the onset of effect was more or less instantaneous. The situation is more complex if the dose is not an i.v. bolus. As can be seen from Figure 4.1, increasing the size of

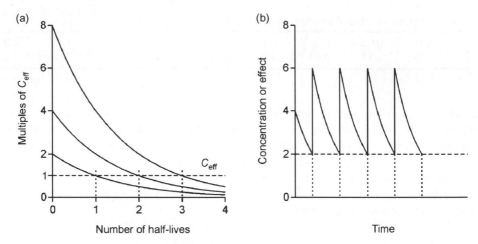

Figure 4.17   (a) Doubling the dose increases the time to reach the minimum effective concentration, C_{eff} by one elimination half-life. (b) Multiple doses given when the plasma concentration or effect has fallen to some arbitrary level. The duration of effect after the first dose is less than that after the second and subsequent doses. After the second dose the maximum concentration/effect is the same.

an oral dose decreases the delay before the onset of effects, but predicting the duration of effect is more difficult.

4.5   Limitations of single-compartment models

The astute reader will have noticed that most of the figures in this chapter use modelled data. This reflects the difficulty in finding convincing biological examples of data which fit a single-compartment model, although pharmacokinetic growth and decay curves are to be found throughout this book. For example, in Figure 4.6, which has been presented as representing decay from a single-compartment, the first sample was not collected until one hour after the injection. Thus, an earlier, faster, decay phase may have been missed. This is supported by the fact that the apparent volume of distribution, calculated using Equation 2.4, is ~50 L. This is larger than the volume of plasma or extracellular fluid, with which drug may be expected to equilibrate rapidly. In fact, it approximates to total body water, including water in tissues, into which a drug will penetrate relatively slowly, so in all probability analysis of earlier samples would have revealed another exponential decay representing distribution of the drug into tissues. Even ethanol concentrations, which are frequently treated as if they conform to single-compartment kinetics, show multi-exponential decay after intravenous infusion. Because it is now recognized that the timing of samples may influence conclusions about the most appropriate compartmental model, more recent studies are usually designed with sampling at sufficiently early collection times to define any distributional phase(s).

A related problem arises because of inhomogeneity of plasma. Thus absorption and metabolism/excretion do not take place in the same location within the single compartment. The drug must pass from one site to another through the very blood that is sampled, usually from the antecubital vein within that compartment, raising a question that is more than theoretical about the validity of the concept of homogeneity in this case. This is considered further in Chapter 7.

There are examples of what appear to be single-compartment models when some drugs are given orally or by i.m. injection. This is because the absorption phase masks more complex kinetics, as with fluphenazine enanthate described above [Figure 4.8(a)]. In such cases there is no alternative but to analyse the data as if from a single-compartment, even if other routes of administration (e.g. intravenous) show otherwise.

These situations may be referred to as apparent or pseudo single-compartment cases. Consequently, it is not unusual to find examples in the literature, for individual drugs, in which biexponential decay equations (two-compartment model) are used for intravenous doses, and biexponential growth and decay equations (one-compartment model) are used for oral doses.

4.6 Summary

The fundamental concepts of pharmacokinetics have been described including half-life, clearance, zero-order and first-order input into a single-compartment model, multiple dosing and attainment of steady-state. However, because single-compartment models are of limited use in describing the plasma concentration–time relationships for the majority of drugs, more complex models are required. These are the subject of the next chapter.

References and further reading

Birkett DJ. *Pharmacokinetics Made Easy.* Sydney: McGraw-Hill, 1998 (Revised 2002).

Clark B, Smith DA. *An Introduction to Pharmacokinetics*, 2nd edn. Oxford: Blackwell, 1986.

Curry SH. *Drug Disposition and Pharmacokinetics*, 3rd edn. Oxford: Blackwell Scientific, 1980.

Curry SH. *Clinical Pharmacokinetics: The MCQ Approach.* New Jersey: Telford Press, 1988.

Curry SH, Whelpton R. *Manual of Laboratory Pharmacokinetics.* Chichester: John Wiley & Sons, 1983.

Curry SH, Whelpton R, de Schepper PJ, Vranckx S, Schiff AA. Kinetics of fluphenazine after fluphenazine dihydrochloride, enanthate and decanoate administration to man. *Br J Clin Pharmacol* 1979; 7: 325–31.

Dost FH. *Der Blutspiegel – Kinetik der Konzentrationsablaufe in der Kreislaufflussigkeit.* Leipzig: Georg Thieme, 1953.

Gambertoglio JG, Amend WJ, Jr., Benet LZ. Pharmacokinetics and bioavailability of prednisone and prednisolone in healthy volunteers and patients: a review. *J Pharmacokinet Biopharm* 1980; 8: 1–52.

Gerber N, Wagner JG. Explanation of dose-dependent decline of diphenylhydantoin plasma levels by fitting to the integrated form of the Michaelis-Menten equation. *Res Commun Chem Pathol Pharmacol* 1972; 3: 455–66.

Gibaldi M. *Biopharmaceutics and Clinical Pharmacokinetics*, 4th edn. Philadelphia: Lea and Febiger, 1991.

Gibaldi M, Perrier D. *Pharmacokinetics*, 2nd edn. New York: Marcel Dekker, 1982.

Kaltenbach ML, Curry SH, Derendorf H. Extent of drug absorption at the time of peak plasma concentration in an open one-compartment body model with first-order absorption. *J Pharm Sci* 1990; 79: 462.

LaDu BN, Mandel HG, Way EL (eds) *Fundamentals of Drug Metabolism and Disposition.* Baltimore: Williams & Wilkins, 1971.

McQueen EG, Wardell WM. Drug displacement from protein binding: isolation of a redistributional drug interaction in vivo. *Br J Pharmacol* 1971; 43: 312–24.

Nelson E. Kinetics of drug absorption, distribution, metabolism, and excretion. *J Pharm Sci* 1961; 50: 181–92.

Ritschel WA. *Handbook of Basic Pharmacokinetics*, 3rd edn. Hamilton: Applied Therapeutics, 1986 (revised 2009).

Ritschel WA. *Graphic Approach to Pharmacokinetics.* Barcelona: J.R.Prous, 1983.

Richens A, Dunlop A. Serum-phenytoin levels in management of epilepsy. *Lancet* 1975; 2: 247–8.

Rowland M, Tozer TN. *Clinical Pharmacokinetics: Concepts and Applications*, 3rd edn. Media, Pennsylvania: Williams & Wilkins.

Schoenwald RD. *Pharmacokinetics in Drug Discovery and Development.* Boca Raton: CRC Press, 2002.

Shargel L, Wu-Pong S, Wu ABC. *Applied Biopharmaceutics and Pharmacokinetics*, 5th edn. New York: McGraw-Hill, 2004.

Teorell T. Kinetics of distribution of substances administered to the body I. The extravascular modes of administration. *Arch Int Pharmacodyn Ther* 1937; 57: 205–25.

Teorell T. Kinetics of distribution of substances administered to the body. II. The intravascular modes of administration. *Arch Int Pharmacodyn Ther* 1937; 57: 226–40.

Tozer TN, Rowland M. *Introduction to Pharmacokinetics and Pharmacodynamics.* Philadelphia: Williams & Wilkins, 2006.

Wagner JG. *Biopharmaceutics and Relevant Pharmacokinetics*. Hamilton: Drug Intelligence Publcations, 1971.

Widmark EMP. *Die theoretischen Grundlagen und die praktische Verwendbarkeit der gerichtlich-medizinischen Alkoholbestimmung*. Berlin: Urban and Schwarzenberg, 1932. Republished as: Widmark EMP. *Principles and applications of medicolegal alcohol determination*. Davis: Biomedical Publications, 1981.

5

More Complex and Model Independent Pharmacokinetic Models

5.1 Introduction

In the previous chapter the fundamental principles of pharmacokinetics were discussed using the simplest of compartmental models. However, it is often the case that the plasma concentration–time data cannot be explained adequately using a single-compartment model, indeed it is difficult to find practical examples of drugs that conform to single-compartment pharmacokinetics after intravenous bolus doses. Generally, more complex, multi-compartment models are required, or it may not be possible to define a compartment model at all and other approaches are required. These are the topics of this chapter.

5.2 Multiple compartment models

5.2.1 Intravenous injections

Quite commonly, when the concentration data are plotted on logarithmic scales, the plot is curvilinear and may be described as the sum of two or more exponential decays. In this case a multiple-compartment model can be used [Figure 5.1(a)]. This is sometimes described as 'the drug conferring on the body the characteristics of a two-(or multi-) compartment model'.

The general equation for the plasma concentration, C, as a function of time in a multiple-compartment model, following an intravenous bolus injection is:

$$C = \sum_{i=1}^{n} C_i \exp(-\lambda_i t) \tag{5.1}$$

This is exemplified in Figure 5.2 which shows the blood concentrations of methylene blue after a bolus intravenous injection of 100 mg. Methylene blue is of some importance in medicine as a contrast agent and for evaluating body fluid volumes. However, it is also a life-saving drug used as an electron donor in the treatment of both congenital and drug-induced methaemoglobinaemia. It shows multiphasic decline of concentrations in plasma (Peter *et al.*, 2000).

Drug Disposition and Pharmacokinetics, By Stephen H. Curry and Robin Whelpton
© 2011 John Wiley & Sons, Ltd

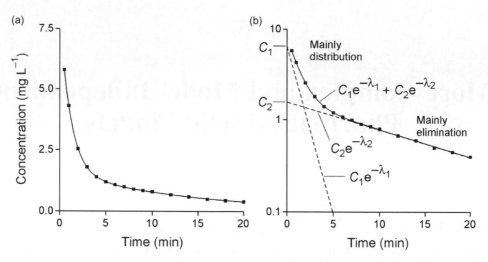

Figure 5.1 Example of data that fits a two-compartment model (a) Cartesian coordinates and (b) semi-logarithmic plot.

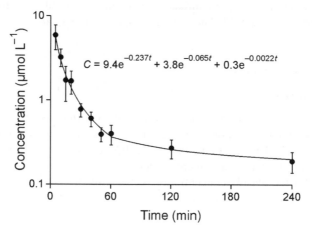

Figure 5.2 Blood concentrations of methylene blue (mean ± SEM, $n = 7$) after 100 mg intravenously. Data fitted using sum of three exponential terms (solid line). (Redrawn from Peter *et al.*, 2000).

For a two-compartment model, $n = 2$, and Equation 5.1 can be written:

$$C = C_1 \exp(-\lambda_1 t) + C_2 \exp(-\lambda_2 t) \tag{5.2}$$

Sometimes (particularly in American literature) Equation 5.2 is written as:

$$C = A \exp(-\alpha t) + B \exp(-\beta t) \tag{5.3}$$

Note that $\lambda_1 > \lambda_2$ ($\alpha > \beta$), as λ_1 (or α) is the rate constant of the steeper part of the decay curve. Indeed this initial steep part may be referred to as the 'α-phase'. The last exponential defines the *terminal* phase, and by convention the rate constant of this phase is λ_z. Thus, for a two-compartment model, $\lambda_2 = \beta = \lambda_z$. An example of a two-compartment model is depicted in Figure 5.3. Rather than the one homogeneous solution of drug of the single-compartment, this model has two solutions, with reversible transfer of drug between them. By definition, plasma is always a component of the central compartment, sometimes called the plasma compartment. The other compartment is the peripheral or tissue compartment. The volumes of these compartments are V_1 and V_2 respectively. The use of other terms such as V_p, can lead to untold confusion – is

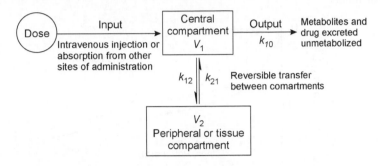

Figure 5.3 Diagrammatic representation of a two-compartment model.

that p for plasma or p for peripheral? The rate constants are labelled according to the direction of movement to which they refer, so the constant relating to transfer from the central (1) to the peripheral compartment (2) is k_{12}. The rate constant for loss from the central compartment by metabolism or excretion of unchanged drug is k_{10} (Figure 5.3) where 0 depicts compartment zero, that is outside the body. Other two-compartment models are possible where drug is metabolized or excreted from the peripheral compartment, or from both central and peripheral compartments. However, the liver and kidneys are often part of the central compartment so the model of Figure 5.3 is usually appropriate.

The shape of the curve of Figure 5.1(b) can be explained as follows:

- Shortly after injection there is rapid transfer of drug from plasma, in the central compartment, to the tissues of the peripheral compartment, with only a small proportion of the dose being eliminated during this time.
- At later times the concentration of drug in the compartments equilibrate and the decay from plasma is chiefly a result of elimination of drug from the body as a whole either by metabolism or excretion of unchanged drug.

5.2.1.1 *Concentrations in the peripheral compartment*

When the drug is injected at $t = 0$, instantaneous mixing in the central compartment is assumed. The initial concentration, $C_0 = C_1 + C_2$, and there is no drug in the peripheral compartment. Consequently, the concentration in the peripheral compartment increases until the forward and backward rates of transfer are equal and, for an instant, there is no net movement of drug. This condition is referred to as steady-state (not to be confused with the concept of steady-state discussed in the previous chapter in which the term was applied to equilibrium concentrations in plasma during long-term dosing), after which net transfer is from the peripheral to the central compartment, because elimination is reducing the concentration in the central compartment. Normally, the amount or concentration in the peripheral compartment is calculated using parameters derived from the plasma concentration–time data:

$$C_{\text{periph}} = \frac{D\,k_{12}}{V_2(\lambda_1 - \lambda_2)}\left[\exp(-\lambda_2 t) - \exp(-\lambda_1 t)\right] \tag{5.4}$$

Unlike the biological situation, it is possible to demonstrate this relationship, using the dyes as models as in the previous chapter. For example, methylene blue (see Figure 5.2) is a phenothiazine dye with an intense blue colour – it was the dye mentioned in Chapter 4. Using the models described earlier, it was possible to generate the data in Figure 5.4 for distribution of methylene blue through a two-compartment model. This emphasizes the fact that the steady-state in question in this context is characterized by there being an instant in time when the concentrations in the two compartments are identical and there is no net movement of drug between the compartments.

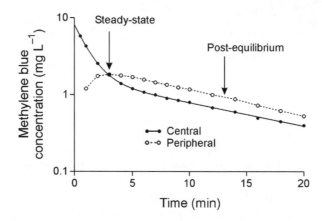

Figure 5.4 Simulation of a two-compartment compartment model showing methylene blue concentrations in the central and peripheral compartments.

The central and peripheral compartment curves cross when the concentration in the peripheral compartment is maximal. This must be the case because this is steady-state and the concentrations are equal. The time of the peak, t_{max} is:

$$t_{max} = \frac{1}{(\lambda_1 - \lambda_2)} \ln\left(\frac{\lambda_1}{\lambda_2}\right) \tag{5.5}$$

At later times (in theory, only at infinite time) the compartments equilibrate and the *ratio* of concentrations remains constant – shown as a parallel decline on a semilogarithmic plot (Figure 5.4).

5.2.1.2 Microconstants

The rate constants of Figure 5.3 are known as microconstants and have to be derived from C_1, C_2 λ_1 and λ_2, which are obtained by resolving the plasma concentration–data into its two exponential terms (Section 5.2). The sum of the microconstants equals the sum of λ_1 and λ_2:

$$\lambda_1 + \lambda_2 = k_{10} + k_{12} + k_{21} \tag{5.6}$$

and

$$k_{21} = \frac{C_1 \lambda_2 + C_2 \lambda_1}{C_1 + C_2} \tag{5.7}$$

$$k_{10} = \frac{\lambda_1 \lambda_2}{k_{21}} \tag{5.8}$$

$$k_{12} = \lambda_1 + \lambda_2 - k_{10} - k_{21} \tag{5.9}$$

Confusion sometimes arises over the relationship between k_{10} and λ_2. The former is the rate constant for loss of drug from the central compartment and is greater than λ_2, which is the rate constant for loss of drug from the body and is derived from the terminal phase of the plasma concentration–time curve once equilibrium has been fully established. The value of λ_2 may be influenced by the other rate constants, particularly k_{21} which controls the rate of return of drug from tissues to the central compartment.

The importance of calculating the microconstants is that they are required to calculate the apparent volumes of distribution. Also, k_{10} and λ_2 are affected differently by diseases and in 'special populations' (Chapters 9, 11 and 12) and this can be very important in understanding requirements for dosing in such populations.

5.2.1.3 Apparent volumes of distribution

Adopting the same approach as that for the single-compartment model, the volume of the central compartment is:

$$V_1 = \frac{D}{C_0} = \frac{D}{C_1 + C_2} \tag{5.10}$$

because at $t = 0$ none of the dose has been transferred to the peripheral compartment, and C_0 is the sum of the constants of the two exponential terms. At steady-state, the concentrations in each compartment are equal, and the forward and backward rates are equal as there is (instantaneously) no net movement of drug:

$$V_1 C_{ss} k_{12} = V_2 C_{ss} k_{21} \tag{5.11}$$

Cancelling C^{ss} and rearranging gives:

$$V_2 = V_1 \frac{k_{12}}{k_{21}} \tag{5.12}$$

The apparent volume of distribution at steady-state, V_{ss} is the sum of apparent volumes of the individual compartments:

$$V_{ss} = V_1 + V_2 \tag{5.13}$$

The volume of distribution throughout the body at steady-state indicates the extent to which the drug is distributed throughout the body. In the methylene blue experiment it is in good agreement with the notional total volume (Table 5.1). However, it is clear from Figure 5.4 that the only time that multiplying V_{ss} by the plasma concentration will give the true amount of drug in the body is that instant at which steady-state conditions occur. Post-equilibrium, using V_{ss} will give an underestimate of the amount of drug in the body

Table 5.1 Results from i.v. injection of methylene blue into a two compartment model

Parameter	Found	Parameter	Found	Parameter	Found	Notional
Dose (mg)	4^a	C_0 (mg mL^{-1})	8.13	V_1 (mL)	492	500
C_1 (mg mL^{-1})	6.56	k_{21} (min^{-1})	0.219	V_2 (mL)	965	1000
λ_1 (min^{-1})	0.847	k_{10} (min^{-1})	0.267	V_{ss} (mL)	1457	1500
C_2 (mg mL^{-1})	1.57	k_{21} (min^{-1})	0.430	V_{Area} (mL)	1902	n.a.^{b}
λ_2 (min^{-1})	0.0691			CL (mL min^{-1})	131.5	130

^{a}Not calculated.
^{b}Not applicable.

because the peripheral concentrations are higher than those in plasma. Consequently a further calculation of apparent volume of distribution is required, V_{Area}:

$$V_{Area} = \frac{D}{AUC\, \lambda_2} \tag{5.14}$$

This apparent volume of distribution is best thought of as a constant of proportionally that allows the amount of drug in the body to be calculated post-equilibrium. Furthermore, V_{area} changes with changes in clearance (which changes λ_2). This must be the case because considering Figure 5.4, as clearance increases the plasma concentrations fall but the concentrations in the peripheral compartment do not fall to the same extent and so V_{area} is greater (Table 5.1). If there was no clearance the drug would not be removed from the body and the concentrations in the central and peripheral compartments would be equal, that is $V_{area} = V_{ss}$.

The term V_{extrap}, which uses a construction line through the terminal phase of the plasma concentration data extrapolated to $t = 0$ may be encountered, particularly in older literature. In the example of Figure 5.3, the intercept is C_2, and $V_{extrap} = D/C_2$, is analogous to the one-compartment case. This approach overestimates the apparent volume of distribution, particularly when V_2 is large relative to V_1. For example using the data of Table 5.1 D/C_2 calculates to be 2550 mL, larger than both V_{ss} and V_{area}.

5.2.1.4   Clearance

The area under the curve is the sum of the areas under the two exponential phases so, by analogy with Equation 4.11, for a two compartment model:

$$AUC = \frac{C_1}{\lambda_1} + \frac{C_2}{\lambda_2} \tag{5.15}$$

Several approaches may be used to calculate clearance. Equation 4.14 is applicable to single- and multiple-compartment models, and if AUC is obtained using the trapezoidal method it is not even necessary to define the number of compartments (Section 5.3):

$$CL = \frac{D}{AUC} \tag{4.14}$$

As the rate of elimination from the body $= C \times CL$, where C is the plasma concentration (or the concentration in V_1) then the rate of elimination from/via the central compartment/plasma is the amount ($V_1 \times C$) multiplied by the elimination rate constant, k_{10}:

$$C \times CL = V_1 \times C \times k_{10} \tag{5.16}$$

Cancelling C from each side gives:

$$CL = V_1 k_{10} \tag{5.17}$$

Post-equilibrium, it has been shown (Gibaldi and Perrier, 1982) that:

$$V_{area}\, \lambda_2 = V_1 k_{10} \tag{5.18}$$

and substituting Equation 5.17 gives:

$$CL = V_{area}\, \lambda_2 \tag{5.19}$$

The amount of drug in the body post-equilibrium is V_{area} multiplied by the plasma concentration so the rate of elimination, *post-equilibrium* is: $(V_{area} \times C)\, \lambda_2$.

5.2.2 Absorption

The addition of an absorption phase to the arithmetic of a two-compartment model leads to an extremely complex situation. A working equation for first-order absorption into a two compartment model is:

$$C = C_1'\exp(-\lambda_1 t) + C_2'\exp(-\lambda_2 t) - (C_1' + C_2')\exp(-k_a t) \tag{5.20}$$

The values C_1' and C_2' are not the same as C_1 and C_2 as they are affected by both the rate and extent of absorption. The situation is further complicated by the relative sizes of the rate constants. If the rate constant of absorption is large then it may be possible to resolve the data and obtain estimates of k_a, λ_1 and λ_2 (Figure 5.4). However, it is often not possible to ascribe values to k_a and λ_1 without investigating the kinetics using an intravenous dose. Furthermore, when $k_a < k_{21}$ the distributional phase is not apparent and the model appears to be that of a single compartment (Figure 5.5). In this situation one could not define a two compartment model without also using an intravenous dose.

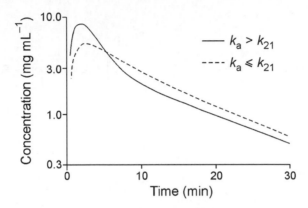

Figure 5.5   Absorption into a two-compartment model.

5.2.3 Infusions

When a drug is infused into a two-compartment model the shape of the rising phase is the inverse of the decay curve at the end of the infusion, just as it was in the single compartment case (Section 4.2.5). Thus the rising phase is composed of two exponential terms, with an initial steep rise. When the infusion is stopped it may be possible to resolve the decay curve into two exponential terms, depending on the relative sizes of C_1 and C_2. The initial steep phase is not as obvious as that after an intravenous bolus injection because during the infusion drug has been transferred to the peripheral compartment and the concentration gradient between the two compartments is not as great, particularly as the concentration approaches steady-state. When the infusion is stopped the exponential curves can be extrapolated to give intercepts, C_1' and C_2' on a vertical construction line drawn at the time the infusion was stopped. If the infusion were continued to steady-state the relationships to C_1 and C_2 are:

$$C_1 = \frac{D\,C_1'\lambda_1}{R_0} \tag{5.21}$$

and

$$C_2 = \frac{D\,C_2'\lambda_2}{R_0} \tag{5.22}$$

where D is the accumulated dose during the infusion (i.e. R_0T). As previously, R_0 is the zero-order infusion rate.

It is possible to calculate a loading dose, R_0/k_{10}, to give an instantaneous steady-state concentration, C^{ss}, in plasma. However, this ignores the transfer of drug from the central to the peripheral compartment and so initially the plasma concentration falls and then rises and asymptotes to C^{ss}. Again this can be demonstrated empirically using methylene blue [Figure 5.6(a)]. To avoid the plasma concentrations during the infusion falling below C^{ss}, a larger loading dose, R_0/λ_2, has to be used, but this may result in the initial concentrations being unacceptably high [Figure 5.6(b)]. This problem is of practical importance for some drugs, for example when lidocaine is used to control ventricular arrhythmias. One solution is to use a 'loading infusion', that is to infuse at a higher rate initially and then reduce the rate to that required for the desired steady-state concentration. An alternative scheme, using intravenous administration of an initial bolus loading dose in conjunction with a constant rate and an exponential intravenous drug infusion has been proposed (Vaughan and Tucker, 1976).

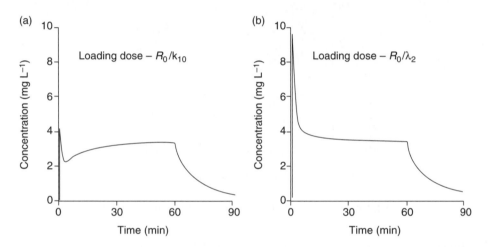

Figure 5.6   Infusion into a two-compartment model with different loading doses. (a) R_0/k_{10} gives instantaneous steady-state concentration, but this falls as drug is transferred to the peripheral compartment. (b) A loading dose R_0/λ_2 ensures the concentrations do not fall below C^{ss}, but the /peak concentration may be unacceptably high.

5.2.4   Multiple oral dosing

The principles outlined in Chapter 4 apply. On repeated dosing the plasma concentrations will increase and tend towards steady-state conditions. As before the situation is simpler at steady-state when following repeated intravenous doses:

$$C = C_1'\exp(-\lambda_1 t) + C_2'\exp(-\lambda_2 t) \tag{5.23}$$

where t is any time during the dosage interval and

$$C_1' = C_1\left(\frac{1}{1-\exp(-\lambda_1\tau)}\right) \tag{5.24}$$

and

$$C_2' = C_2 \left(\frac{1}{1 - \exp(-\lambda_2 \tau)} \right)$$ (5.25)

and τ is the dosing interval. The average plasma concentration at steady-state is given by:

$$C_{av}^{ss} = \frac{AUC}{\tau} = \frac{FD}{V_1 \, k_{10} \, \tau} = \frac{FD}{V_{Area} \lambda \tau}$$ (5.26)

See also Equation 4.38.

5.2.5 Concept of compartments

The concept of pharmacokinetic compartments may be difficult for anyone new to the topic. Understanding is not helped by the fact on some occasions the volume of a 'pharmacokinetic' compartment may be identical to a known anatomical volume such as plasma or total body water, but more often than not the calculated and anatomical volumes of distribution bear no comparison. Then there is the question of which tissues constitute a particular compartment. Generally, well-perfused tissues are components of the central compartment for lipophilic drugs as lipid membranes provide little in the way of a barrier to the movement of such drugs, and tissue and plasma concentrations rapidly equilibrate. Such tissues include liver and kidney and often brain (Figure 5.7). Thus, although the concentrations in individual tissues may be different, the kinetics describing the changes in concentrations are the same, and they decline in parallel on a semi-logarithmic plot (e.g. plasma and liver in Figure 2.13(a)). For a lipophilic drug, the rate of delivery (blood flow) to the tissue is important and so equilibration with the less well-perfused tissue is slower and apparent from the plasma concentration–time plot. In this situation, such tissues constitute the peripheral compartment and might include fat and muscle (Figure 5.7).

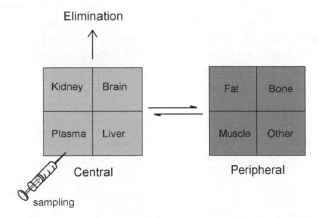

Figure 5.7 Representation of tissues that might constitute the central and peripheral compartments for a lipophilic drug.

It is worth remembering that (i) concentration–time plots, particularly with human subjects, are of plasma or blood, (ii) concentrations in tissues of the central compartment are usually higher than those in plasma and (iii) the concentrations in the peripheral compartment(s) will rise as drug is distributed from the central compartment.

5.2.6   *Relationship between dose and duration of effect*

The relatively simple relationship between dose and duration of effect for a single-compartment model was discussed in Section 4.4. The situation is more complex for multiple-compartment models but these are necessary to explain the duration of action of many drugs. This is best exemplified by thiopental, which clearly exhibits multiple-compartment kinetics (Figure 2.14). Liver concentrations rapidly equilibrate with those in plasma whilst the concentrations in skeletal muscle rise initially and then equilibrate. The concentrations in adipose tissue rise for at least the first 3.5 hours of the experiment. Therefore thiopental needs at least three compartments: a central and two peripheral compartments (Figure 5.8).

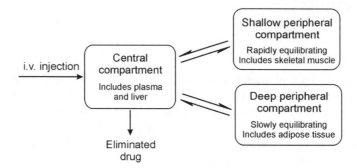

Figure 5.8   Thiopental kinetics require three compartments.

As discussed in Section 2.4.4.1, following an i.v. bolus injection of thiopental the duration of action is short due to uptake of the drug into skeletal muscle and fat. When the drug was originally introduced and larger doses were injected or, more particularly, when doses were repeated, the duration of action was found to be disproportionately long. When dosing was carried out cautiously, noting the patient's response, and using a flexible dosing policy, long-term anaesthesia could be safely maintained. However, incautious administration, generally with a fixed-dose regimen unrelated to patient response led to each successive dose exerting a duration of action longer than the previous one, until an excessively long duration of action occurred, even when administration ceased. It has been incorrectly assumed that this phenomenon is due to saturation of tissue stores leading to increasing proportions of subsequent doses being unable to redistribute from plasma and brain. In fact there is no evidence of tissue saturation (Table 5.2) and the observation can be explained in terms of a multiple-compartment model.

Table 5.2   Duration of action and tissue localization of thiopental at four doses in man (adapted from Curry, 1980)

Dose (g)	Administration time (min)	Duration (h)	Amount in blood at 1 h (g)	Amount in tissues at 1 h (g)	T/B^a
0.4	2	0.25	0.027	0.333	12.3
1	5	0.5–1	0.066	0.834	12.6
2	5	1.5–2.5	0.156	1.744	11.1
3.8	50	4–6	0.288	3.512	12.2

[a]Tissue to blood ratio.

Using composite data for plasma concentrations in human subjects from Brodie's work and *assuming* a two-compartment model (the original work could be fitted to a three-compartment model in agreement with the results for dogs) plasma concentration decay curves for 0.5 and 0.75 g doses were calculated

(Figure 5.9). The subject from whom the data of Figure 5.9 were derived, was anaesthetized with a dose of 0.5 g (i.v.) and regained consciousness after 20 minutes when the plasma concentration was $6.99 \, \text{mg L}^{-1}$. The kinetic model predicts that the subject would be unconscious for 2.8 h had the dose been increased by only 1.5-fold to 0.75 g. This is clearly a different situation from that predicted for intravenous injection into a single-compartment model (Section 4.4).

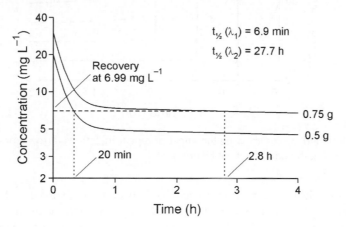

Figure 5.9 Time course of thiopental in plasma calculated from the composite data of Brodie following idealized doses of 0.5 and 0.75 g.

Additionally, the mean terminal half-life of thiopental has been shown to be different in lean and obese subjects, being 6.33 and 27.9 h, respectively, the variations being due to differences in apparent volumes of distribution rather than clearance (Jung *et al.*, 1982). Thus, the curve of Figure 5.9 may represent a more extreme situation than the average, but it does illustrate a general phenomenon applicable to all multiple-compartment models. It is apparent from the figure that the relative duration of effects will vary depending on the concentration at which recovery occurs. If recovery occurred post-equilibrium, during the terminal phase, then the relationship would be the same as that for the single-compartment model. Similarly, if recovery where largely during the distribution phase the increment in duration would not be so great, until of course the dose was increased to the point that recovery occurred in the terminal phase.

The effect of repeating the dose at the time of recovery, and at the same interval subsequently, can be modelled using:

$$C_n = C_1\left(\frac{1-\exp(-n\lambda_1\tau)}{1-\exp(-\lambda_1\tau)}\right)\exp(-\lambda_1 t) + C_2\left(\frac{1-\exp(-n\lambda_2\tau)}{1-\exp(-\lambda_2\tau)}\right)\exp(-\lambda_2 t) \qquad (5.27)$$

Where C_n is the concentration at time t following the nth dose, C_1 and C_2 are the y-intercepts, following the intravenous bolus injection, and τ is the dosage interval (20 min in this example). This is illustrated in Figure 5.10.

While it is the case that with subsequent doses the rate of transfer from the central to the peripheral compartment will decline as the concentration in the peripheral compartment increases, this should not be confused with *saturation* of tissues. Saturation would lead to a change in the rate constants. Furthermore, the clinical situation will be even more complex because of the potential influence of tolerance, both receptor and pharmacokinetic, and the presence of other drugs.

The thiopental research described above was of seminal significance in the pursuit of pharmacokinetic understanding during the development of intravenous anaesthesia and analgesia, and thiopental continues to

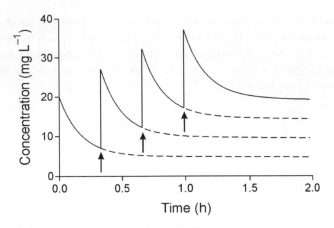

Figure 5.10   Modelled curves showing the effect of repeat injections (arrows) of thiopental at 20 minute intervals using the parameters of Figure 5.9.

be an anaesthetic of significance. Similar pharmacokinetic events are involved in the use of diazepam in status epilepticus, and in delirium tremens of alcohol withdrawal, with the added need in this case to avoid administration that is so rapid that the diazepam comes out of solution in the bloodstream before the injection solution is adequately diluted in the body. However, the most obvious modern example of a drug that follows the properties of the thiopental model is propofol, which is used as an intravenous sedative and anaesthetic, with or without other anaesthetics concurrently, particularly in out-patient procedures such as colonoscopies. Propofol concentrations rise rapidly during the early stages of infusion, then more slowly, even with constant rate infusion (as would be expected for a drug imparting the characteristics of a multiple-compartment model). The drug equilibrates very rapidly with the brain, so the effect is induced quickly, after which it is usually desirable to reduce the rate of infusion to maintain the effect while avoiding the excessive anaesthesia that could develop, relatively slowly, with prolonged administration. A distribution model analogous to that in Figure 5.8 has been devised for this drug. If drug infusion is discontinued after approximately 1 to 24 hours, the concentrations in the plasma decline rapidly and recovery is fast. If administration is for longer, for example for as long as 1 week, in the intensive care unit, the effect declines slowly on cessation of dosing as the terminal half-life is approximately 2 days (Knibbe *et al.*, 2000; Tozer and Rowland, 2006).

5.3   Curve fitting and choice of most appropriate model

Previously, reference has been made to resolving or fitting concentration–time data to obtain estimates of the pharmacokinetic parameters. However, discussion has been left until now because it easier to understand the underlying principles by considering resolution of the decay curve of a two-compartment model into its component exponential terms.

5.3.1   *Graphical solution: method of residuals*

Before the ready availability of personal computers and relatively inexpensive curve-fitting software, pharmacokinetic parameters were often obtained graphically. Although rarely used these days, an understanding of the approach is important when assessing the quality of the data to be analysed. The method of residuals, as it is known, is most easily understood from consideration of an intravenous bolus injection into a two-compartment model. The data are plotted as a semi-logarithmic graph [Figure 5.1(b)]. Because $-\lambda_1$ is the slope of the steeper, initial phase, $\lambda_1 > \lambda_2$ and so the term $C_1\exp(-\lambda_1 t)$ approaches zero faster than $C_2\exp(-\lambda_2 t)$. Therefore at later times the contribution from the first exponential term is negligible and Equation 5.2 approximates to:

$$C \cong C_2 \exp(-\lambda_2 t) \tag{5.28}$$

Provided the plasma concentration–time curve is monitored for long enough, the terminal portion of the $\ln C$ versus t curve, will be a straight line so C_2 and λ_2 can be estimated [Figure 5.1(b)], from the y-intercept and slope of the terminal phase. Values of $C_2\exp(-\lambda_2 t)$ are calculated for earlier time points [i.e. when $C_1\exp(-\lambda_1 t)$ is making a significant contribution to the plasma concentration] and subtracted from the experimental values at those times to give estimates of $C_1\exp(-\lambda_1 t)$, which are referred to as *residuals*. A semilogarithmic plot of the residuals should give a straight line of slope $-\lambda_1$ and intercept C_1. Concave curvature of the residual line indicates that the first estimate of λ_2 is too low, whilst convex curvature indicates that it is too high, assuming that the data fit a two-compartment model.

The method of residuals can be applied to the majority of compartment models, and also first-order input into single- or multiple-compartments, and the post-infusion phase of zero-order infusions. Clearly, the method could not be used for situations where $k_a = \lambda$ as there is no linear phase to which a construction line can be drawn [Figure 4.8(b)]. These data can be fitted iteratively as described below.

5.3.2   *Iterative curve-fitting*

There are several commercially available curve-fitting programs that are sold specifically for pharmacokinetic analyses. However, other packages may be adapted to derive pharmacokinetic parameters. Concentration–time data for an intravenous bolus injection into a single compartment can be transformed to $\ln C$ versus t and solved using linear regression programs available on many hand-held calculators (but see Section 5.2.2.1). Many relationships do not have a mathematical solution and have to be solved iteratively, usually by computing the equation which gives the lowest residual sum of squares. On this occasion 'residual' refers to the difference between the observed value and the value calculated from the derived parameters. The square of the residuals is used because the residuals will be positive or negative. A useful source code for iterative curve fitting is that written in BASIC by Neilsen-Kudsk (1983). This code can be modified to run under different forms of BASIC and it is relatively easy to modify so that it can used to fit a wide variety of equations including calibration curves, most compartmental pharmacokinetic relationships and pH-extraction curves (Whelpton, 1989). With this robust algorithm several sets of data can be fitted to common parameters, for example concentration-time data following oral and i.v. administration of a drug, or after modification, complex (biphasic) dose-response curves in the presence of several concentrations of a competitive antagonist (Patel *et al.*, 1995).

5.3.2.1   *Weighted-regression*

Experimental data will have associated errors. Linear regression assumes that the errors are the same irrespective of the concentration, that is, the data are homoscedastic. However, the size of the error generally increases with concentration – indeed most assays are developed to ensure that the relative standard deviations (RSD) cover a limited range and never exceed a predefined limit, say 10%. Therefore the errors are approximately proportional to the concentration and so the errors associated with high concentrations are higher and may unduly affect the way in which the data are fitted. Because unweighted regression treats all points equally, the fit will be biased towards the higher concentrations at the expense of the lower concentrations. This is particularly a problem with pharmacokinetic data as the concentrations often extend over several orders of magnitude. The answer is to weight the data and to minimize the sum of weighted squares:

$$\text{weighted sum of residual squares} = \sum[(\text{weight})(\text{residual})^2] \tag{5.29}$$

The size of the random errors is assessed by the variance (s^2) and, ideally, the data should be weighted by $1/s^2$, so Equation 5.29 can be written:

$$\text{weighted sum of residual squares} = \sum[1/s^2(\text{residual})^2] \tag{5.30}$$

Thus, the points with the lowest errors assume more importance than those with the largest errors. Ideally, the points should be weighted by 1/variance, but it is unlikely that this will have been measured at every concentration because that would require replicate assays of every sample. However, if the RSD is (approximately) constant over the concentration range, then s is proportional to concentration and the data can be weighted $1/(\text{concentration})^2$. Many commercially-available statistical programs allow the option of weighting data by $1/y^2$.

Weighting data is not some method of manipulating the result to make it appear more acceptable, *it is the correct statistical treatment for heteroscedastic data*. However, this being said, when the errors are small and the concentration range limited, then there may be little difference in the values obtained by using non-weighted or weighted data. It should be noted that if $\ln C$ values are fitted, the transformation weights the data and the results are likely to be different from those when C versus t is fitted.

5.3.2.2   Choice of model

The number of compartments required to fit the data is given by the number of exponential terms that describe the *declining* portion of the curve. The choice of how many compartments to fit should be dictated by the data. Statistical fitting allows the various equations to be compared. Simply choosing the equation which gives the lowest residual sum of squares (*SS*) is unhelpful because this will be the equation with the largest number of parameters. Consequently, most statistical packages compute 'goodness of fit' parameters which take into account the number of parameters in the equation. Some programs will fit two equations simultaneously and compare them using the F-test. Neilsen-Kudsk (1983) used the Akaike information criterion (*AIC*):

$$AIC = N\ln(SS) + 2M \tag{5.31}$$

where N is the number of data points and M is the number of parameters. The equation with the lowest *AIC* is statistically the most appropriate.

The appropriateness of the model, including the weighting, should be tested by plotting the residuals as a function of concentration. These should be randomly distributed about zero. The correlation coefficient, r, is often a poor indicator of the goodness of fit – unless $r^2 = 1$!

5.3.3   Quality of the data

Whether the parameters are derived using the method of residuals or iterative computer fitting the results will be poor if the original data are poor. With modern analytical methods one would expect the concentration data to be reasonably accurate. The chief reasons for poor results are:

• Insufficient number of data points
• Incorrect timing of collection
• Data not collected for sufficient time.

The three bullet points above are related. If the model requires a large number of parameters, for example absorption into a three compartment model will generate seven parameters, then, clearly, the number of points must be sufficient to ensure statistical significance – an absolute minimum of 8 (7 + 1). However, these points must be spaced at appropriate intervals so that each phase is defined, which is very unlikely to be the case with only eight samples. Another issue with timing is that the study design will usually specify the collection times, and these may be printed on the sample tubes prior to collection. It is not always possible to adhere to these times, which need not be a problem provided the correct time is recorded and used in the calculation. Use of uncorrected times will be more significant when the half-life is short. If the duration of the study is too short then the estimate of terminal rate constant, λ_z, is likely to be in error and this will be reflected in the estimates of the other parameters. The 'rule of thumb' is to collect data for at least $4 \times \lambda_z$. If a semi-logarithmic plot of the data does not show a more or less linear terminal phase then the results can be expected to be poor and using iterative fitting rather than the method of residuals will not improve them. It is a case of 'rubbish in = rubbish out'. For this reason data should always be plotted and inspected. In a study of temoporfin kinetics, the terminal half-life in blood was 13.9 h, when samples collected for up to 48 h were analysed; beyond which time the concentrations were too low to be quantified (Whelpton *et al.*, 1995). However, the decay in brain and lung was much slower suggesting that 13.9 h was far too low an estimate. A second study using ^{14}C-labelled temoporfin showed that the terminal half-life in blood was ~10 days (Whelpton *et al.*, 1996), a value that was confirmed from faecal excretion data collected for up to 5 weeks (Section 6.4). Some indication of for how long data should be collected can be obtained by modelling theoretical curves, as has been done for absorption into a two-compartment model (Curry, 1980).

5.4   Model independent approaches

5.4.1   Calculation of V_{Area} and clearance

Some of the pharmacokinetic parameters described above can be obtained without defining a model. Clearly, this applies to systemic availability, *F*, which is obtained by comparison of areas under the curve (Section 8.3). The *AUC*, calculated by the trapezoidal method, can used to derive systemic clearance and V_{Area} using Equations 4.26 and 5.14 or:

$$CL = F\frac{D}{AUC} \tag{4.26}$$

and

$$V_{Area} = F\frac{D}{AUC\,\lambda_z} \tag{5.14}$$

for extravascular doses. The data have to be such that the rate constant of the terminal phase can be derived. The approach can be illustrated using the data of Figure 5.5. Although it is not possible to define the correct number of compartments in one of the cases, the agreement between the parameters is good (Table 5.3). In

Table 5.3   Absorption of methylene blue into a two-compartment model with different rate constants of absorption

Parameter	Rapid absorption	Slow absorption
Terminal rate constant, λ_z (min^{-1})[a]	0.0679	0.0719
$AUC_{(0\text{-}30)}$ (mg min L^{-1})	68.5	67.9
C_{30}/λ_z (mg min L^{-1})	7.2	8.0
$AUC_{(0\text{-}\infty)}$ (mg min L^{-1})	75.7	75.9
CL/F (mL min^{-1})	132.1	131.8
V_{Area}/F (mL)	1946	1833

[a]From last five data points.

this particular instance we know that $F = 1$ and the clearance value is very close to the nominal flow rate of $130\,\text{mL min}^{-1}$.

Also, the average steady-state concentration can be obtained without defining number of compartments, Equation 5.26.

5.4.2 Statistical moment theory

Although statistical moment theory (SMT) is sometimes described as being non-compartmental, it does assume that the drug is measurable in plasma and also that the rate of elimination (flux) of drug is proportional to the plasma concentration (i.e. linear kinetics apply). The parameters that can be assessed are limited, but they are derived from measurements of areas under the curve without having to define compartments and assign, what can be ambiguous, rate constants to them. The method is useful when designing dosage regimens and interspecies extrapolations.

If one considers how long a single drug molecule stays in the plasma after administration, then it could be very short, very long or some intermediate interval, which is not particularly useful in itself. However, if large numbers of molecules are considered then there will be a mean residence time, *MRT*, which according to statistical moment theory is:

$$MRT = \frac{\int_0^\infty Ct\,\mathrm{d}t}{\int_0^\infty C\,\mathrm{d}t} = \frac{AUMC}{AUC} \tag{5.32}$$

where *AUMC* is known as the area under the (first) moment curve. The relationship between these areas is depicted in Figure 5.11.

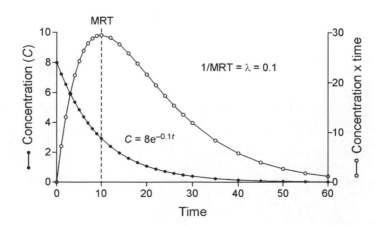

Figure 5.11 Model data for i.v. injection into a single-compartment, showing *C* versus *t* and *Ct* versus *t*.

For a bolus i.v. injection into a single-compartment model:

$$MRT_{i.v} = \frac{1}{\lambda} = \frac{V}{CL} \tag{5.33}$$

where λ is the rate constant of elimination. If comparing the results with a two-compartment model, then $MRT_{i.v.} = V_{ss}/CL$, and the apparent first-order rate constant has a value between λ_1 and λ_2. It follows from Equation 5.33 that MRT is the time when 63.2% of the intravenous dose will have been eliminated. This can be shown by substituting Equation 5.33 into Equation 1.8 when $t = MRT = 1/\lambda$:

$$C = C_0 \exp[-\lambda(1/\lambda)] = C_0 \exp(-1) = 0.368\, C_0 \tag{5.34}$$

Note that $\exp(-1) = 0.368$. As 36.8% remains, 63.2% must have been eliminated. The apparent volume of distribution at steady-state, V_{ss}, can be computed from:

$$V_{ss} = D\frac{AUMC}{AUC^2} = CL \times MRT \tag{5.35}$$

For first-order absorption, for example after an oral dose, the mean arrival time, MAT, can be used:

$$MAT = MRT_{p.o.} - MRT_{i.v.} \tag{5.36}$$

and

$$k_a = \frac{1}{MAT} \tag{5.37}$$

5.4.2.1 Estimating AUMC

The most obvious way of obtaining $AUMC$ is to use the trapezoidal method, analogous to that used for AUC (see Appendix). The area of the nth trapezium, $AUMC_n$, is:

$$AUMC_n = \frac{C_n + C_{n+1}}{2}(t_{n+1} - t_n)t_{n+1} \tag{5.38}$$

However, unlike AUC, the results are weighted by time, and the weight increases as the time increases. Also, Equation 5.38 overestimates the area, while using t_n underestimates $AUMC$. This can be reduced by using the midpoint for time:

$$AUMC_n = \frac{C_n + C_{n+1}}{2}(t_{n+1} - t_n)\frac{(t_n + t_{n+1})}{2} \tag{5.39}$$

Even so, Equation 5.39 overestimates the area for the decay part of the curve. A more accurate equation is:

$$AUMC_n = \frac{C_n(t_{n+1}^2 - t_n^2)}{2} + \frac{C_{n+1} - C_n}{6}(2t_{n+1}^2 - t_n t_{n+1} - t_n^2) \tag{5.40}$$

and, despite its complexity, is easy to use once it has been entered into an Excel spreadsheet. The area beyond the last time point is obtained by extrapolation:

$$AUMC_{(z-\infty)} = \frac{t_z C_z}{\lambda_z} + \frac{C_z}{\lambda_z^2} \tag{5.41}$$

where z denotes the last measured variable. Thus, despite being considered as a non-compartmental approach, it is necessary to calculate the terminal rate constant, which of course means identifying the last exponential phase.

5.4.2.2   *Example of application of SMT*

The data of Figure 5.1 were subjected to SMT (Figure 5.12). The first thing to note is that data have to be collected for a considerable time to define *AUMC* compared with that required to define two compartments from the *C* versus *t* data, as otherwise a large proportion of *AUMC* will be extrapolated. Obviously this has implications regarding the quality of the data because, in all probability, the errors will be highest at low concentrations and *AUMC* is weighted by the high values of *t*. Furthermore, the extrapolation of the remaining area requires accurate assessment of λ_z. Despite this, the results for the data of Figure 5.12 are in good agreement with those derived previously, reflecting the high quality of data (Table 5.4) in this case.

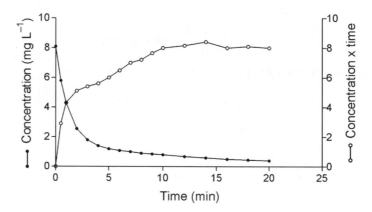

Figure 5.12   Statistical moment approach applied to the methylene blue data of Figure 5.1.

AUMC can be calculated for a compartmental model:

$$AUMC = \sum C_i / \lambda_i^2 \tag{5.42}$$

where *i* is the number of compartments. In the example of Table 5.4, Equation 5.42 gives $AUMC = 338.0\,\mathrm{mg\,min^2\,L^{-1}}$.

Table 5.4   SMT approach to calculating pharmacokinetic parameters

Areas			Derived parameters	
Range (min)	*AUMC* (mg min^2 L^{-1})	*AUC* (mg min L^{-1})		
0–20	141.4[a]	25.0	*MRT* (min)	11.08
20–∞	199.5[b]	5.8	*CL* (mL min^{-1})[c]	130
0–∞	340.9	30.8	V_{ss} (mL)	1440

[a]Using Equation 5.40.
[b]From Equation 5.41.
[c]From Equation 5.35.

Thus the MRT is very easy to calculate, using only the *AUC* and the *AUMC* for a wide variety of pharmacokinetic systems. For one-compartment systems it provides a method for calculating the time for

63.2% of the dose to leave those systems, while for more complex systems it provides a straightforward means of evaluating changes caused by such factors as disease, age, interacting drugs, and many more, on the properties of the drug in question.

5.5 Population pharmacokinetics

This is an analytical process designed to focus on variability and central tendencies in data, and to optimize the use of 'sparse data', which is the type of data most often obtainable during Phase III studies, or after marketing of drugs. Typically, one or two blood samples are occasionally available from any one patient at this stage in the life cycle of a new drug. However, sparse data may be available from large numbers of patients. Thus, in Phase I trials, the investigations yield full pharmacokinetic profiles of investigational drugs in small numbers of healthy volunteers, often only male, and fitting a relatively narrow anthropomorphic profile. Phase II studies may extend this intensity of investigation to the target patient population, thus broadening the scope of the studies in regard to disease factors, interacting drugs, and such factors as age, but still with relatively small numbers of subjects. Later, the scope broadens, the numbers of patients becomes large, but the limitations of sparse data become especially important.

Population pharmacokinetics employs statistical methods based, in part, on a Bayesian feedback algorithm. These methods utilize the gradually accumulating pool of data to draw conclusions about physiological, disease, drug interaction, and other influences on the pharmacokinetic properties of the drug in the human population (Racine-Poon and Smith, 1990). Probability theory is used to facilitate step-by-step prediction of pharmacokinetic properties with ever greater precision, making possible better decisions on the choice of dose for various subpopulations, and eventually for individual patients, in the clinic. Because of this focus on the individual, some authors consider the label 'population statistics' to be unfortunate.

The methods of population statistics were pioneered by Sheiner and Beal (1982) and Whiting *et al.* (1986). Various statistical packages, in particular, and originally, NONMEM (Beal and Sheiner, 1982) have become synonymous with this work, although many others have found application (Aarons, 1991).

A different type of sparse data is obtained when just one sample can be obtained from each patient in a short-term time-dependent study of the type shown in Figure 5.13. This shows the anterior chamber (of the

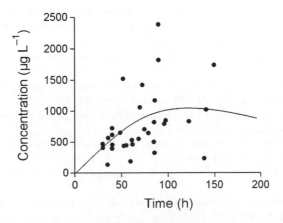

Figure 5.13   Distance-weighted least squares plot of correlation of time after application of ofloxacin and aqueous humour concentration. (Redrawn from Bouchard *et al.*, 1996).

eye – aqueous humour) ofloxacin concentrations in 32 patients each given the same drug regimen, a series of eye drops, at the time of cataract surgery. A statistically significant correlation between time since last dose and concentration was found ($r = 0.39$; $p = 0.025$) using least squares regression analysis, indicating that t_{max} occurred at $\sim$2 h. and that C_{max} was $\sim$1000 μg L^{-1}. While this was at best a crude experiment, and the correlation coefficient indicated that time since dose accounted for less than 16% of the concentration variance, it does provide useful guidance in timing of ofloxacin dosing in future patients of this type.

Experiments of this type have also been useful in laboratory animals, particularly toxicokinetic studies, when often only one blood sample can be obtained from each animal. Such experiments are sometimes described as providing 'descriptive pharmacokinetic data,' as they rarely involve any pharmacokinetic data analysis. While this is acceptable terminology, it is probably unwise to designate studies of this type as population pharmacokinetics.

5.6　Summary

This chapter has illustrated the need for more complex models to explain the kinetics and effects of drugs that cannot be explained using the simpler one-compartment model, and approaches that can be adopted when a model cannot be defined. More complex PK/PD relationships are considered in Chapter 14. The use of other samples (urine and faeces) and metabolites to study pharmacokinetics are considered in the next chapter.

References and further reading

Aarons L. Population pharmacokinetics: theory and practice. *Br J Clin Pharmacol* 1991; 32: 669–70.

Beal SL, Sheiner LB. Estimating population kinetics. *Crit Rev Biomed Eng* 1982; 8: 195–222.

Bouchard CS, King KK, Holmes JM. The kinetics of anterior chamber ofloxacin penetration. *Cornea* 1996; 15: 72–5.

Curry SH. *Drug Disposition and Pharmacokinetics*, 3rd edn. Oxford: Blackwell Scientific, 1980.

Gibaldi M, Perrier D. *Pharmacokinetics*, 2nd edn. New York: Marcel Dekker, 1982.

Haggard HW, Greenberg LA. Studies in the absorption, distribution, and elimination of ethyl alcohol II. The excretion of alcohol in urine and expired air, and the distribution of alcohol between air and water, blood and urine. *J Pharmacol Exp Ther* 1934; 52: 150–166.

Jung D, Mayersohn M, Perrier D, Calkins J, Saunders R. Thiopental disposition in lean and obese patients undergoing surgery. *Anesthesiology* 1982; 56: 269–74.

Knibbe CA, Aarts LP, Kuks PF, Voortman HJ, Lie AHL, Bras LJ, Danhof M. Pharmacokinetics and pharmacodynamics of propofol 6% SAZN versus propofol 1% SAZN and Diprivan-10 for short-term sedation following coronary artery bypass surgery. *Eur J Clin Pharmacol* 2000; 56: 89–95.

Neilsen-Kudsk F. A microcomputer program in BASIC for interative, non-linear data-fitting to pharmacokinetic functions. *Int J Biomed Computing* 1983; 14: 95–107.

Patel J, Trout SJ, Palij P, Whelpton R, Kruk ZL. Biphasic inhibition of stimulated endogenous dopamine release by 7-OH-DPAT in slices of rat nucleus accumbens. *Br J Pharmacol* 1995; 115: 421–6.

Peter C, Hongwan D, Kupfer A, Lauterburg BH. Pharmacokinetics and organ distribution of intravenous and oral methylene blue. *Eur J Clin Pharmacol* 2000; 56: 247–50.

Racine-Poon A, Smith AFM. Population models. In: Berry DA, editor. *Statistical Methodology in the Pharmaceutical Sciences*. New York: Marcel Dekker, 1990: 139–62.

Riviere JE. *Comparative Pharmacokinetics: Principles, Techniques, and Applications*. Chichester: John Wiley & Sons, 2003.

Sheiner LB, Beal SL. Bayesian individualization of pharmacokinetics: simple implementation and comparison with non-Bayesian methods. *J Pharm Sci* 1982; 71: 1344–8.

Tozer TN, Rowland M. *Introduction to Pharmacokinetics and Pharmacodynamics*. Philadelphia: Williams & Wilkins, 2006.

Vaughan DP, Tucker GT. General derivation of the ideal intravenous drug input required to achieve and maintain a constant plasma drug concentration. Theoretical application to lignocaine therapy. *Eur J Clin Pharmacol* 1976; 10: 433–40.

Whelpton R. Iterative least squares fitting of pH-partition data. *Trends Pharmacol Sci* 1989; 10: 182–3.

Whelpton R, Michael-Titus AT, Basra SS, Grahn M. Distribution of temoporfin, a new photosensitizer for the photodynamic therapy of cancer, in a murine tumor model. *Photochem Photobiol* 1995; 61: 397–401.

Whelpton R, Michael-Titus AT, Jamdar RP, Abdillahi K, Grahn MF. Distribution and excretion of radiolabeled temoporfin in a murine tumor model. *Photochem Photobiol* 1996; 63: 885–91.

Whiting B, Kelman AW, Grevel J. Population pharmacokinetics. Theory and clinical application. *Clin Pharmacokinet* 1986; 11: 387–401.

6

Kinetics of Metabolism and Excretion

6.1 Introduction

There are situations where it is necessary to define the kinetics of the metabolites of the drug under investigation. Intellectual curiosity apart, this is obviously the case for prodrugs for which the pharmacological properties reside in the metabolite. In some situations the drug and one or more its metabolites may be active, although not necessarily in the same way. Indeed the metabolite may be responsible for toxicity. The FDA has taken an interest in potentially toxic metabolites, particularly those unique to human beings, and issued its guidelines, Safety Testing of Drug Metabolites, in 2008. Metabolite concentrations may accumulate on repeated drug administration and sometimes it may be more appropriate to define the kinetics of a drug via its metabolite.

A similar situation applies to quantification of a drug in urine or faeces. The investigation may be to determine the proportion of drug that is metabolized as part of a mass-balance study or it may be important that the drug is excreted via the urine, for example in the case of diuretics or antimicrobial drugs intended for treating urinary tract infections. On the other hand the excretion rate may be used to define the kinetics of drug and or its metabolite. As far back as 1929 Gold and DeGraff studied the intensity and duration of the effect of digitalis, which to a large extent reflects urinary excretion of digoxin, and demonstrated that elimination results from a fixed proportion, not a fixed amount, of the body content leaving the body in each 24-hour period.

6.2 Metabolite kinetics

It may not be possible to study the kinetics of all the metabolites of a drug. For example, it is unlikely that *all* of the metabolites of any drug will have been identified. Also, when a metabolite is the product of sequential metabolism in an organ such as the liver, the intermediate metabolites may not be released into the bloodstream. Similarly, a metabolite may be formed in the liver and excreted in the bile without entering the general circulation. Thus, we will be considering metabolites that reach the systemic circulation and will assume, in the main, first-order input and output from a single-compartment.

6.2.1 Basic concepts

The time course of a metabolite or metabolites, and hence its or their kinetics, can be influenced by a number of factors and consequently complex relationships are possible. The simplest case would be when all the

Drug Disposition and Pharmacokinetics, By Stephen H. Curry and Robin Whelpton
© 2011 John Wiley & Sons, Ltd

administered drug, D, is converted to one metabolite, M, which is then eliminated, either by further metabolism or excretion:

$$\text{D} \xrightarrow{k_\text{m}} \text{M} \xrightarrow{k_\text{m·z}} \text{elimination}$$

where k_m is the rate constant of formation and $k_\text{m·z}$ the rate constant of elimination of the metabolite. This situation is analogous to first-order absorption into and elimination from a single compartment model and the equation, which gives the concentration of metabolite, C_m, in plasma at any time, t, takes the same form as Equation 4.17:

$$C_\text{m} = D\frac{k_\text{m}}{V_\text{m}(k_\text{m}-k_\text{m·z}t)}\left[\exp(-k_\text{m·z}t)-\exp(-k_\text{m}t)\right] \tag{6.1}$$

where V_m is the volume of distribution of the metabolite. For Equation 6.1 to be valid the dose, D, should be in moles, or corrected for the differences in the relative molecular masses of the drug and metabolite. If the drug is also eliminated via other pathways then D needs to be corrected for the fraction of metabolite formed, f_m.

The elimination half-life of a metabolite may be longer or shorter than that of the drug. Because metabolites tend be more polar than the parent drug then the volume of distribution and the degree of reabsorption by the kidney are usually less that those of the parent drug. Consequently, the elimination half-life of the metabolite is shorter. Thus, if the drug and metabolite were injected intravenously, on separate occasions then the amount of metabolite in the body would decline more rapidly [Figure 6.1(a)]. Because metabolites are rarely licensed for use in human beings, metabolite kinetics are usually studied after administration of parent drug. When $k_\text{m·z} > k_\text{m}$, the rate of formation is rate-limiting and the drug and

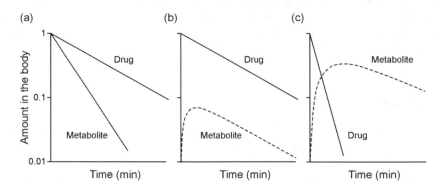

Figure 6.1   (a) Amount of drug and metabolite in the body (as a fraction of dose given) when the half-life of the metabolite is shorter than that of the drug following an i.v. dose of each. (b) Same drug and metabolite as in (a), but when only the parent drug is injected. (c) Situation when the metabolite has a longer half-life than the parent drug.

metabolite concentrations decline in parallel [Figure 6.1(b)]. The term $\exp(-k_\text{m·z}t)$ approaches zero faster than $\exp(-k_\text{m}t)$ so that at later times the slope of the terminal phase of metabolite curve is $-k_\text{m}$, that is the same as that of the parent drug. The situation with regard to the metabolite concentrations is analogous to 'flip-flop' (Section 4.2.4.2) when, without further information, it is not possible to assign the rate constants. When k_m is the larger rate constant, $\exp(-k_\text{m}t)$ approaches zero more quickly and so the slope of terminal phase for the metabolite is $-k_\text{m·z}$ [Figure 6.1(c)].

As already implied, regulations generally prevent human dosing for measurement of the kinetics of drug metabolites except when they are administered as the precursor parent drugs. However, it is possible to learn much from a limited number of examples where a drug and its metabolite(s) are separately approved for medical use. Table 6.1 is a compilation of pharmacokinetic data for ten compounds related in this way. Group I consists of three pairs of drugs, carbamazepine and its epoxide metabolite, codeine and morphine, and *N*-acetyl procainamide and procainamide; in each pair, the half-life of the metabolite administered in its own right is shorter than that of the parent drug. These examples show data of the type in Figure. 6.1(a) after administration of the two drugs separately, but of the type in Figure 6.1(b) after administration of the parent drugs. Note that procainamide and *N*-acetylprocainamide are in fact interconverted. Group II includes three pairs, caffeine and theophylline, amitriptyline and nortriptyline, and imipramine and desipramine; in each pair the half-life of the metabolite is longer than that of the parent drug. These examples show data of the type in Figure 6.1(c) after administration of the parent drugs. The other data in the table is for four benzodiazepines that are related to each other in various ways as precursors and metabolites (Figure 3.7), providing a challenge for the intellectually curious. The half-life relationships in this table show no consistent relationship

Table 6.1   A compilation of pharmacokinetic data for examples of drug pairs used in patients where there is a relation within each pair of precursor and metabolite

Drug example	Protein binding (%)	Half-life (h)	V (l kg^{-1})	CL (ml min^{-1} kg^{-1})	Urinary excretion (%)	Metabolic reaction
Group 1						
Carbamazepine	74	15 (36)[a]	1.4	1.3 (0.36)[a]	<1	
Carbamazepine-10,11-epoxide	50	7.4	1.1	1.7[b]	<1	Epoxide formation
Codeine	7	2.9	2.6	11[b]	0	
Morphine	35	1.9	3.3	24	4–14	*O*-demethylation
N-Acetylprocainamide	10	6	1.4	3.1	81	
Procainamide	16	3	1.9	2.7	67	Hydrolysis
Group II						
Caffeine	36	4.9	0.61	1.4	1.1	
Theophylline	56	9.0	0.5	0.65	18.0	*N*-demethylation
Amitriptyline	94.8	21 (19.5)[c]	15	11.5	<2	
Nortriptyline	92	31 (41.5)[c]	18.4	7.2	2	*N*-demethylation
Imipramine	90.1	12	18	15	<2	
Desipramine	82	22	20	10	2	*N*-demethylation
Benzodiazepines						
1. Diazepam	98.7	43	1.1	0.38	<1	1 to 2, *N*-demethylation
2. Nordazepam	97.5	73	0.78	0.14	<1	1 to 3, hydroxylation
3. Temazepam	97.6	11	0.95	1.0	<1	2 to 4, hydroxylation
4. Oxazepam	98.8	8	0.6	1.05	<1	3 to 4 *N*-demethylation

All data from Goodman and Gilman and so represent interpretations of multiple publications except data in parentheses for amitriptyline.
[a] Data in parentheses from single doses – other data from long-term treatment (carbamazepine is a self-inducer)
[b] Data from CL/F; clearance data for carbamazepine metabolite from renal clearance
[c] Data in parentheses from amitriptyline and nortriptyline measured after amitriptyline doses (Curry *et al.*, 1985, 1986, 1987, 1988)

with other pharmacokinetic parameters of the drugs concerned, with the notable exception that N-demethylation appears to cause a lengthening of the half-life. A cursory search through the pharmacokinetic literature concerning drugs and their metabolites that are not administered in their own right reveals a preponderance of examples of the Figure 6.1(b) type, suggesting that the case exemplified by carbamazepine and its epoxide is the most common.

The examples of Figure 6.1 consider the *amount* of the substances in the body, and when formation is rate-limiting metabolite is removed almost as soon as it has been formed and so the amount of metabolite at any time is less than that of the drug [Figure 6.1(b)]. However, it is usual to measure the *concentrations* of metabolite, and these will be influenced by their apparent volumes of distribution (Equation 6.1). Increased polarity and, possibly, increased plasma protein binding, particularly of acidic metabolites to albumin, reduces the apparent volume of distribution relative to that of the parent drug. Consequently, the *plasma concentrations* of a metabolite can be much higher than those of the drug, even if the *amounts* in the body are considerably less.

The rate of change of amount of metabolite in the body, dA_m/dt, at any time will be the difference in rates of formation and elimination. The rates are given by the plasma concentration multiplied by clearance (Equation 4.3). Thus:

$$\frac{dA_m}{dt} = C \cdot CL_f - C_m \cdot CL_m \tag{6.2}$$

where CL_f is the clearance associated with the formation of metabolite. Integration of Equation 6.2 gives:

$$\frac{AUC_m}{AUC} = \frac{CL_f}{CL_m} \tag{6.3}$$

6.2.2   Fraction of metabolite formed

The amount of metabolite formed, A_m, can be calculated from the area under the plasma concentration–time curve, AUC_m, provided the clearance, CL_m, is known. This will require intravenous administration of the metabolite. By analogy with Equation 4.13:

$$AUC_m = \frac{A_m}{CL_m} \tag{6.4}$$

so

$$A_m = AUC_m \times CL_m \tag{6.5}$$

and the fraction produced is

$$f_m = \frac{A_m}{D} \tag{6.6}$$

An alternative method when it is not possible to administer metabolite is to use excretion rate data. This approach is only applicable when (i) the rate of formation is rate-limiting, (ii) all the metabolite is excreted via the urine, and (iii) there is no metabolism by the kidney. Under these conditions the rate of metabolite excretion approximately equals the rate of metabolism:

$$\text{Rate of renal excretion of metabolite} \approx C \times CL_f \tag{6.7}$$

Measuring the rate of excretion and the plasma drug concentration allows CL_f to be calculated and substituted into Equation 6.3 to solve for CL_m.

6.2.3   More complex situations

In the majority of cases, the drug will have been given extravascularly, probably orally, and more than one metabolite will be formed. This formation may occur in parallel (Figure 6.2), in sequence (Figure 6.3) or a combination of the two. Furthermore, a proportion of dose may be excreted unchanged.

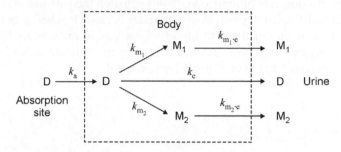

Figure 6.2   A model of metabolism where two metabolites, M_1 and M_2, are produced in parallel. Unchanged drug, D, and the metabolites are excreted into the urine.

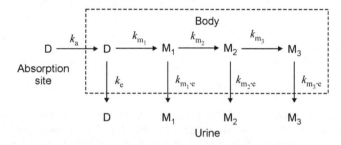

Figure 6.3   A model of excretion where the metabolites, M_1 M_2 and M_3 are produced sequentially. Unchanged drug, D, and metabolites are excreted into urine.

In the case of a single metabolite that is excreted into the urine, the equation relating the amount of metabolite in the body after an i.v. bolus injection of D is:

$$A_m = D \frac{k_m}{(k_{m \cdot e} - \lambda)} [\exp(-\lambda t) - \exp(-k_{m \cdot e} t)] \tag{6.8}$$

where $k_{m \cdot e}$ is the rate constant for urinary excretion of metabolite and the elimination rate constant of the drug is $\lambda = k_m + k_e$. The equation for the amount of metabolite excreted into the urine is:

$$A_{m, \text{urine}} = D \frac{k_m}{\lambda} \left\{ 1 - \frac{1}{(k_{m \cdot e} - \lambda)} [k_{m \cdot e} \exp(-\lambda t) - \lambda \exp(-k_{m \cdot e} t)] \right\} \tag{6.9}$$

In the case depicted in Figure 6.3 any of the steps may be the rate-limiting one, and so the plasma concentrations of any metabolites formed after the rate-limiting step will decline in parallel. In either case, if the rate of absorption is the slowest step then drug and all the metabolites formed from it will decline with a half-life equivalent to that of absorption.

Complex situations arise when a metabolite is further metabolized by more than one route. This is the case with 4′-hydroxypropranolol. The formation of the 4′-hydroxy metabolite of propranolol is rate-limited, and so the propranolol and metabolite concentrations fall in parallel (as illustrated later in Figure 6.6). The major routes of metabolism of 4′-hydroxypropranolol are sulfation or glucuronidation and, at first sight, the half-lives of the conjugates appear to be greater than that of 4′-hydroxypropranolol [Figure 6.4(a)], even though these metabolites would be expected to be more readily excreted. In fact the longer elimination half-lives arise because the kinetics are formation rate-limited and the decline in plasma concentrations are controlled by the relatively slow rate of formation. In this situation the half-life of 4′-hydroxypropranolol is shorter than apparent half-lives of the individual conjugates because it is being metabolized by (at least) two routes. That the apparent elimination half-life of sulfate was formation-rate limited was confirmed by administering the conjugate on a separate occasion, when the half-life was 82 ± 12 min rather than 156 ± 21 min when it was a metabolite of 4′-hydroxypropranolol.

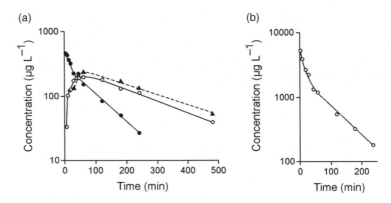

Figure 6.4 Plasma concentrations of 4′-hydroxypropranolol (•) and its sulfate (○) and glucuronide (▲) conjugates (a) after i.v. injection of 4′-hydroxypropranolol, 2 mg kg^{-1} in dogs. (b) Plasma concentrations of 4′-hydroxypropranolol sulfate after i.v. injection of 2 mg kg^{-1} in dogs (adapted from Christ *et al.*, 1990).

6.2.4 *Active metabolites*

The significance of an active metabolite will depend on whether the elimination kinetics of the metabolite are formation dependent or elimination dependent. In the former case the half-life of the metabolite will be the same as that of the parent drug and so it will accumulate at the same rate as does the parent compound. Therefore, dosing can be based upon the disposition parameters for the drug and it is not necessary to be concerned with the kinetic parameters of the metabolite. This even applies to prodrugs where the metabolite is the active species, although in practice a prodrug is likely to have a very short elimination half-life and the rate of formation of the metabolite will not be rate limiting. Under these circumstances the metabolite will be monitored and the dosing will be based on the kinetics of the metabolite.

When the kinetics are elimination dependent, the half-life of the metabolite is greater than the drug and so the metabolite continues to accumulate after the drug has reached steady-state conditions, for example during a continuous infusion or on multiple dosing. Because the metabolite takes longer to reach steady-state, dosing should be determined by the disposition characteristics of the metabolite. The average concentration of metabolite at steady-state can be calculated from the AUC_m after a single i.v. dose:

$$C_{av}^{ss} = \frac{AUC}{\tau} \qquad (6.10)$$

See also Equation 4.38. Furthermore, the frequency of dosing should be based on the half-life of the metabolite, i.e. it is not necessary to give the drug as frequently as the half-life of the drug might suggest.

The *N*-desmethyl metabolites of several centrally acting drugs, including imipramine and amitriptyline, show pharmacological activity similar to that of the parent drug and tend to accumulate to higher concentrations than the parent drug on multiple dosing. Nordazepam is a metabolite of several 7-chloro-benzodiazepine drugs and, because of its long elimination half-life accumulates to concentrations greater than those of the parent drug. In human beings steady-state concentrations of nordazepam are approximately twice those of diazepam after repeated administration of the latter [Figure 6.5(a)].

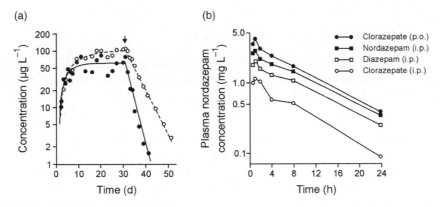

Figure 6.5 (a) Plasma diazepam (•) and nordazepam (o) concentrations in a healthy female volunteer after 2 mg diazepam at night for 30 days (redrawn from Abernethy *et al.*, 1983). (b) Plasma nordazepam concentrations after dosing with clorazepate (4.5 mg kg^{-1}), diazepam and nordazepam (3.0 mg kg^{-1}) (from the data of Curry *et al.*, 1977).

Because the kinetics of nordazepam are not formation dependent then the plasma half-life should be independent of the drug that is administered. This is illustrated in Figure 6.5(b) where diazepam or clorazepate, a prodrug which decomposes to nordazepam in gastric acid, was administered. As predicted the half-life of the metabolite was the same as when nordazepam was injected. The amount of nordazepam produced following i.p. injection of clorazepate was markedly reduced – illustrating the role of gastric acid in the formation of the active moiety [Figure 6.5(b)].

6.2.5 *Effect of presystemic metabolism*

When drugs are taken orally and undergo extensive presystemic metabolism, both drug and metabolite appear in the systemic circulation together. This is like taking a mixture of drug and metabolite. For a metabolite that displays formation rate-limited disposition it would be expected that drug and metabolite peak plasma concentrations would occur at the same time and then decline in parallel (Figure 6.6).

However, the initial concentrations of metabolite in the plasma are not dependent on rate-limited formation from the parent drug and decline according to the disposition kinetics of the metabolite, that is, as if the metabolite had been administered rather than the drug. This results in a bi-exponential decline in metabolite concentrations after oral administration of parent drug (Figure 6.6). It is important not to confuse this bi-exponential decay with that seen for two-compartment models.

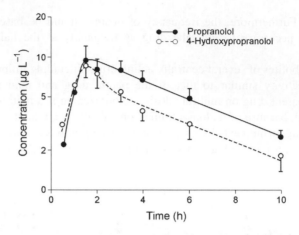

Figure 6.6 Mean plasma concentrations of propranolol and its 4′-hydroxymetabolite after oral administration of propranolol (20 mg) to six normal volunteers. Error bars represent mean ± SEM. (Redrawn from Walle *et al.*, 1980.)

When first-pass metabolism occurs primarily in the liver then the AUC_m values after oral and intravenous doses should be similar, provided, of course, that all the dose of drug is absorbed. This is because the drug can enter the liver whether it is given orally or by injection. Consequently for a drug with low oral bioavailability, the areas under the metabolite plasma concentration–time curves can be used to differentiate poor absorption from extensive hepatic first-pass metabolism. If the AUC_m is higher after an oral dose than when the drug is injected, then this is indicative that presystemic metabolism is occurring in the GI tract.

6.2.6 Interconversion of drug and metabolite

Some metabolites may be converted back to parent drug as depicted below:

$$\text{Drug} \rightleftharpoons \text{Metabolite}$$
$$\downarrow \qquad\qquad \downarrow$$
$$\text{Elimination} \qquad \text{Elimination}$$

The drug and metabolite may also be eliminated via other routes, either further metabolism or excretion. This model is analogous to a two-compartment model, with loss from both the central and peripheral compartments. Initially drug concentrations fall rapidly but as drug and metabolite concentrations equilibrate interconversion has a major influence on the half-life of the drug and metabolite. Several drugs exhibit this phenomenon including: cortisol–cortisone, haloperidol–reduced haloperidol, prednisone–prednisolone, and vitamin K–vitamin K epoxide. The effect of interconversion has been nicely demonstrated for predisone and prenisolone (Figure 6.7). After administration of prednisone, the concentrations of the active metabolite, prednisolone, rose to over 10 times those of prednisone. When prednisolone hemisuccinate was given intravenously the prednisolone concentrations fell rapidly until the prednisolone: prednisone ratio equilibrated at the same value as that after administration of prednisone, demonstrating that interconversion produces the same drug to metabolite ratio irrespective of which drug is given. Prednisolone hemisuccinate is a prodrug ester of prednisolone which is very rapidly hydrolysed to prednisolone.

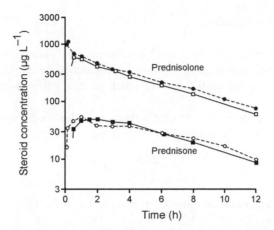

Figure 6.7 Plasma concentrations of prednisone and prednisolone after a single oral dose of prednisone (solid lines) or an i.v. dose of prednisolone hemisuccinate ester to a healthy male volunteer (adapted from Rose *et al.*, 1980).

In the case of sulindac, the sulfide, which is oxidized back to the parent drug, is active whereas the sulfone metabolite is not interconverted and is inactive (Figure 6.8). Thus, sulindac is a prodrug, and it has been suggested that this is why it is less prone to cause gastrointestinal upsets when compared with some other non-steroidal anti-inflammatory drugs as the active form is not produced until after absorption.

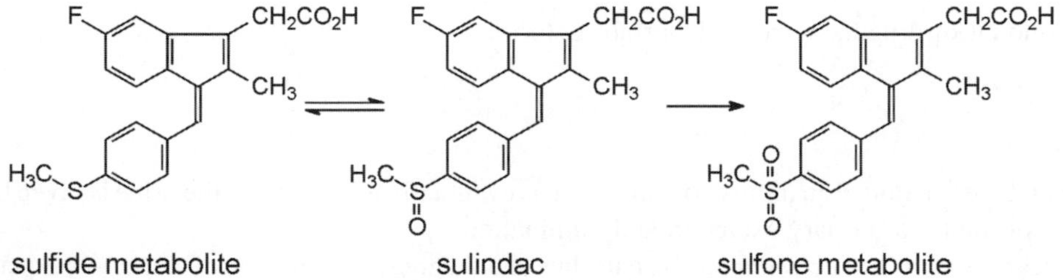

sulfide metabolite   sulindac   sulfone metabolite

Figure 6.8 Sulindac (a sulfoxide) is reversibly reduced to pharmacologically active sulfide or oxidized to sulfone metabolite.

6.3 Renal excretion

The factors contributing to urinary excretion of a compound are:

- Glomerular filtration
- Passive reabsorption from renal tubular fluid
- Pctive secretion into renal tubular fluid.

These were discussed in Section 3.3.1.

6.3.1 Kinetics of urinary excretion

Glomerular filtration and diffusion across the renal tubular epithelium are generally first-order processes so that the rate of transfer of drug is related to the amount of drug in plasma. The excretion of a drug in urine is complicated by further factors including:

- The concentration of drug in renal tubular fluid is influenced by changes in urine volume so there is a tendency for the concentration to increase as water is reabsorbed by the kidney.
- Changes in urinary pH can influence the rates at which weak electrolytes are excreted.

Therefore it is usual to relate the *rate* of renal elimination to the concentration of drug in plasma. The differential equation for the rate of appearance of drug in urine is:

$$\frac{\mathrm{d}Ae}{\mathrm{d}t} = k_{\mathrm{e}}A \tag{6.11}$$

where Ae is the amount of drug excreted in urine at time t and k_{e} is the first-order rate constant for urinary elimination. A is given by:

$$A = D\exp(-\lambda t) \tag{6.12}$$

so

$$\frac{\mathrm{d}Ae}{\mathrm{d}t} = k_{\mathrm{e}}D\exp(-\lambda t) \tag{6.13}$$

Integrating Equation 6.13 gives:

$$Ae = \frac{k_{\mathrm{e}}}{\lambda}D[1 - \exp(-\lambda t)] \tag{6.14}$$

The total amount of drug excreted in the urine, $Ae(\infty)$ is

$$Ae(\infty) = \frac{k_{\mathrm{e}}}{\lambda}D \tag{6.15}$$

indicating that the fraction of an intravenous dose that is eventually excreted into the urine is given by the ratio of the rate constants for urinary excretion and elimination.

A semilogarithmic plot of *excretion rate* versus time should give a straight line of slope $-\lambda$ (Equation 6.13). In practice the excretion rate is calculated from urine samples collected at discrete intervals and the data are either plotted as a histogram or the mid-points of the collection period are used. This approach assumes that there are no changes in the rate of renal excretion as a result of fluctuations in urinary pH, urine volume or unknown factors. To reduce the effects of fluctuations seen in excretion rate plots, an alternative approach is the 'sigma-minus' method. Urine is collected for sufficiently long to allow estimation of $Ae(\infty)$ – this should be for up to about seven elimination half-lives. Substituting Equation 6.15 into Equation 6.14 and rearranging gives:

$$Ae(\infty) - Ae = Ae(\infty)\exp(-\lambda t) \tag{6.16}$$

A semilogarithmic plot of percent of drug remaining to be excreted against time is a straight line of slope $-\lambda$.

Similar approaches can be used to derive equations for urinary excretion following first-order absorption into a single-compartment model, elimination from a two-compartment model and non-linear kinetics.

6.3.1.1   Renal clearance

The rate of elimination of a drug is the systemic clearance multiplied by the plasma concentration (Equation 4.3) The urinary excretion rate is the renal clearance, CL_{R}, multiplied by the plasma concentration, so:

$$CL_R = \frac{dAe/dt}{C} = \frac{k_e A}{C} \qquad (6.17)$$

Note that Equation 6.17 is consistent with the physiologists' definition of renal clearance (Equation 3.6). Because A/C is the apparent volume of distribution, Equation 6.17 can be rewritten:

$$CL_R = k_e V \qquad (6.18)$$

Equation 6.18 gives a route to determining k_e, provided that the apparent volume of distribution of the drug is known (from an intravenous injection) and the renal clearance is measured using Equation 3.6.

6.3.1.2 Effect of urine flow rate

The effect of urine flow rate will depend on whether the plasma and urine concentrations have equilibrated. At equilibrium, U/P (Equation 3.6) will be constant and renal clearance will be directly proportional to urine flow rate. Furthermore, for a neutral molecule that binds to plasma protein the urine concentration and the unbound concentration of drug in plasma will be equal at equilibrium so renal clearance will be:

$$CL_R = f_u \times \text{urine flow rate} \qquad (6.19)$$

where f_u is the fraction of unbound drug in plasma. An alternative way of visualizing the effect of flow rate is that if the urine and plasma concentrations have more or less equilibrated then by whatever proportion the urine flow rate is increased then the *amount* of drug in urine must increase by the same proportion to maintain the required urine concentration. Ethanol, which is not bound to plasma proteins, does not ionize and usually equilibrates rapidly shows a nearly linear relationship between renal clearance and urine flow rate with a slope close to 1 [Figure 6.9(a)]. Correlations between urine flow and clearance have been shown for other drugs, including phenobarbital, sulfafurazole and glutethimide [Figure 6.9(b)]. For glutethimide, the clearance is less than the urine flow rate, which in part can be explained by the fact that drug is approximately 50% bound to plasma proteins. Urine flow will not have much influence on the clearance of drugs for which there is little renal tubular reabsorption.

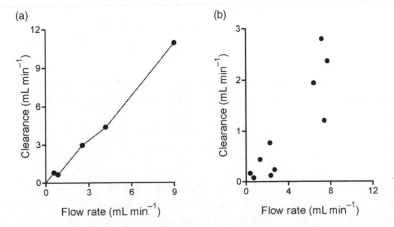

Figure 6.9 Renal clearance as a function of urine flow rate: (a) ethanol, data are five independent estimates in one subject and (b) glutethimide, data are 10 independent estimates.

Despite the increased clearance of some drugs with increased urine flow, forced acidic or alkaline diuresis is no longer employed for treating drug overdose because the changed fluid balance is considered potentially dangerous. Sodium bicarbonate may be used to increase the renal clearance of salicylate and chlorophenoxyacetic acid herbicides, such as 2,4-dichlorophenoxyacetic acid (2,4-D) and 2,3,5-trichlorphenoxyacetic acid (2,3,5-T), but an additional diuretic is not used.

6.3.2   Specific drug examples

6.3.2.1   Amphetamine

The effect of urine volume and pH on the excretion rate of amphetamine is shown in Figure 6.10. There is a clear relationship between the excretion rate and urine pH as would be expected for this weak base ($pK_a \sim 9.8$). There is also an indication that the excretion rate increases with increasing urine volume. Obviously it is not possible to derive pharmacokinetic parameters under the conditions of Figure 6.10, but values have been obtained when the urine pH was adjusted to, and maintained at, $\sim$pH 5 with the administration of ammonium chloride. Under these conditions the renal excretion of amphetamine is markedly increased (c.f. Figure 3.18) which must have an effect on any half-life value calculated.

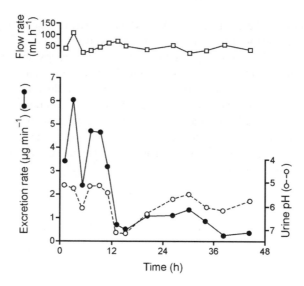

Figure 6.10   Effect of urine pH and volume on the urinary excretion of amphetamine in a single human subject, after oral administration of 10 mg amphetamine sulfate. (Redrawn from Beckett and Rowland 1964.)

6.3.2.2   Ethanol

Ethanol distributes freely with total body water. In spite of wide variations in the volume of urine produced, the concentration of ethanol in urine is closely related to its concentration in blood (Figure 6.11). In the case of ethanol, diffusion of the drug between urine and blood apparently occurs up to a time not long before the urine is removed from the body.

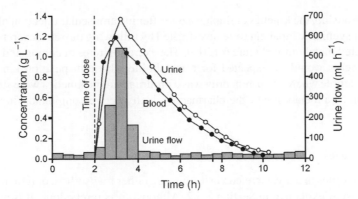

Figure 6.11 Concentrations of ethanol in whole blood (•) and urine (○) after an oral dose of 64 g of ethanol with 116 mL of water. Note that the urine concentrations are plotted at the midpoint of the collection interval (after Haggard *et al.*, 1941).

The near constant ratio of blood to urine concentrations allows urine to be used as an alternative to blood for the purposes of 'drink-driving' laws – $1.07\,\text{g}\,\text{L}^{-1}$ in urine being equivalent to $0.8\,\text{g}\,\text{L}^{-1}$ (0.08%) in blood.

6.3.2.3 Fluphenazine

Fluphenazine is usually administered as an i.m. depot injection of its decanoate ester, a prodrug, which is hydrolysed to fluphenazine and released into the plasma. Before the advent of sufficiently sensitive assays, fluphenazine kinetics could only be studied by administration of radiolabelled drug or from urinary excretion studies. In man, fluphenazine is present in urine as fluphenazine plus conjugates that can be released by hydrolysis with β-glucuronidase. Using ^{14}C-labelled drug the relationship between the plasma concentration and urinary excretion rate of fluphenazine plus conjugates was demonstrated [Figure 6.12(a)]. Having

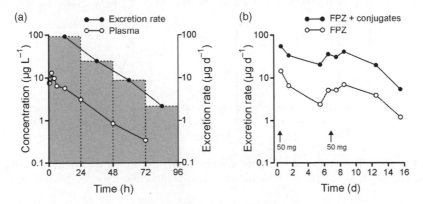

Figure 6.12 (a) Plasma concentrations and urinary excretion rate of fluphenazine after an intramuscular injection of 25 mg. (b) Urinary excretion rate of fluphenazine and conjugated fluphenazine in a patient receiving 50 mg fluphenazine decanoate (i.m.) weekly (after Whelpton and Curry, 1976).

established this relationship, the kinetics of fluphenazine after intramuscular injection of the decanoate ester (50 mg weekly) to a psychiatric inpatient were investigated by following the renal excretion of either the drug or drug plus conjugated metabolites [Figure 6.12(b)]. The excretion rates of conjugated metabolites parallel those of fluphenazine as would be expected for a metabolite whose disposition kinetics are formation-limited. This is an example where the parent drug was a prodrug and the kinetics were studied by monitoring the urinary excretion of a metabolite of the pharmacologically active compound, fluphenazine.

6.4   Excretion in faeces

Drugs and their metabolites that are excreted via the bile are either reabsorbed, or remain in the GI tract to be removed in the faeces, as explained in Section 3.3.2. Although it is rarely done, it is possible to determine drug kinetics from faecal excretion data, as exemplified below for the photodynamic agent, temoporfin. The same arguments as those made for renal excretion apply. Temoporfin was particularly suitable because (i) it is administered by intravenous injection so any drug or metabolite in the faeces is there because it has been excreted, (ii) over 99% is eliminated in faeces and (iii) it has a long elimination half-life. A study with non-radiolabelled drug indicated that the terminal half-life in BALB/c mice was at least 13.9 h but, as it was possible to measure the drug in blood for only 48 h, this was considered to be an underestimate (Whelpton *et al.*, 1995). The slow decline in temoporfin concentrations in some tissues, notably lung and kidney, which were monitored for 96 hours confirmed this view. When ^{14}C-temoporfin became available the study was repeated and the drug monitored for up to 35 days. Faecal excretion rates are shown in Figure 6.13. The decay was bi-exponential

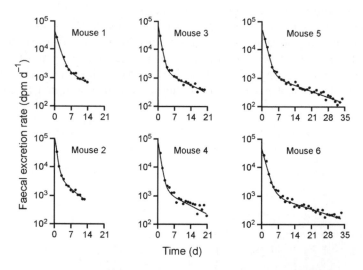

Figure 6.13   Faecal excretion of ^{14}C-label following temoporfin administration to BALB/c mice; solid lines are the least squares fit of the data to a two-compartment model (from Whelpton *et al.*, 1996).

A putative metabolite was detected in liver and faeces but not in blood nor any other tissue. This is in keeping with a metabolite that is produced in, and excreted by, the liver, probably via bile. The metabolite concentrations declined in parallel with those of temoporfin indicating that its disposition was formation rate-limited. Thus, it was possible to define the disposition of temoporfin by monitoring the rates of faecal excretion for up to 7 weeks.

The data of Figure 6.13 illustrate another important point – it is necessary to ensure that the data are collected for sufficient time to derive a reliable estimate of the half-life of the terminal phase. Failure to do so will result in an underestimate (Table 6.2).

Table 6.2   Elimination half-lives derived from faecal excretion data of Figure 6.13

Mouse	Duration 0–14 days[a]		Duration 0–21 days		Duration 0–28 days	
	$t_{1/2}$ (λ_1) (h)	$t_{1/2}$ (λ_2) (days)	$t_{1/2}$ (λ_1) (h)	$t_{1/2}$ (λ_2) (days)	$t_{1/2}$ (λ_1) (h)	$t_{1/2}$ (λ_2) (days)
1	22.7	9.0				
2	13.0	5.0				
3	15.4	8.7	14.6	7.3		
4	14.9	6.2	15.4	7.5		
5	18.0	6.5	19.9	10.5	20.2	11.0
6	18.2	5.8	21.2	10.2	21.1	10.0
Mean	17.0	6.9	17.8	8.9	20.7	10.5

[a] Time over which kinetic parameters were calculated.

References and further reading

Abernethy DR, Greenblatt DJ, Divoll M, Shader RI. Prolonged accumulation of diazepam in obesity. *J Clin Pharmacol* 1983; 23: 369–76.

Beckett AH, Rowland M. Rhythmic urinary excretion of amphetamine in man. *Nature* 1964; 204: 1203–4.

Christ DD, Walle UK, Oatis JE, Jr., Walle T. Pharmacokinetics and metabolism of the pharmacologically active 4'-hydroxylated metabolite of propranolol in the dog. *Drug Metab Dispos* 1990; 18: 1–4.

Curry SH, Whelpton R, Nicholson AN, Wright CM. Behavioural and pharmacokinetic studies in the monkey (*Macaca mulatta*) with diazepam, nordiazepam and related 1,4-benzodiazepines. *Br J Pharmacol* 1977; 61: 325–30.

Curry SH, DeVane CL, Wolfe MM. Cimetidine interaction with amitriptyline. *Eur J Clin Pharmacol* 1985; 29: 429–433.

Curry SH, DeVane CL, Wolfe MM. Pharmacology of combined antidepressant/H$_2$-blocking drug therapy. *Psychopharmacol Bull* 1986; 22: 220–222.

Curry SH, DeVane CL, Wolfe MM. Lack of interaction of ranitidine with amitriptyline. *Eur J Clin Pharmacol* 1987; 32: 317–320.

Curry SH, DeVane CL, Wolfe MM. Hypotension and bradycardia induced by amitriptyline in healthy volunteers. *Human Psychopharmacology* 1988; 3: 47–52.

Duggan DE. Sulindac: therapeutic implications of the prodrug/pharmacophore equilibrium. *Drug Metab Rev* 1981; 12: 325–37.

http://www.fda.gov/downloads/Drugs/GuidanceComplianceRegulatoryInformation/Guidances/ucm079266.pdf (accessed 28th July 2009).

Gold H, Degraff AC. Studies on digitalis in abulatory cardic patients: II. The elimination of digitalis in man. *J Clin Invest* 1929; 6: 613–26.

Haggard HW, Greenberg LA, Carroll RP. Studies in the absorption, distribution and elimination of alcohol: VIII. The diruesis from alcohol and its influence on the elimination of alcohol in the urine. *J Pharmacol Exp Ther* 1941; 71: 349–357.

Rose JQ, Yurchak AM, Jusko WJ, Powell D. Bioavailability and disposition of prednisolone tablets. *Biopharm Drug Dispos* 1980; 1: 247–58.

Walle T, Conradi EC, Walle UK, Fagan TC, Gaffney TE. 4-Hydroxypropranolol and its glucuronide after single and long-term doses of propranolol. *Clin Pharmacol Ther* 1980; 27: 22–31.

Whelpton R, Curry SH. Methods for study of fluphenazine kinetics in man. *J Pharm Pharmacol* 1976; 28: 869–73.

Whelpton R, Michael-Titus AT, Basra SS, Grahn M. Distribution of temoporfin, a new photosensitizer for the photodynamic therapy of cancer, in a murine tumor model. *Photochem Photobiol* 1995; 61: 397–401.

Whelpton R, Michael-Titus AT, Jamdar RP, Abdillahi K, Grahn MF. Distribution and excretion of radiolabeled temoporfin in a murine tumor model. *Photochem Photobiol* 1996; 63: 885–91.

7

Further Consideration of Clearance, and Physiological Modelling

7.1 Introduction

Previous chapters have introduced the concept of clearance and its importance, along with apparent volume of distribution, as a determinant of the elimination half-life of a drug. The term 'clearance' is sometimes used to describe the phenomenon of removal of a drug from the body as a whole, when the term 'elimination' would be better. It should be remembered that clearance always refers to a volume of fluid from which a substance is removed in unit time, and thus will always have units of flow, such as mL min^{-1} or L h^{-1}. However, there are times when it is normalized, for example to body weight, concentration of microsomal protein, or hepatocyte concentration. Clearance can be used to describe the behaviour of a drug *in vitro*, as well as in *in vivo* systems. The term can be applied to individual organs, for example, renal clearance, hepatic clearance, etc., or to the whole body when it may be referred to as systemic (or plasma) clearance.

7.2 Clearance *in vitro* (metabolic stability)

The study of renal clearance dates from the 1930s, when pioneering renal physiologists discovered that kidney function could be assessed in terms of the removal of drugs from the blood in the renal artery (Section 3.3.1.4). Pharmacokineticists have extended this to embrace all processes of drug elimination, and, in the case of the liver, to model it experimentally *in vitro*, in an experiment sometimes called 'metabolic stability'.

7.2.1 Microsomes

Microsomal intrinsic clearance, CL_{mic}, provides an assessment of the ability of the microsomal fraction of the liver to remove the drug from the biophase surrounding the enzyme surface in the absence of any delivery (by blood) or availability (e.g. restrictions imposed by protein binding) influences. The experimental measurement of microsomal intrinsic clearance *in vitro* involves incubation of drug in a fixed volume of fluid in which is suspended a known quantity of liver microsomes. The decay of drug concentration is monitored using a suitable analytical method. First-order decay is ensured by using an appropriately low drug concentration and an appropriately high microsome concentration. The first-order rate constant, k, is obtained from the

Drug Disposition and Pharmacokinetics, By Stephen H. Curry and Robin Whelpton
© 2011 John Wiley & Sons, Ltd

slope of a semilogarithmic concentration–time plot:

$$CL_{mic} = -\text{slope} \times \text{volume of incubation}$$
$$= V.k \tag{7.1}$$

Microsomes are considered to be 100% viable, and so the activity can be expressed in terms of the microsomal protein concentration. Normalizing CL_{mic} to 1 mg of protein, gives units of $mL\,min^{-1}\,mg$ $protein^{-1}$. The rate of metabolism is $CL_{mic} \times C$ (see Equation 4.3)

At relatively high drug concentrations, when non-linear kinetics are seen, Equation 7.1 can be written in terms of Michaelis–Menten enzyme kinetics, as rate of reaction $= V_{max} \times C/(Km + C)$ (see Equation 4.42):

$$CL_{mic} = \frac{V_{max}}{km + C} \tag{7.2}$$

where C is the concentration in the biophase at the enzyme surface (in this case of *in vitro* work, in the fluid). At very high concentrations when the enzymes are saturated with drug, the rate of reaction is V_{max}, so:

$$V_{max} = CL_{mic} \times C \tag{7.3}$$

and

$$CL_{mic} = V_{max}/C \tag{7.4}$$

which is analogous to the zero-order case, Equation 4.43. Obviously, the microsomes contain liver enzyme systems in which only microsomally catalysed chemical change occurs. However, microsomal reactions include oxidations, reductions, hydrolyses and some phase 2 reactions, so multiple chemical changes can occur. Only by measuring the concentrations of the different products can pure, single reaction kinetics be studied. This is not commonly done, as pharmacokineticists have, historically, measured disappearance of substrate, rather than appearance of products because interest was primarily in the disappearance of pharmacologically active molecules. Also, until metabolites have been identified, it is not possible to develop assays for them.

7.2.2   Hepatocytes

Analogous experiments can be performed using hepatocytes instead of microsomes. The clearance is expressed in terms of the numbers of cells: $mL\,min^{-1}\,million\,cells^{-1}$. Because hepatocytes are not necessarily 100% viable, a viability correction determined in a separate experiment with a compound whose properties are known may be needed. Also, hepatocytes reproduce a somewhat larger collection of metabolic reactions, microsomal and otherwise, so that the result with hepatocytes is a kinetic constant assessing a somewhat larger collection of product-formation reactions. Again, separate assays of products are needed if the kinetics of single reactions are to be studied.

Human hepatocytes contain an average of 52.5 mg of microsomal protein per g of liver and there are $\sim 120 \times 10^6$ hepatocytes per g of liver, so there is 0.44 mg of microsomal protein per 1 million hepatocytes *in vitro* and *in vivo*. The corresponding figure for the rat is 0.34 mg of microsomal protein per million hepatocytes. With *in vitro* work, the hepatocyte concentration is limited by the physical properties of the suspension – it is inconvenient if the hepatocyte concentration is such that it is difficult to achieve adequate mixing of the suspension without damaging the cells. Consequently, the suspension of hepatocytes must be relatively dilute. In contrast, microsomal suspensions can contain higher protein concentrations than is the case with the hepatocyte suspension. So, experimentally, drug half-life values are often shorter in the conditions of the microsomal suspensions than in those of the hepatocyte suspensions, in spite of the fact that more reactions take place in the hepatocyte incubations. Normalization to hepatocyte concentration and to the microsomal protein concentration overcomes this small experimental difficulty, and:

- The clearance normalized for protein calculated in the hepatocyte experiment will usually be higher than that in the microsomal experiment.
- The half-life in a rat hepatocyte experiment may be longer than that in a human experiment with the same hepatocyte concentration, but, because *in vivo* the rat has more liver mass than the human per kg of body weight, the clearance when scaled up to *in vivo* expectations will be higher in the rat.

7.3   Clearance *in vivo*

A non-eliminating organ can remove drug molecules from the blood passing through it until equilibrium between the tissue and plasma concentrations is reached, after which elimination in the liver and kidney reduces the concentrations in both blood and tissue. In this situation:

$$\text{Rate of removal from plasma} = QC_a - QC_v$$
$$= Q(C_a - C_v) \tag{7.5}$$

where Q is blood flow, C_a is the afferent arterial concentration and C_v is the efferent venous concentration (see Figure 4.4). In this situation, the extraction ratio, E, can be viewed as assessing organ uptake.

$$E = \frac{C_a - C_v}{C_a} \tag{7.6}$$

This approach has been used in the search for an understanding of brain uptake in particular, where, in an appropriately designed experiment, carotid artery and jugular vein concentrations can be measured. For an eliminating organ, the organ clearance is the elimination rate divided by C_a, and this provides the basis for assessment of renal clearance in particular (see Section 3.3.1.4). The extraction ratio concept is of major importance in relation to the liver.

The concept of systemic clearance was introduced in earlier chapters, particularly Chapters 4 and 5, because of the importance of clearance as a determinant of the elimination half-life. However, it is worth emphasizing some key points. The relationship between elimination rate constant, apparent volume of distribution and clearance in a single-compartment model was demonstrated in Section 4.2.1, resulting in Equation 4.5:

$$CL = \lambda V \tag{4.5}$$

but it is rare that single-compartment models are applicable and the more useful equation is Equation 4.14:

$$CL = \frac{D}{AUC} \tag{4.14}$$

which has the advantage that it is generally applicable. It can be applied to multiple-compartment models (Section 5.1.1.4) or when a model has not been defined (Section 5.3.1). *AUC* is obtained using the trapezoidal method (see Appendix). It is, of course, necessary to define the terminal decay constant, λ_z, in order to extrapolate the area from the last time point to infinity.

7.3.1   Apparent oral clearance

Following oral dosing the equation equivalent to Equation 4.14 is Equation 4.26:

$$CL = F\frac{D}{AUC} \tag{4.26}$$

Sometimes the value given by Equation 4.26 is referred to as 'oral clearance' or '*apparent* oral clearance' or sometimes even just 'clearance', a potential cause of confusion. Obviously, a value for *CL* cannot be derived without knowing the proportion of the dose which reaches the systemic circulation, *F*. Because an accurate value of *F* cannot be obtained without the use of i.v. doses, it would seem to be better to use the data from i.v. studies to obtain systemic clearance. To avoid ambiguity, any value obtained for *D/AUC* from extravascular doses should be referred to as *CL/F* or CL_{oral}. Apparent oral clearance is commonly used in studies of special populations, using literature values of *F*.

7.3.2   Two-compartment models

Any of the appropriate equations in Chapter 5 may be used to calculate systemic clearance, including Equations 5.16 and 5.17 as they are mathematically related. However, Equation 4.14 is equally applicable, where *AUC* is obtained from the trapezoidal method plus extrapolation to $t = \infty$. This obviates the need to (i) define the model and (ii) calculate the values of the microconstants. It is important that sufficient data are collected so that the extrapolation is not more that 5–10% of the total area.

7.3.3   Systemic clearance at steady-state

For a drug infused at a constant rate, R_0, into a single compartment model until steady-state conditions apply (approximately $5 \times t_{1/2}$), *CL* can be substituted for λV in Equation 4.34:

$$R_0 = C^{ss}\, CL \tag{7.7}$$

Rearrangement gives:

$$CL = \frac{R_0}{C^{ss}} \tag{7.8}$$

Similarly, for repeated i.v. doses, substitution into and rearrangement of Equation 4.38 results in:

$$CL = \frac{D}{C_{av}^{ss}\tau} = \frac{\text{Dosing rate}}{C_{av}^{ss}} \tag{7.9}$$

where dosing rate = dose/dosage interval (D/τ). The usual problem arises with oral doses, if systemic availability is unknown then the clearance will be the apparent oral clearance, *CL/F*.

It might appear from Equation 7.7 that systemic clearance can be calculated from a single blood sample taken to measure the average steady-state concentration. While this is laudable from the point of view of generation of the maximum amount of information from minimal data, it should be remembered that an accurate value of C_{av}^{ss} requires determination of the *AUC* following a single dose (Section 4.2.6). Further, practical issues to be considered when calculating *CL* after multiple doses, include:

- The assumption that *F* does not change from single to multiple doses.
- That no enzyme induction or inhibition occurs.

- That linear kinetics apply after single and multiple dosing.
- That the subject is compliant in terms of dosage taking, including timing of doses.
- That enough multiples of the (unknown) half-life of the drug have elapsed to ensure that steady-state has been reached.

In reality, this approach is more likely to be successful with a drug for which the single dose clearance is known, and in testing to see if any of the changes bulleted above have indeed occurred.

7.3.4   *Additivity of clearance*

One of the features of clearance is its 'additivity'. Thus, if a drug is eliminated only by the liver and kidney, systemic clearance must be the sum of the two:

$$CL = CL_R + CL_H \tag{4.2}$$

When this is the case, renal clearance can be obtained as described previously (Equation 3.6) and CL_H obtained by difference. Note that Equation 3.6:

$$CL_R = \frac{U}{P} \times \text{urine flow rate} \tag{3.6}$$

is in keeping with the concepts of clearance discussed in this chapter because the urine concentration multiplied by the urine flow rate is the rate of elimination of the drug (e.g. mg min^{-1}) and P is the plasma concentration at the midpoint of the period of collection.

   If there are other mechanisms by which drug is being eliminated, pulmonary clearance, decomposition and metabolism by plasma esterases, for example, then the difference between CL and CL_R can only be described as *non-renal* clearance, CL_{NR}.

7.4   Hepatic intrinsic clearance

Hepatic clearance, CL_H can be defined as:

$$CL_H = Q_H E \tag{7.10}$$

where Q_H is blood flow through the liver and E is the extraction ratio. The equations which follow arise from a concept of the liver behaving as a single homogeneous pool (the 'well-stirred' model), which is obviously an over-simplification of a complex body organ. There are several other models discussed in the literature, for example the 'parallel-tube model,' and the 'dispersion model,' and it is also possible to invoke multiple plate ideas as in chromatographic columns The homogeneous well-stirred pool concept has a simplicity that facilitates our understanding of a broad range of pharmacokinetic observations, and it is the one most generally used. It assumes that the drug metabolizing enzymes are distributed evenly throughout the liver, and that the hepatic portal vein and the hepatic artery are equivalent in providing blood flow and therefore drug delivery to the liver. There are also differences in blood pressure between the hepatic artery and the hepatic portal vein, and these physiological differences could affect the interaction between substrates and enzyme surfaces. Clearly, during drug absorption, the drug concentrations are much higher in the hepatic portal vein than they can ever be in the hepatic artery, and this would be expected to affect the drug concentrations at the enzyme surfaces, which could in turn reduce the likelihood of first-order metabolism occurring. Also, if there is any product inhibition in the mechanism, then there may be differences in the extent to which this can occur in the two cases. Nevertheless, according to this model, the following holds true:

$$E = 1 - F \tag{7.11}$$

The applicability of this equation is dependent on there being no metabolism by the gastrointestinal mucosa. Equation 7.11 can be useful in scaling up from *in vitro* to *in vivo* (see later), and in understanding clearance calculations with data from oral doses, which are commonly exposed to the liver before they reach the remainder of the body.

Rane *et al.* (1971) predicted the hepatic extraction ratio from V_{max} and Km estimates *in vitro* using rat liver homogenates [Figure 7.1(a)].

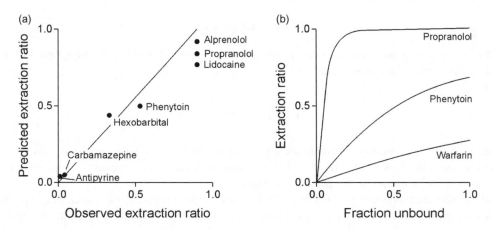

Figure 7.1   (a) Relationship between observed extraction ratios in perfused rat liver and the values predicted using V_{max} and Km values from metabolism in rat liver homogenates. The solid line is the line of idenity, slope = 1 (after Rane *et al.*, 1977). (b) Effect of plasma protein binding on the extraction of a highly extracted drug (propranolol), a poorly extracted drug (warfarin) and one with intermediate extraction (phenytoin) (from Shand *et al.*, 1976).

Hepatic intrinsic clearance, CL_{int}, is considered to be the maximal ability of the liver to remove drug irreversibly without any restrictions due to flow limitations or binding to proteins and so takes the form of Equation 7.2. However, when the substrate concentration is very low compared with Km, the equation can be written:

$$CL_{int} = \frac{V_{max}}{K\mathrm{m}} \tag{7.12}$$

Note that the components of Equation 7.12 are measured with different units in different situations, most obviously in the case of V_{max} which can have either mass/time or concentration/time units. Thus the expression V_{max}/Km is shown as identifying a first-order rate constant k (when the units are reciprocal hours) in Chapter 3. Values will be 'real' or 'apparent' depending on whether purified enzymes are used. This becomes especially important when V_{max} and Km concepts are applied to plasma concentrations of phenytoin (Chapters 5 and 19), and Equation 7.12 is the equation that is usually used for intrinsic clearance of a drug exhibiting first-order elimination kinetics. As hepatic blood flow increases, hepatic clearance increases to a maximum, the value of which depends on CL_{int}:

$$CL_H = Q_H \frac{CL_{int}}{(Q_H + CL_{int})} \qquad (7.13)$$

Comparing Equations 7.10 and 7.12, it follows that:

$$E = \frac{CL_{int}}{Q_H + CL_{int}} \qquad (7.14)$$

Thus, the extraction ratio is a function of flow rate; the larger the blood flow, the smaller the extraction ratio. This relationship has been validated using compartmental and perfusion models (Perrier and Gibaldi, 1974; Rowland *et al.*, 1973). However, if the intrinsic clearance is small relative to the flow rate, then the denominator in Equation 7.13 approximates Q_H, so:

$$CL_H \approx CL_{int} \qquad (7.15)$$

Such drugs are referred to as 'capacity-limited' or 'restricted'. For a drug with $CL_{int} > Q_H$, Equation 7.13 reduces to:

$$CL_H \approx Q_H \qquad (7.16)$$

These drugs are referred to as 'flow rate-limited' drugs. Note that if a drug is entirely removed from the body by hepatic clearance, then $CL_H = CL$.

7.4.1 Effect of plasma protein binding on elimination kinetics

There is little doubt that binding to plasma proteins can affect the rate of elimination of a drug; but in what way, and to what extent, may be difficult to predict. The equations in Section 2.4.3 apply to equilibrium between bound and unbound drug, and because plasma protein binding influences the apparent volume of distribution, it would be expected to affect elimination as:

$$t_{1/2} = \frac{0.693V}{CL} \qquad (4.7)$$

provided that protein binding does not influence CL, and that is where there have been misunderstandings as to the influence of protein binding. At one time, a widely held belief was that plasma protein binding inevitably delayed elimination because less drug was available to the drug metabolizing enzymes. This erroneous generalization was 'supported' by some studies that demonstrated a negative correlation between percent bound and degree of metabolism. However, it was pointed out that metabolism is dynamic and when unbound drug is metabolized, bound drug dissociates to maintain the equilibrium:

$$\text{Drug–protein complex} \underset{k_1}{\overset{k_{-1}}{\rightleftharpoons}} \begin{array}{c}\text{Protein} \\ + \\ \text{Drug}\end{array} \overset{k_m}{\longrightarrow} \text{Metabolite}$$

so for binding to delay metabolism, k_{-1} would have to be smaller than k_m (Curry, 1977, 1980). In a series of theoretical calculations, Gillette (1973) reasoned that 'it seems probable that the rate of dissociation of the drug-protein complex seldom becomes rate limiting in the metabolism of drugs'; indeed he demonstrated that it is possible for plasma protein binding to hasten metabolism by the efficient transport of drug to the liver.

7.4.1.1   *Influence of protein binding on hepatic clearance*

Wilkinson and Shand (1975) showed that there is a delay only if protein binding is high and intrinsic clearance is low. They did this by a modification of the original equations (7.13 and 7.14):

$$E = \frac{f_u CL'_{int}}{Q_H + f_u CL'_{int}} \tag{7.17}$$

where f_u is the fraction unbound and

$$CL'_{int} = \frac{E Q_H}{f_u(1-E)} \tag{7.18}$$

CL'_{int} is the intrinsic clearance of the unbound drug. (Note that in some older literature the symbol f_B was sometimes used for fraction unbound.) Modifying Equation 7.13 to take account of protein binding gives:

$$CL_H = Q_H \left(\frac{f_u CL'_{int}}{Q_H + f_u CL'_{int}} \right) \tag{7.19}$$

If the intrinsic unbound clearance is very small compared to the flow, Q_H, then Equation 7.19 approximates to:

$$CL_H = f_u CL'_{int} \tag{7.20}$$

Drugs with a low intrinsic clearance (capacity-limited) include warfarin and diazepam and, as predicted by Equation 7.20, the elimination of these drugs is affected by the degree of plasma protein binding. If the liver is the major route of elimination for these drugs, changes in CL'_{int}, resulting from enzyme induction or inhibition may markedly affect their elimination half-lives.

Some drugs such as propranolol and lidocaine have intrinsic clearances greater than liver blood flow and when the intrinsic unbound clearance is very large compared with the hepatic flow,

$$CL_H = Q_H \tag{7.21}$$

According to Equation 7.21 the clearance of these drugs will be unaffected by changes in plasma protein binding, but will be affected by changes in hepatic blood flow, as might occur with heart or liver disease or drugs that affect cardiac output. Enzyme induction or inhibition should have less impact on the kinetics of these drugs. However, for a constant rate infusion the steady-state concentrations will be:

$$C^{ss} = \frac{R_0}{Q_H} \tag{7.22}$$

by rearrangement of Equation 7.6 and substitution of Equation 7.21. A clinically important point is that Equation 7.22 predicts that the steady-state total concentrations will be unaffected by alterations in protein binding. There may be some change due to redistribution between plasma and tissue concentrations, but it is possible that total concentrations may remain reasonably constant when the unbound concentration increases. Thus dosing should be based on the unbound concentrations. Most drugs fall between the extremes of capacity-limited and flow-limited [Figure 7.1(b)].

Capacity-limited and flow rate-limited drugs may be referred to as lowly and highly extracted drugs, respectively, indicating the relationship with E. This is the case for the equations presented so far which consider the liver to be a homogeneous solution of the drug. An alternative model is the parallel-tube model of the liver which assumes an exponential gradient exists between arterial and venous blood such that:

$$CL_H = Q_H[1-\exp(-f_u CL'_{int}/Q_H)] \tag{7.23}$$

Now, rather than E, the critical term is $f_u CL'_{int}$. For example when this is larger than Q_H, Equation 7.23 approximates to Equation 7.21.

7.4.1.2 Influence of protein binding and volume of distribution on half-life

Wilkinson and Shand also examined the significance of tissue distribution and protein binding on half-life by using the following definition of volume of distribution:

$$V = V_b + V_t \frac{f_u}{f_t} \tag{7.24}$$

where V_b is the blood volume, V_t is the apparent volume of distribution made up of other tissues of the body, and f_u and f_t are the fractions of unbound drug in the blood and tissues, respectively. It was shown that the half-life is a function of volume of distribution, hepatic blood flow, fraction unbound, and unbound intrinsic clearance:

$$t_{1/2} = 0.693\left(\frac{V}{Q_H} + \frac{V}{f_u CL'_{int}}\right) \tag{7.25}$$

Increases in the left-hand term within the brackets (increased volume of distribution or reduced liver blood flow) will tend to increase $t_{1/2}$, as might be expected. An increase in intrinsic clearance will decrease $t_{1/2}$. The effect of binding is more complex, depending on whether the drug has a high or low intrinsic clearance. Basically however, with increased binding (i.e. decreased f_u) the right hand term of the part of Equation 7.25 in brackets will tend to increase $t_{1/2}$. For a drug with low intrinsic clearance, as f_u decreases from unity, half-life increases to become very long as f_u approaches zero whereas for a drug with high intrinsic clearance, the increase in $t_{1/2}$ is very much less marked (Figure 7.2). Taking the effect of tissue

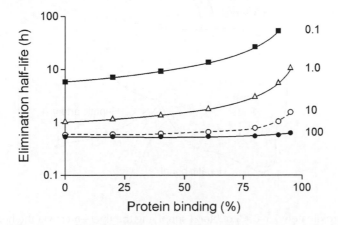

Figure 7.2 Effect of plasma protein binding and elimination half-life for four values of intrinsic clearance. Calculated from Equation 7.25, using $Q_H = 1.5\,\text{L}\,\text{min}^{-1}$, $V = 70\,\text{L}$ and intrinsic clearance $= 100, 10, 1$ and 0.1 times Q_H.

binding into account gives:

$$t_{1/2} = 0.693 \left(\frac{V_b}{f_u CL'_{int}} + \frac{Vt}{f_t CL'_{int}} \right) \qquad (7.26)$$

The precise effect on half-life will be determined in each case by the interplay of binding, intrinsic clearance, hepatic blood flow, and tissue binding. At one extreme, the half-life of propranolol was shown to be relatively short in the presence of high protein binding. In contrast, for drugs at the other extreme (e.g. warfarin and tolbutamide), the consequence of high plasma protein binding will be a long half-life. It should be noted that the combination of a low value for CL'_{int} and V, and a high value for percent plasma protein binding, is likely to be rare, because of the significance of the physical properties of drugs leading to the expectation that high V, low f_u, and high CL'_{int} will occur in parallel. The combination of low V, low f_u, and low CL'_{int} occurs with tolbutamide, warfarin, and non-steroidal anti-inflammatory drugs, and it is only with such drugs that protein binding effects on elimination are recorded.

7.4.2   *First-pass metabolism*

For a drug taken orally that is *completely absorbed* from the gastrointestinal tract and *only metabolized by the liver*, the fraction of the oral dose that reaches the systemic circulation, F, is given by rearranging Equation 7.11:

$$F = 1 - E \qquad (7.27)$$

In this case F can be considered to be the fraction of the dose that escapes first-pass metabolism. It is possible to show that:

$$CL_{oral} = CL_{int} \qquad (7.28)$$

It must be noted that this only applies when the strict caveats stated above apply. The ratio of AUC values is commonly used to evaluate F and Equation 7.27 used to calculate E.

However, oral and i.v. doses reach the liver by different routes. Oral doses more or less completely pass through the hepatic portal vein (which thus behaves like an artery), while i.v. doses pass through the hepatic artery (Figure 7.3). Once both doses are fully equilibrated within the body the hepatic artery becomes the

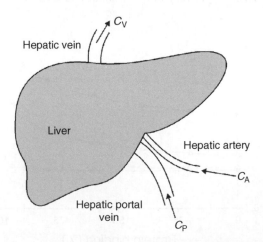

Figure 7.3   Approximately 75–80% of blood supplying the liver enters via the hepatic portal vein which carries deoxygenated blood containing substances that have been absorbed from the GI tract. Oxygenated blood from the heart enters via the hepatic artery. Blood leaves via the hepatic vein which drains into the vena cava.

major route for both. One can envisage two values of E, one for the hepatic artery/hepatic vein transfer, and another for the hepatic portal vein/hepatic vein transfer. Because the products of these two transfers intermingle in the hepatic vein, these two Es are not readily accessible (they can be assessed in heroic experiments in which radioactive doses are used, and different isotopes are incorporated into the i.v. and oral doses which are given together – and these experiments are most successful if the various blood vessels are separately sampled).

Failure to recognize that Equation 7.27 only applies when the liver is the only site of metabolism will result in erroneous conclusions as illustrated in Figure 7.4 which shows data relating CL_{mic} and F for a series of compounds. There is no useful correlation. In spite of this, $1 - F$ has been used for estimating E, and hence CL_H, and then back calculating from CL to intrinsic microsomal clearance.

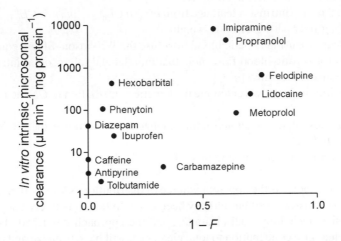

Figure 7.4 Comparison of *in vitro* microsomal intrinsic clearance from a variety of literature sources with assessment of bioavailability (F) for 13 representative drugs, showing only the most slender of relationships. Bioavailability data are from Goodman and Gilman (1996, 2001 and 2005), and *in vitro* data from 13 individual papers located by means of a literature search.

The literature concerned with intrinsic clearance creates potential for confusion concerning what is what. This occurs because the concepts were created on the basis of protein-free incubations *in vitro*, and assays of plasma concentrations (including protein-bound material) *in vivo*. There can be only one *in vivo* or hepatic 'intrinsic clearance' (the ability of the liver to metabolize the drug in the absence of delivery or availability restrictions). This should be given the symbol CL'_{int}. There is however also microsomal intrinsic clearance, which may or may not (usually not) reproduce hepatic intrinsic clearance – the symbol CL_{mic} has been used for this. This is measured experimentally *in vitro* using microsomes.

The calculation of clearance in hepatocytes can be of CL'_{int}, from the experimentally measured clearance in the incubation making use of the data on the hepatocyte concentration. This experiment may or may not reproduce the intrinsic clearance that occurs *in vivo*, as it involves hepatocyte reactions in the absence of blood flow or availability restrictions.

In vivo data starting with the experimental observation of systemic clearance, CL (using plasma assays that include protein-bound drug) is first used to calculate hepatic clearance (probably less than CL because of the renal, and other, contributions). Hepatic clearance, CL_H, can then be used to 'back-calculate' an equivalent of microsomal intrinsic clearance (using data for the number of hepatocytes per gram of liver, and again using the data for microsomal protein concentration per gram of liver) but it should be recognized that this does not calculate CL_{mic}. Rather, it calculates CL_{int}, which is equal to $f_u CL'_{int}$. Note that, because f_u is a fraction, $CL'_{int} > CL_{int}$, in keeping with the definition that CL'_{int} is the maximum hepatic activity in the absence of availability, that is protein binding and delivery, restrictions.

7.5   *In vitro* to *in vivo* extrapolation

It is desirable to use intrinsic clearance to help determine the expected systemic clearance, and hence the half-life, in human investigations. The strategy for this *in vitro/in vivo* scaling is relatively straightforward:

- Measure the half-life of metabolism of the drug *in vitro*.
- Calculate microsomal *in vitro* intrinsic clearance.
- Calculate microsomal *in vivo* intrinsic clearance from *in vivo* CL_{mic} = *in vitro* CL_{mic} × microsomal protein (mg per g of liver) × g liver per kg of body weight.
- Calculate hepatic clearance using hepatic blood flow (use the Wilkinson–Shand equation, Equation 7.13, and literature values for hepatic blood flow; note that this 'labels' microsomal intrinsic clearance as the only contributor to hepatic clearance).
- If the volume of distribution is known calculate the hepatic contribution to CL and hence the contribution to the half-life.
- If the percent of the dose that is excreted unmetabolized is known use the additivity of clearance to calculate the anticipated CL.
- Correct the calculation for protein binding using $CL_{int} = f_u CL'_{int}$.

In regard to the fifth point above, if the apparent volume of distribution is known then an *in vivo* experiment and assessment of half-life assessment has already been done. In fact, it is likely that the *in vivo* kinetics in a suitable animal species, including renal clearance, and the approach described above to determine the microsomal intrinsic clearance contribution to total clearance, will have been carried out. This collection of data can be used to make predictions for human beings, in combination with allometric scaling approaches (Chapter 15). A selection of scaling factors for the rat is given in Table 7.1.

Table 7.1   Scaling factors for the rat

Property	Value	Scaling values for standard weight rat (250 mg)
Liver weight	45 g kg body weight^{-1}	11 g
Liver blood flow	1.8 mL min^{-1} g liver^{-1}	20 mL min^{-1}
Hepatocyte number[a]	1.35×10^8 cells g liver^{-1}	1.5×10^9 cells
Microsomal protein yield[a]	45 mg protein g liver^{-1}	500 mg protein

[a] Literature averages.

Ideally, the intrinsic clearance obtained *in vitro* would equal that observed *in vivo*. Various investigators have studied correlations between *in vitro* and *in vivo* values, for both rats and humans. Typically, in quite complex studies, *in vitro* microsomal intrinsic clearance accounted for, on average, only about one-fifth of *in vivo* intrinsic clearance in humans (Naritomi *et al.*, 2001). Similarly, *in vitro* hepatocyte intrinsic clearance accounted for, on average, one fifth to one quarter of *in vivo* intrinsic clearance in rats (Lavé *et al.*, 1997). Further, *in vitro* hepatocyte intrinsic clearance accounted for, on average, approximately one fifth of *in vivo* intrinsic clearance in humans (Lavé *et al.*, 1997). Among the possible reasons for these results were:

- Intrinsic clearance *in vivo* includes all processes of elimination, including renal excretion, and non-hepatocyte, non-microsomal metabolism.
- While the calculations can allow for blood flow and protein binding effects, they do not allow for variations in fine detail of liver perfusion local to enzyme surfaces.
- Non-specific binding effects could reduce the actual drug concentrations at enzyme surfaces.
- There could be lack of homogeneity of distribution of the enzymes through the liver.

As the result, *in vivo* intrinsic clearance is found in fact to correlate quite well with extraction ratio. The best *in vitro* predictor of human *in vivo* data appears to be human hepatocytes.

7.6 Limiting values of clearance

Conceptually, it seems to be obvious that clearance numbers will relate to blood flow properties of organs, and will have upper limits, such as

- CL cannot exceed cardiac output: $5.3\,\text{L}\,\text{min}^{-1}$ (or $75\,\text{mL}\,\text{min}^{-1}\,\text{kg}^{-1}$).
- CL_H cannot exceed hepatic blood flow: $1.5\,\text{L}\,\text{min}^{-1}$.
- CL_R cannot exceed renal blood flow: $1.5\,\text{L}\,\text{min}^{-1}$.
- CL_R for drugs for which there is no renal tubular membrane transfer cannot exceed glomerular filtration rate ($125\,\text{mL}\,\text{min}^{-1}$ plasma $\cong 230\,\text{mL}\,\text{min}^{-1}$ blood).

Many measurements of clearance conform to these concepts and indeed the clearance of some compounds may be used to estimate plasma/blood flows, for example *p*-aminohippuric acid to measure renal plasma flow (Section 3.3.1.4). However, Table 7.2 shows a selection of clearance values, together with data for apparent volume of distribution and half-life, which show that in certain cases systemic clearance can exceed cardiac output. Although it might be expected, there is no obvious correlation between the values of clearance, volume of distribution, and half-life among the compounds in Table 7.2.

Table 7.2 Values of systemic clearance, apparent volume of distribution and elimination half-life for selected drugs

| Drug | Systemic clearance, CL | | V | Half-life |
	($\text{mL}\,\text{min}^{-1}\,\text{kg}^{-1}$)	($\text{L}\,\text{min}^{-1}$)	($\text{L}\,\text{kg}^{-1}$)	(h)
Glyceryl trinitrate	230	16.1	3.3	2.3 (min)
Prazepam	140	9.8	14.4	1.3
Triametrine	63	4.4	13.4	4.2
Azathioprine	57	4.0	0.81	0.16
Hydralazine	56	3.9	1.5	0.96
Isosorbide	45	3.2	1.5	0.8
Cocaine	35	2.5	2.1	0.71
Desipramine	30	2.1	34	18.0
Nicotine	18.5	1.3	2.6	2.0
Propranolol	12	0.84	3.9	3.9
Diltiazem	11.5	0.81	5.3	3.2
Chlorpromazine	8.6	0.60	21.0	30.7

Examples of very high clearance drugs also include physostigmine, esmolol, loratidine, misoprostol, spironolactone, and, according to some reports, selegiline. Some of the explanations as to why *CL* can appear to, or actually, exceed cardiac output could include:

- Experimental errors in measurement of *CL* or calculation errors resulting from use of inappropriate models.
- Widespread non-enzymatic chemical degradation of the drugs throughout the body.
- A major contribution from non-hepatic and non-renal elimination. This seems to occur with glyceryl trinitrate, which is extensively metabolized in blood vessel walls and something similar could occur with drugs metabolized by plasma esterases, e.g. physostigmine. In these cases metabolism or chemical degradation occurs continuously, independent of blood flow to any particular organ of the body.
- Also, we should not lose sight of the fact that, when clearance is calculated from *D/AUC* for a drug with a very high apparent volume of distribution and therefore a very low *AUC*, the errors in the result of the calculation will be relatively high.

A major consequence of the risk that exists of an erroneously calculated value for systemic clearance is that use of Equation 4.2 could lead to seriously incorrect estimates of hepatic clearance as it is based on the difference between systemic clearance and renal clearance.

7.7   Safe and effective use of clearance

- *In vitro* work can and should involve the measurement of microsomal intrinsic clearance for all drugs of interest. This is the core process of drug metabolism reduced to fundamentals free of blood flow and protein-binding influences.
- *In vitro* work can and should involve the measurement of hepatocyte clearance, and calculation of intrinsic clearance from the data involved, so that the contribution to clearance from the presence of enzymes in the intact liver cells can be evaluated.
- Use of perfused liver (*in vitro* or *in vivo*) is a valuable research technique, which can provide data on hepatic activity towards the drug with physiological and biochemical processes intact but with delivery controlled; the same could be said of perfused kidney, or renal artery/renal vein sampling, but this is not commonly attempted.
- Systemic clearance from *D/AUC* should be calculated in all *in vivo* work, but care must be taken in its interpretation, in particular it is important not to view it as an assessment of hepatic activity.
- Comparison of members of a series of compounds in drug development or clinical pharmacology (including Phase I) can use clearance values as operational numbers (for comparative purposes) provided due allowance is made for the potential 'noise' in such work.
- Dosing adjustments with digoxin and other drugs can be based on clearance rather than half-life.
- Clinical dosing must respect the phase of the kinetics that is in effect when dosing regimens or adjustments are made. For example, gentamicin dosing regimens are based on the distribution phase of a biexponential decay, while most dosing regimens are based on the terminal phase of a bi- or multi-exponential decay – clearances measured appropriate to the clinical objective must be used. This is considered in more detail in Chapter 19.

7.8   Physiological modelling

The concepts of organ clearances can be applied to what are known as 'physiological models'. In the previous chapters most of the pharmacokinetic models have been based on the concept of the body as one or more 'compartments' or 'pools' which are treated as if they contain homogeneous solutions of the drug. These models are useful for such things as deriving dosing schedules; however these compartments may have little or no relationship to anatomical spaces or organs. Furthermore, tissues in the same compartment can have markedly different concentrations, for example, the plasma and liver concentrations depicted in Figure 2.14(a). In physiologically based pharmacokinetic modelling (PBPK), the tissues and organs that play a role in the disposition of the drug being investigated are included in the model. Thus, there is no general physiological model; individual models will be dictated by the nature of the drug and to some extent by the route of administration. The various organs are connected by arterial and venous blood flows. Because of their importance in drug disposition the liver and kidney are usually included. If the lungs are included in the model then they are placed in series with the right and left heart (Figure 7.5).

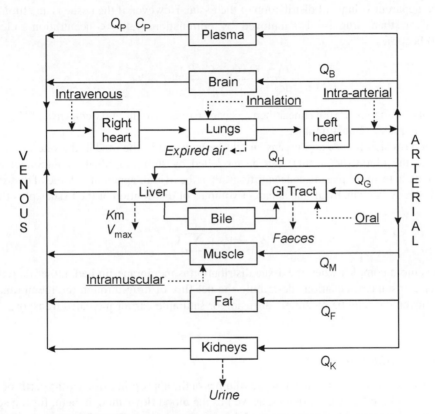

Figure 7.5 Hypothetical physiological model demonstrating how relevant organs are connected by arterial and venous blood flows and how various routes of administration can be depicted as required by the model.

Each organ or tissue type has an associated blood or plasma flow, Q_t and volume V_t. A further complexity is that each organ is modelled as consisting of plasma, interstitial and intracellular components. Thus many physiological factors can be incorporated in the model including the effects of plasma and tissue binding and the effects of the drug on blood flows to the organs, should that be appropriate; for example the effect of propranolol on cardiac output. *In vitro* data such as Km and $V_{\max}$ values from metabolism studies and partitioning between the components of a tissue can be incorporated.

In many situations the distribution of drug between tissue and blood is flow-limited. As blood flows through a tissue, drug is extracted so that at equilibrium the tissue to blood concentration is given by a partition coefficient, R_t:

$$R_t = C_t/C_B \tag{7.29}$$

The differential equation describing the rate of change in the tissue is:

$$\frac{dC_t}{dt} = \frac{Q_t C_B - Q_t C_t/R_t}{V_t}$$
$$= \frac{Q_t(C_B - C_t/R_t)}{V_t} \tag{7.30}$$

where V_t is the apparent volume of distribution of the tissue. However, if the tissue is an eliminating organ, such as the liver, then drug is also removed by elimination (rate = concentration × clearance) and Equation 7.30 becomes:

$$\frac{dC_t}{dt} = \frac{Q_t(C_B - C_t/R_t)}{V_t} - C_t.CL_t \tag{7.31}$$

If the kinetics of elimination are non-linear then CL_t can be related to the apparent Km and V_{max} values using an equation analogous to Equation 4.42.

Having decided which tissues and organs should be included in the model, the rate of change of drug concentration in the blood (or plasma) can be modelled by summing all the component terms. If the modelling is done in terms of plasma concentrations, then the overall apparent volume of distribution of the drug at steady-steady is given by all the apparent volumes of distribution of the tissues plus the volume of plasma, V_P:

$$V_{SS} = V_P + \sum V_t R_t \tag{7.32}$$

The situation is more complex when the tissue distribution is membrane-limited, rather than flow-limited. Under these circumstances, equations describing the net flux of drug through the membrane have to be derived. The movement may be by simple diffusion or saturable carrier-mediated transport.

7.8.1   *Practical considerations*

To utilize PBPK models is necessary to know the volumes of the appropriate tissues, the partition coefficients of the drug between blood and those tissues, as well as the blood flows though them, for the species under investigation. It may be possible to use published data for these values (Table 7.3) or it may be necessary to measure them. Sometimes allometric scaling is used (Chapter 15). Blood flows may be determined using such techniques as microsphere, laser Doppler velocimetry or tracer dilution techniques. It should be remembered that the total blood flow through the tissues cannot exceed cardiac output. Values of R_t can be obtained by infusing drug to steady-state conditions, after which the animals are killed for analysis of tissue concentrations so that Equation 7.29 can be used. Non-linearity of R_t values with increasing drug doses indicates binding or complex diffusion in the tissue being studied.

Once all the parameters have been obtained these can be used in the model. The plasma concentration data are not fitted statistically as in other models but the physiological parameters are adjusted to obtain the most appropriate model. Tissues with large blood flows and volumes will have the greatest influence on the model while smaller tissues may have little influence on the overall quality of 'fit'. Thus models, unsurprisingly, are likely to be heavily dependant on the liver and kidney.

Table 7.3 Physiological parameters for several species[a]

Parameter	Mouse	Rat	Monkey	Dog	Man
Body weight (g)	22	200	5000	17 000	70 000
Volume (mL)					
Plasma	1.0	9.0	220	650	3000
Muscle	10	100	2500	7500	35 000
Kidney	0.34	1.9	30	76	280
Liver	1.3	8.3	135	360	1350
Gut	1.5	11	230	640	2100
Plasma flow rate (mL min^{-1})					
Muscle	0.5	3.0	50	140	420
Kidney	0.8	5.0	74	190	700
Liver	1.1	6.5	92	220	800
Gut	0.9	5.3	75	190	700

[a]Bischoff *et al.*, 1971.

7.9 Inhomogeneity of plasma

Plasma is not necessary homogeneous with regard to drug concentration while absorption of oral or intramuscular doses is continuing. There will be a concentration gradient through the blood stream with the highest concentration just beyond the absorption site and the lowest concentration in blood arriving at the absorption site. This has been demonstrated for ethanol (Table 7.4). Blood ethanol concentrations were measured at five locations after oral administration. During absorption highest concentrations occurred in arteries and lowest in veins but when absorption was more or less complete (90–150 min) blood concentrations were almost homogeneous. It should be recognized, therefore, that pharmacokinetic analysis and concentration–effect studies may be markedly affected by time and site of sampling.

Table 7.4 Ethanol content of blood drawn simultaneously from various parts of the body: dog (15 kg) given 3 g kg^{-1} into stomach

Time (min)	Concentrations in blood (g L^{-1})				
	Artery Left heart	Jugular vein	Femoral vein	Right heart	Skin capillary
30	2.31	2.00	1.09		2.13
60	2.86	2.65	2.10	2.85	
90	2.91	2.80	2.60		2.74
150	2.58	2.50	2.46		
210	2.22	2.17	2.09	2.21	2.14
270	1.9	1.88	1.82		
330	1.65	1.63	1.6	1.65	1.62

Haggard and Greenberg, 1934.

The issues associated with arteriovenous differences in drug concentrations have been discussed in a two-part review by Chiou (1989a,b), which he described as 'critical or even provocative'. Generally, when a drug is administered, whether it be injected or taken orally, it is transported to the heart and enters the arteriolar circulation (Figure 7.5). In the early phases uptake by tissues reduces the venous concentrations relative to the afferent arterial ones [Figure 7.6(a)]. There comes a time when the arteriolar and venous concentrations become equal. At later times, when the plasma concentrations have been reduced by elimination, the tissues release drug into the blood flowing through them so that the venous concentrations are now higher than the arterial ones [Figure 7.6(b)].

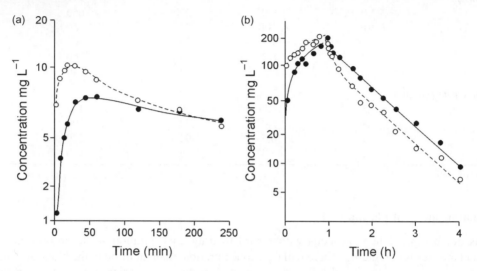

Figure 7.6 (a) Mean arterial (○) and peripheral venous (•) plasma concentrations of lidocaine in 10 patients after extradural injection of 400 mg. (b) Femoral arterial (○) and venous (•) plasma concentrations of propranolol in a 18.3 kg dog given a constant intravenous infusion (15.3 mg h^{-1}) for 60 minutes. (Redrawn from Chiou, 1989a.)

Using a physiology-based principle, the ratio of arterial to venous concentrations can be calculated:

$$\text{ratio} = \frac{1-\lambda R}{\dot{Q}} \tag{7.33}$$

where λ is the first-order elimination rate constant, R is the apparent partition coefficient and $\dot{Q}$ is the blood flow per unit weight of tissue. Therefore, short half-lives (larger λ) and extensive tissue binding will result in a large arteriovenous difference, whereas an increase in blood flow will tend to reduce the difference. Not only will there be differences in the arterial and venous concentrations but the concentrations in blood from different sampling sites may be different. Capillary blood concentrations should lie between those of arterial and venous blood, but may be affected by blood flow to the sampling site, for example finger-tip samples had greater ethanol concentrations than those collected from the big toe.

The consequences of inhomogeneity and sampling site on derived pharmacokinetic parameters have been largely ignored. The terminal half-lives may be the same [Figure 7.6(b)] but volume of distribution data based on intercepts on a concentration axis and even *AUC* estimations are likely to be in error. (Chiou, 1989a,b), throwing into doubt some of the fundamental concepts of the subject.

References and further reading

Bischoff KB, Dedrick RL, Zaharko DS, Longstreth JA. Methotrexate pharmacokinetics. *J Pharm Sci* 1971; 60: 1128–33.

Chiou WL. The phenomenon and rationale of marked dependence of drug concentration on blood sampling site. Implications in pharmacokinetics, pharmacodynamics, toxicology and therapeutics (Part I). *Clin Pharmacokinet* 1989a; 17: 175–99.

Chiou WL. The phenomenon and rationale of marked dependence of drug concentration on blood sampling site. Implications in pharmacokinetics, pharmacodynamics, toxicology and therapeutics (Part II). *Clin Pharmacokinet* 1989b; 17: 275–90.

Curry SH. *Drug Disposition and Pharmacokinetics*, 2nd edn. Oxford: Blackwell, 1977.

Curry SH. *Drug Disposition and Pharmacokinetics*, 3rd edn. Oxford: Blackwell, 1980.

Gillette JR. Overview of drug-protein binding. *Ann N Y Acad Sci* 1973; 226: 6–17.

Goodman & Gilman's The Pharmacological Basis of Therapeutics (Bruton L, Lazo J, and Parker K,eds) 11th. edn.; 2005, New York, McGraw-Hill. See also 10th edn (2001) Hardman JG, Limbard LE, and Gilman AG (eds) and 9th edn (1996) Hardman JG and Limbard LE (eds).

Lavé T, Dupin S, Schmitt C, Chou RC, Jaeck D, Coassolo P. Integration of in vitro data into allometric scaling to predict hepatic metabolic clearance in man: application to 10 extensively metabolized drugs. *J Pharm Sci* 1997; 86: 584–90.

Naritomi Y, Terashita S, Kimura S, Suzuki A, Kagayama A, Sugiyama Y. Prediction of human hepatic clearance from in vivo animal experiments and in vitro metabolic studies with liver microsomes from animals and humans. *Drug Metab Dispos* 2001; 29: 1316–24.

Perrier D, Gibaldi M. Clearance and biologic half-life as indices of intrinsic hepatic metabolism. *J Pharmacol Exp Ther* 1974; 191: 17–24.

Rane A, Wilkinson GR, Shand DG. Prediction of hepatic extraction ratio from in vitro measurement of intrinsic clearance. *J Pharmacol Exp Ther* 1977; 200: 420–4.

Rowland M, Benet LZ, Graham GG. Clearance concepts in pharmacokinetics. *J Pharmacokinet Biopharm* 1973; 1: 123–36.

Shand DG, Cotham RH, Wilkinson GR. Perfusion-limited of plasma drug binding on hepatic drug extraction. *Life Sci* 1976; 19: 125–30.

Wilkinson GR, Shand DG. Commentary: a physiological approach to hepatic drug clearance. *Clin Pharmacol Ther* 1975; 18: 377–90.

8

Drug Formulation: Bioavailability, Bioequivalence and Controlled-Release Preparations

8.1 Introduction

There was a time when it was believed that if a tablet contained its labelled quantity of active drug, then it was as clinically effective as a pure solution of that drug. Also, chemical equivalence was historically equated with clinical equivalence. However, it is now well established that differences in product formulation can lead to large differences in speed of onset, intensity and duration of drug response.

The study of formulation factors in pharmacological response is described as the science of 'biopharmaceutics'. The word 'bioavailability' may be used to describe the extent to which a drug is released from its pharmaceutical dosage form to be available to exert an effect. Regulatory authorities have defined bioavailability as:

> The rate and extent to which the therapeutic moiety is absorbed and becomes available to the site of drug action (Chen *et al.*, 2001).

Clearly, no single pharmacokinetic assessment measures both rate and extent, and some authorities would prefer the definition to refer to only extent. Of course, what is normally studied is *systemic availability*, that is, the appearance of the drug in the *general* circulation, which has the potential to vary in both rate and extent (Section 8.3). This is a composite result of pharmaceutical and biological factors. Note the potential for conflict between 'absorption' and 'bioavailability'. Also, the term 'bioequivalence' is commonly used, implying that two or more products are comparable, to some standard, in their release of active medication *into the blood* (see later).

Variations in bioavailability were first observed in the 1960s and 1970s as the result of:

- Therapeutic failures on changing to new suppliers of certain drugs (presumably as the result of lesser availability of the drugs in the second preparations).
- Increased incidence of unwanted effects and toxicity on changing to new suppliers of certain drugs (presumably as the result of greater availability of the drug in the second preparation).
- Observation of other differences, such as in clinical response.
- Observation of differences in drug concentrations in blood and blood fractions.

Drug Disposition and Pharmacokinetics, By Stephen H. Curry and Robin Whelpton
© 2011 John Wiley & Sons, Ltd

Scientific investigation is not straightforward. There is a hierarchy of testing methods for bioavailability. In descending order of accuracy, sensitivity and reproducibility, the methods are:

- *In vivo* tests in humans with plasma concentration measurements.
- *In vitro* tests (e.g. dissolution) that have been correlated with human *in vivo* data from the first bullet point.
- *In vivo* tests in animals that have been correlated with human *in vivo* data from bullet point 1.
- An *in vivo* test in humans based on urinary excretion of the active drug substance (not metabolites).
- An *in vivo* test in humans using measurement of pharmacological effec.
- A well-controlled clinical trial in humans testing therapeutic outcome, specifically conducted to test bioavailability, involving comparison of two products.
- A validated *in vitro* test, without the support of *in vitro/in vivo* correlation studies.

By far the most satisfactory bioavailability investigations are conducted *in vivo* in humans by the study of drug concentrations in blood and blood fractions. In practice, studies of the first bullet point are required for new products and for new formulations of existing drugs, and they should be supported and extended by batch testing with dissolution tests (second bullet point). Only when such evaluations are not possible will other methods be acceptable.

8.2 Dissolution

Strictly speaking dissolution is the process of active medicament dissolving in the fluid around it. However, it has long been recognized that the release of a drug from a tablet involves at least five steps in sequence. These are wetting of the dosage form, penetration of the dissolution medium into the dosage form, disintegration of the tablet, deaggregation of the dosage form and dislodgement of the drug-containing granules (see later) almost universally necessary in tablet production, and, finally, dissolution of the active medicament (as defined above). Pharmaceutical dissolution tests, such as those using Wood's Apparatus, which is a rotating disc system, attempt to reproduce this sequence of events. They do not reproduce the biological factors that play a part in transfer of drug to the systemic circulation. In bioavailability testing a correlation will be sought between *in vivo* (plasma concentration) data and dissolution data. This correlation may be seen as the same rank order of a number of different formulations in the *in vivo* and *in vitro* test results, and/or reproducibility of *in vivo* and *in vitro* data, such that the dissolution test can be applied to future batches in the quality control process, in the expectation that reproducible dissolution test data can be taken as assurance that the *in vivo* data will be reproducible. It is not practicable to test every batch *in vivo*.

8.3 Systemic availability

Bioavailability assessments using plasma concentrations rely on three fundamental descriptive pharmacokinetic observations: the maximum concentration, C_{max}, the time of the maximum concentration, t_{max}, and the area under the plasma concentration–time curve, AUC. C_{max} and AUC evaluate the extent to which drug becomes bioavailable, whilst t_{max} evaluates the rate at which a drug becomes available.

Measuring the proportion or fraction of a dose of drug which appears in the general circulation is relatively simple to assess by comparing AUC following the test dose with the AUC following an intravenous dose ($AUC_{i.v.}$). For a drug that is eliminated according to first-order kinetics, the area under the blood concentration curve–time curve from $t=0$ to $t=\infty$, is directly proportional to the amount of drug that enters the systemic circulation. However, plasma concentrations are usually measured on the assumption that there is also a direct relationship between blood and plasma concentrations of the drug. When a drug is injected intravenously, the entire dose enters the circulation and so the $AUC_{i.v.}$ can be used to estimate what fraction, F, of a dose given by an alternative route, reaches the systemic circulation. For an oral dose:

$$F = \frac{AUC_{\text{p.o.}}}{AUC_{\text{i.v.}}} \times \frac{Dose_{\text{i.v.}}}{Dose_{\text{p.o.}}} \qquad (8.1)$$

where $AUC_{\text{p.o.}}$ is the area under the curve flowing an oral dose. If equal sized doses are given then Equation 8.1 simplifies to:

$$F = \frac{AUC_{\text{p.o.}}}{AUC_{\text{i.v.}}} \qquad (8.2)$$

In sequential experimental designs, the doses of drugs to be compared are given to the same subjects with a suitable time interval between the doses to ensure that all of the first dose has been removed before the second is given. The AUC values are usually calculated using the trapezoidal method (see Appendix).

8.3.1 Effect of bioavailability on plasma concentration–time curves

It should be obvious that a reduction in bioavailability, say for an orally administered dose, will reduce the plasma concentrations relative to those of an equal size intravenous dose. However it is worth considering how the concentration–time curves are affected. Figure 8.1(a) shows typical curves for an intravenous injection and an oral administration for which the systemic availability, $F = 1$. Note that the concentration

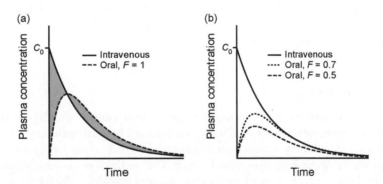

Figure 8.1 Comparison of plasma concentration–time curves for various values of F. (a) $F = 1$. At later times the concentrations after the oral dose must be higher than the i.v. and the areas of shaded areas must be equal. (b) $F = 0.7$ concentrations at later times are superimposed, whereas $F = 0.5$ all the oral concentrations are less that the i.v. ones. C_0 is the initial plasma concentration after the i.v. injection.

following the i.v. injection declines very rapidly, as would be expected for a drug eliminated according to first-order kinetics. The concentrations after the oral dose increase from zero to the maximum concentration (t_{max}) – the point at which the rate of absorption equals the rate of elimination – and then decline because from this point the rate of elimination is greater than the rate of absorption. The $AUC_{(0-\infty)}$ is the same for each route of administration (although of course infinite time cannot be shown on the figure). Part of the area under the curves is common to both routes but the shaded areas represent areas that are only under the i.v. or the oral curve. These areas must be equal. It is, at least in theory, possible for the concentrations to be superimposable at later times. This occurs when $F \sim 0.7$ [Figure 8.1(b)] whereas for lower values *all* the oral concentrations are less than those after the i.v. injection.

Although the curves in Figure 8.1 are labelled 'oral' the principles apply equally to other extravascular routes of administration such as intramuscular, subcutaneous and inhalation.

8.4 Formulation factors affecting bioavailability

8.4.1 *Origins of variation*

As implied earlier, the simplest oral preparation would be a pure solution of a drug in water, and oral solutions are sometimes used in comparisons with tablets in studies of 'relative bioavailability'. Solution preparations are used in therapeutics, but the bitter taste, instability and insolubility of many compounds necessitate complex formulations with solubilizing agents, flavouring agents and antioxidants. These difficulties could be overcome by packing the dry powdered drugs in gelatin capsules. However, machine filling of gelatin capsules without additives is difficult because of the small size of many drug doses. Hence the use of tablets, but as tablets have no outer gelatin shell, they must be made durable by compression. Various other inactive constituents (excipients) are needed for a variety of purposes, and they can lead to a considerable range of bioavailability problems. Apart from the active constituent it is generally necessary to include the following:

8.4.1.1 *Diluents*

As already mentioned, the weight of medicament is often too small for easy handling. This problem is commonly overcome by dilution of the active material with an inert material, such as lactose, starch, calcium phosphate or calcium sulfate, to increase the bulk. These substances can form complexes with the active ingredient, and affect solution of the latter in biological fluids once the preparation has disintegrated. They can also directly affect the drug absorptive process. The use of lactose in this context is waning because of the risk of lactose intolerance in some patients.

8.4.1.2 *Granulating and binding agents*

The drug, or drug–diluent mixture, cannot usually be pressed into a tablet of sufficient strength to survive buffeting in bottles. It is more satisfactory to first prepare granules by mixing the dry powder with a natural gum or mucilage, such as solution of acacia or tragacanth, or with syrup (sucrose), gelatin, povidone, various cellulose products or partially hydrolysed starch. The particles of powder are moistened with the granulating agent until they aggregate in relatively large angular granules, which have a very large surface area. The next stage is sieving, to control the granule size, and thorough drying of the now uniform granules.

The dry granules are commonly at this stage given a further external coating of the granulating solution, this time as a binding agent. The combined effect of the large surface area, the angular properties of the granules, and the adhesive properties of the binding agent then leads to a cohesive tablet when the mixture is divided into quantities of the required size and compressed into tablets in the tablet-making machine. These materials are present to preserve the structural integrity of the tablet, and inevitably retard drug release.

8.4.1.3 *Lubricants and surfactants*

The use of machinery in tablet making necessitates the use of lubricants. It is desirable that the granules should adhere to themselves but not to the tablet punches, and this ideal is achieved by the incorporation of a small amount of talc or other dry consumable lubricant in the powder. Materials used include stearic acid and various stearates, hydrogenated vegetable oils, polyethylene glycol and sodium lauryl sulfate. The term 'glidant' is sometimes used in this context. Some of the materials may be water repellent and so affect the 'wettability' of the disintegrating tablet. Equally, surfactants may be included to increase dissolution.

8.4.1.4 Disintegrating agents

The manufacturer may be content with a robust tablet, but the patient wants all of the manufacturers' work undone in order to effect rapid disintegration. Three types of agent can be incorporated into the tablet for this purpose.

- Substances (e.g. starch) that swell up on contact with moisture.
- Substances (e.g. cocoa butter) that melt at body temperature.
- Substances (e.g. a dry mixture of sodium bicarbonate and tartaric acid) that effervesce on contact with water.

Of these, the last is the most popular. Also used to aid disintegration are alginic acid, microcrystalline cellulose and colloidal silicates.

8.4.1.5 Miscellaneous

Apart from the above, it is sometimes necessary to include antioxidants to prevent decomposition, or substances (such as potassium carbonate) to control the pH of the tablet in the event of attack by moisture. Dyes will be included if coloured tablets are required, and dyes may be adsorbed on to aluminium hydroxide, providing another adsorbing surface for the drug and, in this case, another opportunity for a pH influence. If appropriate, flavouring may be included.

8.4.1.6 Coated tablets

Coating of tablets is carried out for both cosmetic and practical purposes. Sugar-coated tablets can be produced in bright colours and polished to a high degree for the purpose of making them attractive to the eye (a dubious virtue). Coating involves solutions of sucrose, as well as various coloured materials. The sugar and colours are dried on to the compressed tablets and the dry residue is mechanically polished. A practical purpose of coating is to ensure that the tablet reaches the intestine before disintegrating. This is achieved by including materials in the coating which are not attacked by the acid in the stomach, but which are attacked in the less acidic intestine. Control of tablet disintegration in this particular way *can* involve further additives. Finally, coated tablets are often marked with an identification symbol. This involves a small amount of an edible ink. Coating can involve the use of sugars, starch, calcium carbonate, talc, titanium dioxide, acacia, gelatin, wax, shellac, cellulose acetate, other cellulose materials and polyethylene glycol.

8.4.1.7 Capsules

Some of the above additives are required in capsules, although obviously not granulating agents. Particularly with capsules, colours, opaquants (e.g. titanium dioxide), dispersants, hardening agents (e.g. sucrose), fillers, lubricants and glidants are needed. There is of course, also the gelatin, which is available in a variety of types based on bone and skin waste from other industries. In particular, successful products using soft gelatin capsules are now common. There is some concern that gelatin will have to be phased out because its use is not acceptable to all patients.

8.4.2 Examples of drugs showing bioavailability variations

A number of important bioavailability examples have been studied. Some were potentially life-threatening. Some of the drugs that were reported as having problems of bioavailability are no longer available, or rarely prescribed, but they serve to illustrate the issues that needed to be considered. The features of a compound, or those materials mixed with it, which will lead to a bioavailability influence include:

- Particle size, in that smaller particles of drugs when released from tablets dissolve more quickly (phenacetin – now obsolete, nitrofurantoin, griseofulvin, sulfadiazine, spironolactone and aspirin).
- Crystalline form, salt form and complexing with tablet constituents, affecting the rate of solution in a similar way (chloramphenicol existed in various polymorphic forms).
- Low aqueous solubility leading to slow solution at the best of times.
- Wetting agents and other materials in the tablet affecting the interaction of constituent drug and aqueous media.
- Variations in tablet making consequent on granule compression, humidity, etc.

It is obvious that solubility in aqueous media is the key factor and it has been suggested that this is especially so when one of the following apply:

- A sparingly soluble drug is used in relatively high doses.
- Absorption only occurs in the upper part of the gastrointestinal tract.
- A steady and prolonged release of the drug is required.
- High early plasma levels are required for the desired action.
- The gastrointestinal contents or gut flora exert a destructive effect on the drug.
- The drug is absorbed by carrier-mediated mechanism.

Solubility, and particularly the rate at which solids dissolve, will be influenced by the particle size of the active ingredient (Figure 8.2).

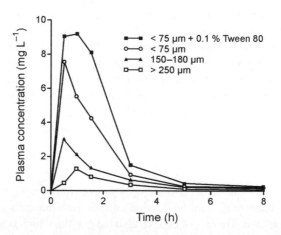

Figure 8.2   Mean plasma concentrations of phenacetin (acetophenetidin) in six human subjects after 1.5 g doses in suspensions of different particle sizes and with and without Tween 80 (after Prescott *et al.*, 1970).

Apparently simple differences such as using capsules rather than tablets, or vice versa, may result in marked changes in bioavailability and consequently clinical effects. For example, the urinary excretion rate of triamterene was greater following administration of tablets rather than capsules and this was accompanied by corresponding increases in sodium excretion (Figure 8.3).

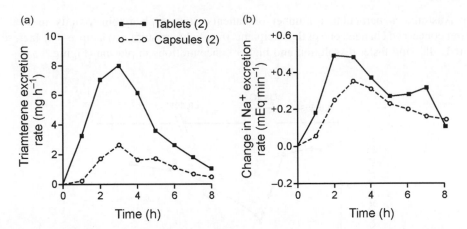

Figure 8.3 (a) Mean rate of excretion of triamterene in eight subjects following oral doses as capsules and tablets. (b) Mean rate of sodium excretion. (From the data of Tannanbaum *et al.*, 1968).

Considering bioavailability from the clinical point of view, Turner (1974) emphasized the facts that the possibility of a bioavailability problem achieving clinical significance would be enhanced with:

- Sparingly soluble drugs for which there is a close relationship between dissolution rate and plasma levels, and where different formulations with similar disintegration times show marked differences in dissolution times.
- In replacement therapy, such as for thyroid and adrenal cortical deficiency, and in diabetes mellitus. The clinical effects of small changes in the bioavailability of replacement drugs in conditions such as hypothyroidism and Addison's disease may develop only slowly and insidiously, and may not, therefore, be easily recognized until a serious condition has developed.
- In the control of serious clinical conditions in which there is a narrow range of optimum plasma concentrations of drugs for correct therapy.
- Therapies requiring the use of drugs with a very small therapeutic ratio, so that relatively small changes in plasma concentrations may lead to the development of signs of toxicity.

Unsurprisingly, the most dramatic bioavailability demonstrations have been with drugs with small therapeutic windows and marked toxicity. Problems with digoxin toxicity were explained by differences in tablet dissolution (Figure 8.4).

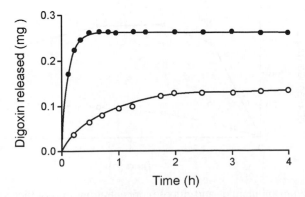

Figure 8.4 Dissolution *in vitro* of two formulations of digoxin (redrawn from Fraser *et al.*, 1972).

Also, in Australia in particular, a number of patients showed phenytoin toxicity in 1968 when the manufacturer concerned changed one of the excipients from calcium sulfate to lactose. The lactose was more easily wetted, allowing faster dissolution and higher concentrations in plasma (Figure 8.5).

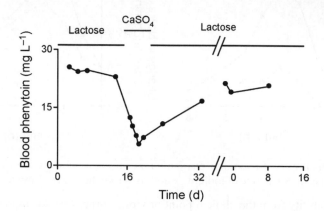

Figure 8.5   Influence of lactose and calcium sulfate as excepients on the concentrations of phenytoin in blood in a patient taking 400 mg per day. (Redrawn after Tyrer *et al.*, 1970.)

Similarly major differences in serum glucose concentrations after tolbutamide were shown to be due to formulation differences (Figure 8.6)

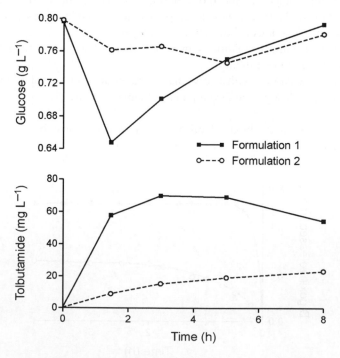

Figure 8.6   Mean serum tolbutamide and glucose concentrations in 10 subjects given identical dose in two different formulations (after Varley, 1968).

8.5   Bioequivalence

Bioequivalence studies are initiated to investigate differences between products; usually so called 'generics' (an unfortunate term) are compared with established preparations. For example, a generic diazepam might be compared with a proprietary brand such as Valium. The aim is not to show that the test compound is better than the established one but to show equivalence to it. Thus, if the innovator product (the proprietary brand) has low bioavailability then the new generic product must also have low bioavailability. In the United States the Food and Drug Administration (FDA) has defined bioequivalence as, 'the absence of a significant difference in the rate and extent to which the active ingredient or active moiety in pharmaceutical equivalents or pharmaceutical alternatives becomes available at the site of drug action when administered at the same molar dose under similar conditions in an appropriately designed study.'

Typically a bioequivalence study will involve *in vivo* testing of the generic drug against the standard drug in a cross-over design using 24 to 36 healthy, normal volunteers. Sometimes the study will call for the experiments to be conducted after meals but usually the subjects are fasted before they are given the drugs. Sufficient blood samples must be collected so that the C_{max}, t_{max} and AUC can be measured. The FDA usually considers two products bioequivalent if the 90% confidence intervals (CI) of the relative mean C_{max}, $AUC_{(0-t_z)}$ and $AUC_{(0-\infty)}$ of the generic formulation to reference is within 80% to 125% in the fasting state. Similar requirements were designed to apply in Australia, but the 90% CI values are based on log-transformed data. This is because $\ln(AUC)$ data are usually normally distributed. Protocols for bioequivalence evaluation will be designed with strict statistical control so that they adequately test for, say, $\pm 20\%$ differences between pairs of products. Much of the modern bioavailability literature is devoted to the design of such protocols, including detailed statistical control and analysis, as well as to debate on such issues as crossover and sequential study designs. The exact standard in any particular case will depend on what is practicable and desirable, as well as such considerations as the likely clinical result of, say, a 10% difference between products, and whether therapeutic objectives would be best met by emphasis on one, two or all three of the pharmacokinetic assessments commonly made, or on other criteria. Thus, there are commonly tighter requirements for drugs with a narrow therapeutic window (e.g. thyroxine and digoxin) and/or those with saturable metabolism (phenytoin, for instance).

In addition to requirements for new products, bioequivalence studies are required when a manufacturer changes the formulation of an existing product. Bioequivalence concepts are rarely applied to competing controlled-release preparations, or to immediate-release and controlled-release products with the same active constituent. Controlled-release products face very little generic competition.

8.6   Controlled-release preparations

Controlled-release preparations are either those that provide a sustained-release of drug or a delayed-release. The latter are usually enteric-coated oral preparations, not *designed* for delayed-release, but to avoid breakdown in the acid environment of the stomach, which of course will delay release to some extent depending on gastric emptying.

The principle of sustained-release is to ensure the rate constant of release and, hence, absorption (k_a) is less than the elimination rate constant (k) (Figure 8.7). As with any sequential reaction, the rate constant of the slowest step is rate limiting (Section 1.5.1.1). Therefore the half-life of elimination of the drug from plasma is determined by k_a not k, i.e. $t_{1/2} = 0.693/k_a$ (i.e. flip-flop, Section 4.2.4.2). This results in a longer duration of action which in turn means less frequent dosing, more convenient dosing and better patient compliance.

Sustained-release preparations are available for most routes of administration, including sublingual, oral, subcutaneous, intramuscular, transdermal and rectal. Oral preparations make use of wax matrices or tablets with different layers disintegrating at different rates, and capsules containing hundreds of pellets of different types (and often of different colours), each type disintegrating at a different rate. Transdermal preparations

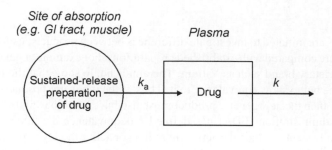

Figure 8.7   Principle of sustained-release: $k_a \ll k$ so the overall rate is determined by k_a.

include glyceryl trinitrate, hyoscine and nicotine patches where the patch includes a rate-limiting membrane to ensure a steady release of drug. Several preparations of insulin are available for subcutaneous injection, each giving different rates of release. Sustained-release preparations for intramuscular injection include microcrystalline salts of penicillin G and esters of the antischizophrenic drug, fluphenazine. In these preparations, the drug is slowly released from the injection site and so these injections are sometimes referred to as depot injections. The way in which the rate of release controls the time course of the drug in plasma is illustrated in Figure 8.8. Fluphenazine was quickly absorbed after intramuscular injection and the plasma half-life was approximately 12 h. However, when fluphenazine enanthate, dissolved in sesame oil, was injected fluphenazine was only slowly released, resulting in low but sustained plasma concentrations with a half-life of ~3.5 days. Note that although the *rate constant* of elimination is some seven times greater than the *rate constant* of absorption, initially all the dose is in the muscle and none in the plasma, so the *rate* of absorption is greater than the rate of elimination. It is only later, when the plasma concentration is greater, that the rate of elimination becomes greater than the rate of absorption.

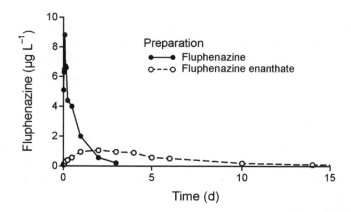

Figure 8.8   Fluphenazine concentrations in plasma after i.m. injections of fluphenazine, as the hydrochloride salt, and as the enanthate ester (after Curry *et al.*, 1979).

Fluphenazine enanthate and fluphenazine decanoate, which is even longer acting ($t_{1/2}$ of release ~12 days), are prodrugs, as the esters are hydrolysed to release the active drug. Long-acting i.m. preparations of penicillin G are microcrystalline salts; procaine penicillin G acts for approximately 3 days, whereas benzathine penicillin G acts for up to 7 days.

Enteric-coated preparations are used to reduce gastric disturbances, such as bleeding after oral administration of non-steroidal anti-inflammatory dugs such as aspirin and diclofenac, or ferrous sulfate. The delay in onset is not a problem when drugs are being used chronically. Coated tablets are also useful for drugs that are unstable in gastric acid, for example pancreatic enzymes given to sufferers of cystic fibrosis.

8.7   Conclusions

There is no doubt that bioavailability is a real and important factor in drug response, but it can now be considered to be under control. Thanks to legislation that has required higher standards of quality control, improved analytical methods, easier availability of human volunteer research facilities, a greater sense of responsibility within the industry, better scientific data on excipient factors in product performance, and more diligent use of dissolution testing, there should never again be therapeutic failures, product variations or toxicity induction on the scale seen in the 1960s and 1970s. It should be noted that bioavailability problems were most obvious with phenytoin, digoxin, thyroxine and tolbutamide, drugs with low therapeutic indices, and for which analytical methods were available at the time for the purpose of therapeutic monitoring. Modern research is designed to bring to the market drugs with better margins of safety.

There have, however, been two long-term social consequences of the enlightenment initiated by the bioavailability scares of the 1960s and 1970s, and of the improvements in the relevant science that ensued:

- A long-term distrust of generic drugs.
- A long-term practice of physicians, patients and pharmacists on insisting that thyroid hormone products, anticonvulsants (especially phenytoin), and, to some extent, digoxin, continue to be dispensed as the brand-named drug.

As noted earlier, both phenacetin and tolbutamide, used as examples here, and also chloramphenicol (which was the subject of extensive bioavailability research at one time) are now virtually obsolete, although not, primarily for bioavailability reasons. The other key examples, thyroxine, phenytoin and digoxin remain just too important in medicine for bioavailability considerations to adversely affect their positions in the pharmaceutical armamentarium.

The consequences continue to fuel a vigorous debate concerning the effectiveness and safety of *all* generic prescribing with different interest groups promoting their particular philosophies to the confusion of a distrustful consumer population. It can now be presumed that, unless proved otherwise for a specific case, that products that have been adequately tested for their bioequivalence, and are thus certified to be bioequivalent, are clinically equivalent.

References and further reading

Chen ML, Shah V, Patnaik R, Adams W, Hussain A, Conner D, *et al*. Bioavailability and bioequivalence: an FDA regulatory overview. *Pharm Res* 2001; 18: 1645–50.

Curry SH, Whelpton R, de Schepper PJ, Vranckx S, Schiff AA. Kinetics of fluphenazine after fluphenazine dihydrochloride, enanthate and decanoate administration to man. *Br J Clin Pharmacol* 1979; 7: 325–31.

Ding X, Alani AWG, Robinson JR. Extended-release and targeted drug delivery systems. In: Remington G, editors. *The Science and Practice of Pharmacy*. Philadelphia: Williams & Wilkins, 2006.

Fraser EJ, Leach RH, Poston JW. Bioavailability of digoxin. *Lancet* 1972; 2: 541.

Malinowski H, Johnson SB, Bioavailability and dissolution testing. In: Remington G, editors. *The Science and Practice of Pharmacy*. Philadelphia: Williams & Wilkins, 2006.

Prescott LF, Steel RF, Ferrier WR. The effects of particle size on the absorption of phenacetin in man. A correlation between plasma concentration of phenacetin and effects on the central nervous system. *Clin Pharmacol Ther* 1970; 11: 496–504.

Tannenbaum PJ, Rosen E, Flanagan T, Crosley AP Jr. The influence of dosage form on the activity of a diuretic agent. *Clin Pharmacol Ther* 1968; 9: 598–604.

Turner P. Trade names or approved names. Point II. Points of view. The clinical pharmacologist. *Postgrad Med J* 1974; 50: 93–5.

Tyrer JH, Eadie MJ, Sutherland JM, Hooper WD. Outbreak of anticonvulsant intoxication in an Australian city. *Br Med J* 1970; 4: 271–3.

Varley AB. The generic inequivalence of drugs. *JAMA* 1968; 206: 1745–8.

9

Factors Affecting Plasma Concentrations

9.1 Introduction

The factors that affect plasma concentrations of drugs include:

- Those associated with the disposition and fate such as route of administration (Chapter 2) and rate of elimination (Chapter 3).
- Pharmaceutical factors such as tablet and capsule properties (Chapter 8).
- General physiological factors such as time of administration of dose and food, diet and nutritional state, weight, sex, hormone balance including pregnancy, circadian rhythms, genetics (Chapter 10) and age (Chapter 11).
- Pathological state, especially diseases of the organs involved in disposition and fate (Chapter 12) and.
- Drug interactions (Chapter 17).

This chapter is concerned with those general physiological factors that are not the subjects of individual chapters.

In considering factors that affect plasma concentrations of drugs following single doses it is essential to refer to the standard rising and falling patterns discussed at the beginning of Chapter 4 and shown in model form in Figures 4.1 and 4.2. For single doses (Figure 4.1), we need to recall that with regard to plasma concentrations:

- A change in the *extent* of absorption will change the concentrations at all times, leading to a larger or a smaller area under the concentration–time curve (*AUC*).
- A change in the *rate* of absorption, with no changes in the extent, will lead to changes in one direction at early time points, with changes in the opposite direction at later time points, and no change in *AUC*.
- A change in the rates of metabolism and/or excretion will lead to changes in the concentrations in the same direction at all points, and to changes in *AUC*.
- Changes in tissue localization will lead to changes in concentrations in the opposite direction, e.g. increased tissue binding leading to reduced plasma concentrations.
- Any one factor can change one or more of the pharmacokinetic properties of the drug.

In regard to multiple-dosing, it is essential to refer to the standard pattern shown in Figure 4.2. With i.v. infusions, we note the constant pharmacokinetic steady-state concentration, and the time to reach that steady-state. With oral dosing, in addition to the average concentration within each dosage period and the time to reach that average, and the pharmacokinetic pseudo steady-state that it represents, it is necessary to consider

Drug Disposition and Pharmacokinetics,   By Stephen H. Curry and Robin Whelpton
© 2011 John Wiley & Sons, Ltd

the fluctuation between peaks and troughs. Thus factors affecting the rate and extent of absorption of a drug in different ways, and also factors affecting elimination, will cause complex changes in the time to reach a plateau and in the height of that plateau.

By applying the concepts discussed above, it should be possible to deduce the influence of any factor when it is described in terms of rate and extent of absorption, metabolism, excretion, or tissue localization. The factors affecting plasma concentrations are considered in this light in the following sections of this chapter.

9.2   Time of administration of dose

9.2.1   *Time of day, and association of dosing with meal times*

The dosing of drugs 'three times a day with meals' was at one time deeply ingrained in the practice of medicine. This was at a time when there was little pharmacokinetic knowledge, when most drugs were given according to this regimen, either because they were antacids or antispasmodics designed to reduce the unwanted discomfort experienced after the wanted enjoyment of a fine meal, or because the frequency of dosing was unimportant in comparison with the need to provide a reminder that it was time for a dose, and so increase compliance. At one time, new drugs were introduced with this regimen for no reason other than that this was standard practice.

Combining drug dosing with eating is not always good science, and the influence of food on drug absorption is considered in the next section. Also, a lunchtime dose is notoriously subject to being forgotten, partly because people are erratic in their lunch habits, and partly because people are at work at lunchtime and do not wish to be seen taking tablets and capsules by their workmates, or are just too busy to remember. The dose in the middle of the day is particularly difficult in the case of schoolchildren who need medication, so there is a powerful drive to design drugs with pharmacokinetic properties such that doses can be given twice a day or, better still, once a day. Additionally, there are variations in the receptivity of the body to doses given in the morning rather than the evening. This is considered in the circadian rhythms section. This is quite separate from the obvious fact that we want sleep-inducing drugs to exert their effects only at night, or when we want to sleep.

Generally speaking, daily doses are scheduled to be taken in the morning. However, pharmacological needs in this context can be quite subtle. For example, cholesterol synthesis is greater at night. The 'statin' drugs, which inhibit cholesterol synthesis, vary in first-pass effects, which can in turn be affected by diurnal rhythms. They also vary in that some of them are prodrugs, and also in drug interactions. In some cases, these drugs are recommended to be taken at bedtime, as a way of achieving greater efficacy. This would emerge as more cholesterol reduction per unit of dose during therapy. This timing of dose recommendation is not always followed – it might be most achievable by using once a day controlled-release technology. Controlled-release products have the potential to be less affected by daily rhythms.

An example of a dosing schedule based on a combination of pharmacological and pharmacokinetic knowledge leads to the recommendation that sedative antihistamines used in the treatment of allergic disorders be taken at bedtime, so that the sedative effect is experienced at night, leaving the residual antihistamine effect, sometimes exerted by metabolites with relatively long half-life values, to prevail the following day. Another example based on pharmacology is that some antihypertensive drugs are commonly recommended not to be taken at bedtime, as, if patients under the influence of these drugs arise during the night they are at considerable risk of orthostatic hypotension causing them to fall over at a time when their defensive reflexes are slow. A further example is the complex situation that can arise in patients taking several drugs, with the potential for absorption interactions. In this case it can be necessary to space out the various doses during the day, to avoid the effects of the various drugs on each other. Drug interactions are the subject of Chapter 17.

9.3 Food, diet and nutrition

9.3.1 Physical interaction with food

The presence of food in the gastrointestinal tract, particularly the stomach:

- Provides *adsorbing* surfaces to which drugs can adhere, in competition with sites of *absorption*.
- Prevents free access of drug molecules to the absorbing surface by reducing the efficiency of mixing.
- May provide lipid layers in which drug molecules will dissolve, mostly reducing their availability for absorption.
- Stimulates gastric acid secretion, which affects the ionization of weak electrolytes, and can cause chemical decomposition of some drugs.
- Modifies gastric emptying time, and hence changes the rate of movement of drugs from the stomach to the intestine.

It is thus not surprising that food can markedly reduce the rate, and sometimes also the extent, of absorption. This can be turned to good effect if a relatively low, prolonged effect is wanted, or if, as with some of the non-steroidal anti-inflammatory drugs reduction in gastrointestinal irritation and bleeding is required. It can be a nuisance if the highest possible peak concentration, or the earliest possible effect is required, or if bioavailability (Chapter 8) is reduced, so prescribing instructions sometimes include advice on timing of doses in relation to food, depending on the clinical need.

Food in the stomach both stimulates and delays gastric emptying. In the fasting state, the stomach experiences periodic waves of peristalsis which cause any accumulated fluid to pass through the pyloric sphincter. When solid or semisolid food enters the empty stomach, the pyloric sphincter closes or remains closed. A slurry (called 'chyme') is then created by the combined effect of stomach acid and the grinding effect of stomach muscle, so that virtually no solids enter the intestine. Once the slurry is formed, the pyloric sphincter opens to allow the slurry through. Fatty food, especially, slows gastric emptying. Drug particles become caught up in this process when taken with food. There are drugs, both therapeutic and non-therapeutic, that shorten or lengthen the gastric emptying time. Alcohol is one of the more important non-therapeutic drugs that slows this process, and when alcoholic beverages are combined with fatty meals late at night, and followed quickly by sleep, quite prolonged gastric emptying can occur, preserving the experience of a full stomach into the next morning.

Griseofulvin is an interesting example of a drug where absorption is faster and more complete when the dose is accompanied by dietary lipid – this drug is absorbed along with the fat molecules in the course of normal lipid absorption. Another interesting case arises with the anti-obesity drug orlistat, which exerts its effect by inhibiting gastric and pancreatic lipases, slowing the conversion of dietary fat into absorbable products, and hence reducing the absorption of those products. At the same time it can reduce the absorption of dietary fat-soluble vitamins, such as vitamins A and E, and β-carotene, and of fat-soluble drugs, such as warfarin and ciclosporin.

9.3.2 Macronutrients

The role of nutrition in the drug-metabolizing enzyme system in animals was reviewed by Campbell and Hayes in 1974. No specific role for carbohydrates had been shown, although sugar intake (glucose, sucrose and fructose) had been shown to decrease enzyme activity and to prolong hexobarbital sleeping times in mice (a pharmacological method of evaluating microsomal oxidation). A fat-free diet had been shown to depress activity, with decreased V_{max} for various substrates. Protein lack had been shown to reduce activity, as had vitamin deficiency, particularly with vitamins B, C and E and, to a lesser extent, vitamin A. Calcium, magnesium and iron deficiencies had been shown to reduce activity, as did starvation.

Early human studies supported these observations, with a highly significant partial correlation between protein intake and dosage requirements for digoxin. A lesser correlation was shown with fat intake, and this was ascribed to influences on renal clearance. Antipyrine and theophylline oxidation were early study topics, with half-life reductions being shown when subjects changed to a low carbohydrate/high protein diet (antipyrine from 16.2 to 9.5 h; theophylline from 8.1 to 5.2 h). A change to a high carbohydrate/low protein diet reversed these effects. These changes were ascribed to metabolic differences. Early studies on fasting in obese subjects showed no effect on antipyrine half-life, or on that of tolbutamide, but did cause changes in apparent volumes of distribution, in part because of greater fluid volumes, as well as fat masses, in obese people. Reduced dietary protein intake reduces synthesis of plasma proteins such as albumin.

More recent data support the above conclusions. Long-term consumption of a high-protein diet has been shown to increase the clearance of propranolol and antipyrine, a high-carbohydrate diet has been shown to reduce the clearance of theophylline, but a high-fat diet has been shown to increase the clearance of ciclosporin after i.v. injection and the fraction absorbed after oral dosing. Restriction of calorific intake has been shown to reduce the clearance of aminopyrine, as has intravenous nutrition with glucose. In regard to glucuronidation, a lack of sensitivity to diet has been shown with paracetamol and oxazepam. The changes that occur with alterations in dietary protein intake are supported by studies in patients on total parenteral nutrition (TPN). They can be detected within a few days of dietary changes.

At one time it was believed that dietary protein could change the activity of the P-450 system in an acute way, as it was shown that the extent of systemic availability of propranolol could be increased by as much as 70% within 5 minutes of consumption of a high-protein meal. The effect lasted for about 30 minutes. This occurred with both i.v. and oral (immediate-release) but not controlled-release propranolol. This could only be explained by a meal-related reduction in presystemic drug elimination, as the apparent volume of distribution, protein binding, plasma half-life and oral clearance of the enantiomers of propranolol were unaffected in these circumstances, as was the metabolic pattern. The explanation was that the meal caused a dramatic increase in hepatic blood flow, and as such, similar effects might be expected with all meals of any composition. However, it was shown that protein caused a particularly large increase in hepatic blood flow, raising the portal vein concentrations of propranolol after both i.v. and oral doses to levels that partially saturate the enzymes responsible for presystemic elimination. These levels were not reached with controlled-release propranolol. Similar observations have been made with several other high extraction ratio drugs, notably oral metoprolol, labetalol and intravenous lidocaine, and also with hydralazine, which is metabolized by acetylation. Similarly, although it was discovered in a long-term study, the observation with ciclosporin has been at least in part ascribed to an absorption influence in the belief that the ciclosporin complexed with fat passes membranes more rapidly.

9.3.3   Micronutrients

Pyridoxine supplementation can decrease the systemic availability of phenytoin and phenobarbital in epileptic patients. Levodopa can be affected similarly. This appears to be an effect on intrinsic clearance. Folic acid supplementation has a similar effect on phenytoin metabolism. Ascorbic acid can compete with ethinyloestradiol for sulfate conjugation, and dietary supplements can cause enhanced bioavailability of the steroid. Similar results have been observed with oral contraceptives, with an enhanced effect on blood clotting being observed. Also, high dose supplementation with ascorbic acid can impair antipyrine metabolism. Enriched vitamin K in certain plant foods can cause resistance to warfarin. However, the same plant foods can stimulate drug metabolism, and this result may have been caused by an increase rate of warfarin metabolism.

A variety of effects related to intake of particular dietary constituents, food additives, and herbal supplements have received intense study. These include cruciferous vegetables (such as cabbage, broccoli, cauliflower and spinach), which contain indoles that have the ability to increase the activity of the P-450 system. For example, the now-obsolete analgesic phenacetin (acetophenetidin) was shown to have 50% reduced bioavailability under such influences, and paracetamol metabolism is also similarly enhanced. The enzyme-inducing influence of polycyclic hydrocarbons in barbequed (charcoal-broiled) and smoked food has received much attention. In contrast, food additives, particularly butylated hydroxyanisole and butylated hydroxytoluene (BHT), which are approved preservatives in processed food and beverages, have been shown not to affect the pharmacokinetic properties of antipyrine and paracetamol. St. John's wort and grapefruit juice, a medicinal herb and a popular beverage, respectively, interact with a variety of drugs to a considerable extent, and are considered in Chapter 17.

9.4 Smoking

The properties of approximately one third of the one hundred most prescribed drugs are known to be affected by, or have the potential to be affected by, cigarette smoking. The most common influence is through enzyme induction, exemplified in clearance studies, but there are also influences on absorption. The main enzyme inducing agents are polycyclic aromatic hydrocarbon (PAH) compounds, among the thousands of constituents of cigarette smoke. These compounds induce the activity of CYP1A1, CYP1A2 and possibly CYP2E1. One model PAH is 3-methylcholanthrene, which was one of the first inducers of drug metabolism to be discovered. This compound is still used as a tool in drug metabolism and carcinogenicity research. Nicotine is primarily metabolized by CYP2A6, and it is also an inducer, increasing the activity of the same enzymes as those affected by the PAH compounds, and also inducing CYP2B1 and CYP2B2, but this effect is probably not clinically important. Examples of drugs affected by smoking include theophylline, propranolol, diazepam and chlordiazepoxide. The induction of propranolol metabolism is associated with a decreased effect on blood pressure and heart rate reduction. More recent drug examples include caffeine, imipramine, haloperidol, pentazocine, flecainide and oestradiol. Effects on oestrogen kinetics can negate the efficacy of oral contraception. When smokers give up smoking, this induction dissipates at a rate related to the turnover of microsomal protein. This includes past-smokers who use nicotine replacement therapy to help them achieve their 'cure'. Cigarette smoking also results in relatively fast clearance of heparin, possibly related to enhanced heparin binding to antithrombin III.

Insulin absorption from subcutaneous sites of injection can be relatively slow in smokers, because of the vasoconstriction caused by nicotine. However, and in contrast, absorption of inhaled insulin can be enhanced, leading to earlier and higher maximum insulin levels. Inhaled corticosteroids can also be affected leading to reduced effects in asthmatics who smoke. There is also the potential for a pharmacodynamic interaction involving sedative effects of central depressant drugs, such as the benzodiazepines and opioid analgesics, and the stimulant effect of nicotine, with the one effect competing with the other. This is analogous to the interaction between amphetamine and amylobarbital, given as a single regimen, in which each is thought to negate the effect of the other. This combination was once widely prescribed as an antidepressant despite the lack of evidence as to its efficacy.

9.5 Circadian rhythms

Circadian rhythm refers to a cycle in biochemical, physiological and behavioural processes of approximately 24 hours. The term, from the Latin *circa* ('around') and *diem* or *dies* ('day'), is attributed to Franz Hallberg. The formal study of such temporal rhythms is chronobiology and more recently the term 'chronopharmacokinetics' has been coined. Other related terms, such as 'chronokinetics' and 'chronopharmacodynamics' are to be found in the literature, together with 'diurnal variation,' and 'chronopharmacology'. Circadian rhythms are internally generated, but they can be reset by external cues, such as daylight, and even by conscious decisions, such as eating schedule. Functionally, they facilitate adaptation by the body to environmental changes and maintenance of homeostasis within a particular 'daily round'.

Central control is through the master 'biological clock' – the suprachiasmatic nucleus (SCN), or nuclei, a pair of cell groups in the hypothalamus. The SCN receives input from the retina and relays it to the pineal gland, which secretes melatonin, with production peaking at night. In the absence of input on the light/dark cycle, the body adjusts to a day varying from about 23.5 to 24.65 hours. The SCN processes information on, and also influences, hunger/satiety cycles, body temperature, and many other functions. In fact, food is a very important synchronizer (or 'zeitgeber') for this mechanism of homeostasis. There are believed to be other, independent peripheral 'oscillators' in the body with similar functions in relation to specific organs, including the liver. These clocks can take from 1–7 days to reset when influences are imposed voluntarily, such as with travel across time zones, and changes in the timing of meals, leading to the familiar total body disruption associated with long-distance air travel.

Diurnal variations occur in all of the physiological processes of importance in pharmacokinetics, such as gastric emptying time, gastric and urinary pH, and blood flow to the liver and kidneys. Consequently, diurnal variations are observable in at least drug absorption, metabolism and excretion. However, it should be noted that such variations occur in parallel with, for example, physical activity, and it can be difficult to determine the component contributions to pharmacokinetic variation in any particular case. Cardiovascular and non-steroidal antiinflammatory drugs have received special consideration in this context.

9.5.1 Absorption

Several studies have shown nifedipine and propranolol to have increased C_{max} and earlier t_{max} values after morning compared with evening administration. Because these differences occurred only with immediate-release dosage forms, and not with sustained-release dosage forms, and as these drugs are lipophilic, these differences have been attributed to the relatively short gastric emptying time and high gastrointestinal perfusion in the morning. This is presumed to result in faster absorption, and it is probably a general phenomenon. It is also thought that this would not be the case with less lipophilic drugs, such as atenolol, which has been shown to be less susceptible to this effect. However, it is important to note that drugs in this class are used for the treatment of hypertension. Blood pressure itself shows a circadian rhythm, thus introducing a pharmacodynamic sensitivity component. Time-dependent drug absorption is almost always a consequence of time-dependent gastric emptying, which is affected most obviously by physical activity, rather than the biological clock. However, physical activity can reset the biological clock. Thus the desire to go to the gymnasium may be triggered by the biological clock, or cause the biological clock to reset. Similar considerations apply to the effects of food, which are considered elsewhere in this chapter.

Studies with non-steroidal antiinflammatory drugs have revealed a number of observations, including, for example with diclofenac, highest C_{max} and highest AUC, but no differences in t_{max} and $t_{1/2}$, with morning doses. With indomethacin (Figure 9.1) and ketoprofen, similar absorption differences were noted, but additionally the terminal half-lives were higher in the evening. Additionally, with ketoprofen given i.v., there was an indication of slower elimination in the evening. In the case of salicylic acid, urinary elimination is apparently relatively slow in the morning.

Circadian rhythms in protein binding have been reported, dependent on temporal changes in plasma protein concentrations. This has been noted as significant with such highly protein bound drugs as carbamazepine, diazepam, phenytoin, valproic acid and cisplatin. This would be pharmacologically important only with highly protein-bound drugs that have low apparent volumes of distribution.

9.5.2 The liver

Animal investigations have shown circadian rhythms in hepatic blood flow and liver enzyme activity including expression of the cytochrome P-450 enzymes. Cytochrome P-450 enzymes metabolize melatonin

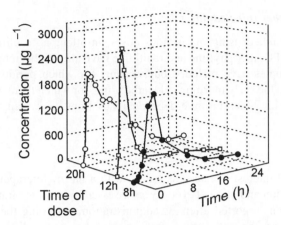

Figure 9.1 Indomethacin plasma concentrations according to time of dose, after oral doses of a sustained-release preparation, 75 mg h^{-1} (adapted from Bruguerolle, 1998).

in humans. Several examples of model compounds have been studied in rats, including, in particular, the *O*-dealkylase enzymes that detoxify various coumarin compounds. Higher levels of activity were shown during the dark period of the 24-hour cycle, but only in males. Similarly, erythromycin *N*-demethylase activity is relatively high in rats during the dark period, in this case in the second third of that period. While emphasis has been placed on the biological clock influencing P-450 activity, it should be noted that there is significant feedback with activity of CYP proteins affecting circadian rhythms – the pregnane X receptor (PXR) has a role in this – and, of course, the rat is a nocturnal animal.

In humans, plasma concentrations of 5-fluorouracil (5-FU) are subject to circadian variation in dihydropyridine dehydrogenase, which has relatively low activity at night. During constant rate i.v. infusion equilibrium concentrations were highest at 1 am, indicating relatively slow metabolism at that time. However, several studies in humans have failed to show a significant correlation between diurnal changes in enzyme activity and pharmacokinetics, including studies with tacrolimus, a substrate for CYP3A4, in spite of changes being noted in the hydrocortisol/cortisol ratio, an established biomarker for CYP3A4 activity. Using the constant rate infusion technique with theophylline, no diurnal variations were observed, although with ranitidine, relatively high concentrations were seen at 10 pm. With diclofenac there was no diurnal variation in half-life, while indomethacin and ketoprofen showed relatively high half-life values at 7 and 8 pm, respectively, but the highest salicylic acid half-life was at 6 am. The half-life of paracetamol was 15% longer at 6 am compared with 2 pm. Thus while the general picture prevails of relatively slow metabolism of drugs in humans at night, the opposite of the situation in rats, no general rule can be applied. It could be that, with all of the other influences at work in any particular case, the diurnal variation in enzyme activity is a minor factor. Hepatic blood flow may be more influential, for example, one study estimated hepatic blood flow in supine healthy individuals was greatest at 8 am. As a result, it might not be clear whether a difference seen with a drug with hepatic blood flow dependent properties was an effect of posture or a biological clock.

9.5.3 The kidney

Glomerular filtration, renal blood flow, urinary pH and tubular reabsorption have all been shown to have higher values during the daytime in humans. pH-dependent effects have been shown with salicylic acid and sulfasymazine, with faster excretion in the evening when the pH is relatively high. This is probably a food effect, with food choices made in the evening causing slight alkalinization of the urine.

9.5.4   Intravenous and other injected doses

Several drugs have been studied using intravenous doses, in attempts to eliminate the drug absorption component. Ketoprofen, 5-FU, theophylline and ranitidine have already been mentioned (see for example, Section 9.5.1). Other examples have included bupivacaine, for which peridural infusion over 36 hours revealed higher clearance in the early morning, and terbutaline, for which there were highest concentrations at 11 pm, and etoposide, with which no time-dependent variations were seen.

9.5.5   Pharmacodynamics

Theophylline is of especial interest because, as a drug with a narrow therapeutic index, the relationship between plasma concentration and clinical efficacy, as assessed by peak expiratory flow rate is important. Notwithstanding observations reported after i.v. administration, regarding the rate of metabolism being relatively slow in the morning, there is a contrasting report of therapeutically significant concentrations in patients being found after evening, but not morning, administration.

9.6   Weight and obesity

9.6.1   General principles

To standardize exposure between experimental animal species, drugs may be administered on a weight-corrected basis (usually $mg\,kg^{-1}$). Laboratory animals are obviously considerably smaller than humans and this is considered in detail in Chapter 15. Clinically, intravenous dosing apart, doses are mostly given in multiples of unit doses, as represented by a number of tablets or capsules, and while higher doses may be given to heavier people, exact proportionality in dosing is rarely attempted. Within a group of individuals, changes in dose on a $mg\,kg^{-1}$ basis mostly lead to proportionate changes in concentrations in plasma and tissues, and it might be reasonable to suppose that changes in weights of individuals will lead to proportionate changes in concentrations. However, it must be remembered that there can be different reasons for differences in body weight. Clearly, sumo wrestlers, ballet dancers, long distance cyclists and competitive swimmers, for example have differences in both weight and in the proportions of the various constituents of their weight. It is necessary to consider the *proportion* of bone, fat, muscle, and fluid as there may be different pharmacokinetic consequences of changes in the relative amounts of these components, when weight increases or decreases. This in turn leads us to take note of the lipophilicity or otherwise of any compound under investigation for influences of weight changes and differences, and especially consider localization of drugs in lipid deposits as well as their binding to muscle. Because tissue localization reflects equilibrium distribution between plasma water and binding or other sites in tissues, binding to plasma protein is also a part of this equation (Figure 2.12). It is relatively easy to make seemingly logical predictions concerning the likely effect of lipophilicity differences among a group of anaesthetics, based on data considered in Chapter 2, and to make similar predictions on the likely response rate of a lean person compared with a fat person in the case of thiopental anaesthesia. However, such predictions rarely match experimental data.

Human pharmacokinetic studies are, generally speaking, conducted in subjects and patients in the 'normal' range with regard to their weight and age. Rarely does a pharmacokinetic protocol involve tight control of the weight of the subjects, even less frequently are objective measures of weight used as selection criteria for inclusion in a study. There are in fact many objective measures of weight. For example, body mass index (BMI), historically known as Quetelet's number (or index), provides a convenient and useful indicator of body fat. BMI is the body weight in kilograms divided by the square of the height in metres. Individuals with values of 25–30 are considered to be overweight, and those with values > 30 are designated as obese. Obesity is further classified as moderate (BMI 30–35), severe (35–40), and morbid (> 40). Obviously, it is possible to have a very muscular body without excess fat, and thus have a relatively high BMI, as this index does not differentiate adipose tissue and muscle mass. Muscle has a higher density than fat, so the small framed but muscular ballet dancer mentioned earlier will have a higher specific gravity than the sumo wrestler. Nevertheless, generally speaking, a high BMI indicates an excessive body content of fat.

Ideal body weight (IBW) is a concept derived from data collected by the Metropolitan Life Insurance Company of New York. It relates weight to mortality data. It is an estimate of desirable weight corrected for sex, height and frame size. IBW is considered relevant only to life expectancy, and not to pharmacokinetics. However, there is a concept of lean body weight (LBW):

$$\text{Males}: \text{LBW} = 1.1 \times \text{TBW} - 0.0128 \times \text{BMI} \times \text{TBW}$$
$$\text{Females}: \text{LBW} = 1.07 \times \text{TBW} - 0.0148 \times \text{BMI} \times \text{TBW}$$

where in this instance TBW is total body weight (not to be confused with total body water). Some investigators use lean body *mass* (LBM), which has been derived by substituting the equation for BMI into the equations above:

$$\text{Males}: \text{LBM in kg} = 1.1 \times \text{TBW in kg} - 128(\text{TBW/height in cm})^2$$
$$\text{Females}: \text{LBM in kg} = 1.07 \times \text{TBW in kg} - 148(\text{TBW/height in cm})^2$$

It has been suggested that dosing of different drugs should be based on TBW, depending on their lipophilicity. Lipophilic drugs would be adjusted based on TBW, and hydrophilic drugs on LBW.

9.6.2   Obesity

Differences in distribution of drugs associated with obesity, for reasons connected with body size and composition might be expected, and because obese people have reduced cardiac output and slower tissue perfusion. A broad selection of drugs has been studied in the search for differences in their pharmacokinetic properties that can be related to obesity. Generally speaking, no consistent or important influences of obesity on drug absorption have been identified. However, there is no shortage of differences in apparent volume of distribution, both when expressed using volume units and also when weight corrected volume units are used. There is also no shortage of examples of obesity affecting clearance and half-life, but there is no consistent pattern of any obesity-induced changes in plasma protein binding of drugs or in renal excretion. There is a close association with cardiac output and obesity, with the potential to explain the changes seen in clearance when it is affected by, for example, hepatic blood flow. Table 9.1 shows a representative selection of examples of pharmacokinetic observations in obesity (Cheymol, 2000).

Table 9.1   Selection of pharmacokinetic observations in obesity. Each pair of numbers separated by / is the mean from control/obese patients

Drug	Therapeutic group	V (L)[a]	V (L kg^{-1})	CL (L h^{-1})	$t_{1/2}$ (h)
Ciprofloxacin	Anti-infective	219/269[b]	3.08/2.46[b]	44.6/53.8[b]	4.0/4.2
Ifosfamide	Anticancer	33.7/42.8[b]	0.53/0.55	4.33/4.56	4.9/6.4
Carbamazepine	Anticonvulsant	69.7/98.4[b]	0.96/0.87[b]	1.38/1.19	31.0/59.4[b]
Propofol	Sedative/anaesthetic	13.0/17.9	2.09/1.8	1.70/1.46	4.1/4.05
Dexfenfluramine	Appetite suppressant	668.7/969.7[b]	11.3/10.2	37.3/43.9	13.5/17.8
Propranolol	β-Blocker	180.0/226.8	3.1/2.4	41.6/46.2	3.4/3.9

[a]V_{ss} for propranolol.
[b]Significantly different.

The only consistent theme evident from these data is a totally expected increase in apparent volume of distribution in obesity. However, quite commonly, this is accompanied by a decrease in the value corrected for body weight, indicating that the drugs tend not to penetrate the excess weight proportionately. It would be expected that lipophilic drugs would show relatively high apparent volumes of distribution when the excess weight is primarily fat, and that hydrophilic drugs would show relatively high apparent volumes of distribution when the excess weight is primarily water, but this proposed pattern is not always confirmed by data. It is also possible that excess fat in particular, given the relatively low perfusion rate of fat, could require longer times for equilibration of drug concentrations and that apparent volumes of distribution in fatty obese people have been underestimated. Livers of obese people show fatty infiltration, and clearance can be lower, higher or no different in obesity, although studies with antipyrine have suggested that there are no specific effects on hepatic intrinsic clearance. However, the erythromycin breath test (Section 10.2.2) has shown a strong negative correlation with the percentage of IBW. Other examples, not included in Table 9.1, support this conclusion, and the examples in the table are typical of their groups. However, among the anti-infectives, vancomycin differs from ciprofloxacin by showing a shorter half-life in obesity. The anaesthesia group are of particular interest in not showing dramatic effects of retention in lipid in obese patients, in spite of their lipophilicity. The renal clearance of lithium, which, as an inorganic cation that is excreted unchanged, was found to be relatively fast in obese people in spite of there being no difference in the creatinine clearance in the patient groups studied. Finally, some oncologists dose anticancer drugs on the basis of IBW rather than TBW, as these drugs tend to be lipophobic.

9.7 Sex

There has been for many years a general belief that females are more sensitive to drugs than males. This has been based on observations with alcohol, on extrapolations from knowledge of glomerular filtration rates in males and females, on early studies that showed that male rats tend to metabolize drugs relatively rapidly, and on clinical observations, to some extent made by anaesthetists. However, differences in human beings have never been as prominent as those observed in laboratory rats.

9.7.1 Absorption and bioavailability

Gastric emptying time is relatively long in females, and this is thought to reflect effects of oestrogen. This can be expected to cause delays in absorption of drugs, with the same AUC, but longer lag times and lower rate constants of absorption, and hence lower C_{max} and later t_{max} values. There is no particular abundance of data supporting this proposed general rule. In fact, relatively fast absorption of salicylate has been shown in females, and a population study with mizolastine, an orally administered antihistamine drug, also demonstrated relatively slow absorption in males. However, absorption of ferrous sulfate has been shown to be relatively fast in prepubertal girls, apparently attributed to a hormonal effect, which would have to be on the carrier mechanism for iron.

Gastric alcohol dehydrogenase levels are relatively low in females. This leads to lesser presystemic metabolic losses, and so to relatively high blood alcohol concnetrations. This accounts, at least in part, for the observations of sensitivity differences with this drug. In contrast, intestinal concentrations of CYP3A4 do not show a similar, or indeed, any consistent pattern. One positive observation of a difference was made with oral verapamil, which is cleared relatively quickly in men, a difference not observed after i.v. doses.

9.7.2 Distribution

On average, body weight and BMI are relatively low in females, who also have a higher percentage of fat, and a relatively large plasma volume – and organ blood flow is also relatively high in women. On the basis of general principles, theories can be generated that these differences would be expected to cause tissue distribution differences, depending on the lipophilicity of the drug. A limited number of examples either do or do not support such theories. For example, vecuronium, a skeletal muscle relaxant, shows relatively fast onset of effect and relatively long duration of action in females, not immediately explained by assumptions about lipohilicity, tissue distribution and duration of effect. However, diazepam shows a relatively high apparent volume of distribution in females, and metronidazole, which has low lipophilicity shows a relatively low apparent volume of distribution in females. Also, metronidazole shows higher clearance in females supported by the lower *AUC*. Protein binding may make a contribution in this context, but observations made to date are inconsistent.

9.7.3   Metabolism

Although there are more data on metabolism, the pattern is no more consistent than in regard to absorption and distribution. There does seem to be a general trend of relatively fast rates of metabolism of CYP3A4 substrates in females, and males seem to have relatively high P-gp levels. In relevant physiology, cardiac output and hepatic blood flow are relatively low in females.

In vitro studies have mostly shown relatively high CYP3A4 concentrations in female tissue samples, and erythromycin and isofsamide have been shown to be metabolized more rapidly. However, it should be noted that *in vitro* experiments do not necessarily reproduce the hormonal differences that occur *in vivo*, and this could lead to inconsistent observations. However, *in vivo*, i.v. studies have shown relatively fast metabolism of erythromycin in females. With midazolam, the overall weight of evidence is that there are basically no differences between the sexes with either i.v. or oral doses. However, oral, but not i.v., verapamil is cleared more rapidly in males. This is reflected in blood pressure changes. The midazolam and verapamil data together seem to provide insight into the relative significance of the enzyme exposure and P-gp transport. To reach the enzymes, the substrate needs to penetrate the hepatocytes, and to do this it must by-pass the effect of P-gp. Midazolam is a substrate for CYP3A4 but not for P-gp-mediated efflux, whereas verapamil is a substrate for both. Thus verapamil shows sex differences because of differences in P-gp affecting the exposure of oral doses to CYP3A4, with relatively low intracellular concentrations of verapamil in males and faster presystemic metabolism. Generally-speaking, drugs that are substrates for both CYP3A4 and P-gp show sex-related differences, but those that are only substrates for CYP3A4 do not. Most of the work in this context has been with CYP3A4. In regard to CYP2D6 and CYP1A2 there is evidence of metabolism by males being relatively fast. In the case of CYP2C19 there seems to be no difference.

Hormonal changes *per se* seem to have little effect. Studies during the menstrual cycle with midazolam after oral and i.v. doses, of eletriptan, a migraine treatment, and with dextromethorphan have shown no effects. Premenopausal patients apparently metabolize midazolam faster than do postmenopausal patients, but this is not reversed by hormone replacement treatment. Analogous studies with erythromycin have shown no differences.

9.7.4   Excretion

The differences between males and females in glomerular filtration rate are relatively small and in proportion to body weight, so there are sex differences that are really weight differences. There is a small amount of evidence of sex differences in active tubular secretion of drugs, mainly obtained from studies with frusemide (furosemide), which is a substrate for a renal tubular transporter – the clearance in females is lower than that in males. Similarly, amantidine, which is a substrate for the organic cation transporter, shows faster clearance in males. The renal clearance of acetylsulfadimidine, the metabolite of sulfadimidine (sulfamethazine), is reduced in females (Table 9.2).

Table 9.2   Mean urinary clearance values for sulfadimidine and *N*-acetylsulfadimidine in 54 subjects (Curry 1980)

Compound	Renal clearance (mL min^{-1})	
	Male ($n = 35$)	Female ($n = 19$)
Sulfadimidine	3.82	3.93
N-Acetylsulfadimidine	21.16*	22.20^a

^{a}Sex difference significant $p < 0.5$ (*t*-test).

9.7.5   Effects

Links between sex differences in pharmacokinetics and effect have been sought with prednisolone, for which there are pharmacokinetic differences but no pharmacodynamic differences. In contrast, vecuronium shows a relatively high effect in females because of differences in the apparent volume of distribution. Verapamil sex differences, with which there is a correlation between effect and pharmacokinetic properties, have already been mentioned. In regard to centrally-acting drugs, a limited number of studies has confirmed that there are cases of relatively high pharmacological sensitivity in females. Undoubtedly, part of the sex difference in alcohol response is the result of pharmacokinetic influences, but there seems to be a pharmacodynamic contribution to this. With morphine, there is a relatively narrow therapeutic index in females, with a 60% higher incidence of nausea and vomiting associated with a comparable analgesic effect than that in males. With diazepam and some antidepressants, there is evidence of relatively high pharmacodynamic sensitivity in females. With propofol, there is about 30–40% higher effect in females. This is not attributable to differences in pharmacokinetics.

9.8   Pregnancy

Pregnancy leads to a wide variety of anatomical, physiological and biochemical changes, and all of them have the capacity to modify the pharmacokinetic properties of drugs. While there has always been a tendency to discourage the use of medication during pregnancy, because of the risk of teratogenic effects, many patients have to continue with chronic medication, such as with antiepileptics, antiasthmatics and antidepressants, during pregnancy. There is also an ongoing need for acute treatments, such as with anti-infective agents during pregnancy, and it has been estimated that pregnant women receive an average of 1.3 prescriptions per clinic visit.

9.8.1   Physiological and biochemical changes

The cardiovascular system shows profound changes in pregnancy. Cardiac output, heart rate and stroke volume increase, and peripheral resistance and blood pressure (except in abnormal situations) decrease. Plasma volume can also increase. Total hepatic blood flow can increase by over 50% above non-pregnant rates, especially in the third trimester. Renal blood flow and glomerular filtration rate also increase by as much as 50%, as the result of renal vasodilatation. Thus changes in drug absorption, tissue distribution, metabolism and excretion can all be proposed as likely. However, the pharmacokinetics of the majority of drugs remain to be studied in this condition (Hodge and Tracy, 2007). Renal excretion has been investigated the most. For example, in one study the renal clearance of atenolol, a drug commonly studied for its renal elimination because of its near dependence on the kidney for its removal from the body, was 12% above the postpartum level during the third trimester. Similarly, the renal clearance of digoxin, which is 80% excreted unchanged, increased by 21% and the clearance of lithium doubled during pregnancy.

The isoforms of the P-450 system show variable changes during pregnancy. For example, CYP1A2 and CYP2C19 show decreases in activity, and the half-life of theophylline, metabolized by CYP1A2, has been shown to be increased sufficiently to lead to a need for dosage reduction. Proguanil, which is converted to an active metabolite, cycloguanil, by CYP2C19, shows sufficiently decreased conversion to require an increase in dosage in pregnancy. Other isoforms show increased activity, so that, for example, fluoxetine (CYP2D6) shows relatively fast metabolism and relatively low plasma concentrations during pregnancy, and dosage increases may be needed. Phenytoin (CYP2C9) also shows increased clearance, and dosage adjustment based on maintenance of the optimum plasma concentration may be needed. There is also increased clearance of nicotine (CYP2A6) and methadone (CYP3A4). Conjugating enzymes also show a variety of effects. For example, the clearances of lorazepam (UGT2B7) and paracetamol (UGT1A1) are increased while that of caffeine (NAT2) and lamotrigine (UGT1A4) are decreased. The anticipated changes in half-life and *AUC* occur.

9.8.2   Hormonal effects

The significance of hormonal changes *per se* is not clear. Studies in pregnancy *per se* are in short supply, but much can be learned from studies during oral contraceptive use. There is considerable evidence of a component of hormonal control over the activity of CYP1A2, the activity of which is reduced among women taking oral contraceptives, although studies have shown no correlation between either oestrogen or progesterone levels and the activity of this isoform. An analogous situation exists with CYP2A6, the activity of which is increased during oral contraceptive use. The metabolism of omeprazole (CYP2C19) is decreased during oral contraceptive use, while that of dextromethorphan (CYP2D6) is apparently unchanged. Nifedipine and midazolam (CYP3A4) show decreased metabolism in users of oral contraceptives, and it is theorized that this may be due to inactivation of the CYP3A4 enzyme by oestrogen. However, this apparently does not occur in pregnancy, as neither oestrogen nor progesterone has been shown to inhibit, nor, for that matter, induce the enzyme *in vitro* or *in vivo*, although medroxyprogesterone has been shown to have an inhibitory effect on the human enzyme *in vitro*. Similar incomplete information is to be found for the conjugating enzymes.

9.8.3   Transporters

Studies of transporters, especially P-gp, multi-drug resistance associated protein, and breast cancer-resistance protein have been mostly restricted to the role of transporters in the placenta. There is some evidence that hormonal changes may induce or inhibit the expression of transporter proteins affecting intestinal uptake and efflux, and renal excretion and reabsorption. For example, *in vitro* accumulation studies with digoxin appear to indicate potential for changes in the disposition of this drug in the intestine and kidney during pregnancy, although details remain unclear. In the placenta, although approximately twenty transporter proteins have been identified, few have been linked to xenobiotic transport. For example, P-gp is expressed on the maternal side of the placenta, and the use of knock-out mice has shown that the experimental teratogen, avermectin is at least in part prevented from reaching the foetus by this efflux protein. Also, measurement of foetal levels of digoxin, paclitaxel and saquinovar, all P-gp substrates, has been used to demonstrate a similar exclusion. Measurement of P-gp levels in the human placenta have been considered to provide evidence of similar exclusions in humans. P-gp expression is relatively high in the earlier stages of pregnancy, the time when the foetus is most susceptible to teratogenic damage. Analogous, but less complete information, has been obtained for multi-drug resistance associated protein and breast cancer-resistance protein.

9.8.4   *The foetus*

Some of the enzymes involved in drug metabolism are present as early as the sixth week of gestation, although appreciable levels are mostly not reached until after birth. CYP3A7 is the predominant enzyme in the foetus, actually declining in activity after the end of the first week from birth. Data from *in vitro* studies support the belief that this enzyme has a role in detoxifying certain endogenous compounds, notably dehydroepiandrosterone-3-sulfate, and also the potentially toxic metabolites of retinoic acid. Enzymes detected as present also include CYP2C9, CYP2C19 and CYP2D6, and it can be presumed that these enzymes are present to metabolize exogenous molecules that fail to be excluded by transporters such as P-gp. There is relatively little expression of UGT activity in the foetus, although *in vitro* studies have shown measurable glucuronidation of morphine in cells from foetal livers.

9.9   Ambulation, posture and exercise

In the course of any day, humans stand up and walk, sit, lie down for rest and sleep, and deliberately exercise their bodies. Exercise can be described as 'acute' (relatively short-lived as in a 'work-out') or 'chronic' (a programme of training continued over a relatively long period of time). These activities obviously involve changes in posture among other body features. The distinction between acute and chronic exercise is important, as, for example, during acute exercise blood is shunted to the working organs, primarily the muscles, at the expense of other organs, some of them crucial in drug disposition and pharmacokinetics, such as the gastrointestinal (GI) tract and the liver. In acute exercise, heart rate, cardiac output, systolic blood pressure, and pulmonary ventilation all increase, while hepatic blood flow is decreased. In chronic exercise, cardiac output is relatively high, although resting heart rate is lowered, so perfusion of all tissues, including the GI tract and the liver, the brain, the kidneys, and tissues in which drug molecules are stored such as voluntary muscle and fat deposits is increased. This is assessed as increased regional blood flow. Also, blood volume and activity of oxidative enzymes is increased, while fat mass is reduced. Thus there is potential in both acute and chronic exercise for all of the drug disposition sites to operate differently. Maximal oxygen uptake ($VO_{2_{max}}$) expressed as $mL\,min^{-1}$ or $mL\,min^{-1}\,kg^{-1}$, is commonly used as the best measure of aerobic (cardiorespiratory) fitness, higher values being associated with higher fitness levels. Maximal oxygen uptake is considered to be reached when oxygen consumption shows no further increase with increased exercise workload, and represents the maximal capacity of the system to extract, deliver and utilize oxygen.

9.9.1   *The gastrointestinal tract*

Drug absorption is affected in several different ways by posture and body movement. For example, alcohol absorption is relatively slow when the body is in the supine position, for example during alcohol-induced sleep, as opposed to when standing and/or walking, and this has been attributed to relatively slow gastric emptying consequent on the adoption of the supine position, and also to a specific pharmacological effect of alcohol slowing gastric emptying. Exercise has little effect on gastric emptying until the intensity exceeds 70% of maximum oxygen uptake. Both acute and chronic exercise then have the ability to both shorten and lengthen gastric emptying time depending on the type of stomach contents involved. Intestinal transit time shows only minor changes with exercise, and then only with chronic exercise, when it is shortened. Intestinal blood flow is reduced in acute exercise, as the blood is shunted to the working organs, principally voluntary muscle, but it is increased by chronic exercise. The absorption of quinidine, salicylate and sulfadimidine (sulfamethazine) is unaffected by the level of acute exercise. In contrast, sulfamethazole, tetracycline and doxycycline showed increases in C_{max} on acute exercise, but the comparison was with bed rest (as opposed to upright but not exercising), which is a different posture. The *rate* of absorption of digoxin has been shown to be higher during acute exercise, again in contrast with the body in the supine position. One solitary chronic exercise absorption study with asozemide in rats, showed no effect.

9.9.2 Transdermal absorption

Exercise increases skin temperature, and can cause vasodilation, or vasoconstriction if compensatory mechanisms of the vasomotor centre prevail. In this context, glyceryl trinitrate (GTN, nitroglycerin) plasma concentrations obtained from skin patches have been shown to increase three-fold in exercise, with a weak correlation with skin temperature. However, GTN concentrations in plasma are very sensitive to posture, and to hepatic blood flow, and observations attributed to exercise may well be related to changes in the metabolism of the drug in blood vessel walls, caused by changes in cardiac output (Figure 9.2).

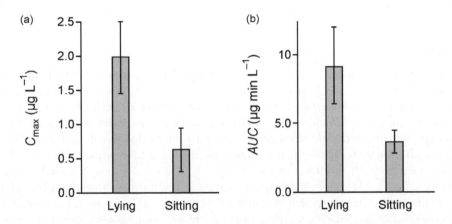

Figure 9.2 (a) Maximum concentration and (b) area under the curve for glyceryl nitrate in two matched groups of healthy volunteers, one group lying down (face up) and the other sitting. Data are mean ± SEM. (Adapted from Curry and Kwon, 1985.)

9.9.3 Tissue distribution

Because blood is shunted to the active tissues in acute exercise, changes in apparent volume of distribution are to be expected, and animal studies with atropine, theophylline and antipyrine, in which reductions in apparent volume of distribution were observed, support this.

There were no changes in plasma protein binding of verapamil and propranolol during acute exercise. However, digoxin concentrations in skeletal muscle have been shown to be increased during exercise, with corresponding decreases in concentrations in erythrocytes. Thus a shunting from red cells to active tissues occurs, but most likely with no impact on myocardial concentrations.

Chronic exercise is likely to be associated with reduction in fat mass and an increase in lean mass, with potential for effects on tissue distribution related to muscle binding and fat uptake. This has been simulated for thiopental, using the kind of information discussed in Section 2.4.4.1. In this work, it was shown that the apparent volume of distribution would increase by as much as 76% in overweight individuals (physically inactive) when compared with lean individuals (exercising regularly). Chronic exercise has been shown to increase the apparent volume of distribution of antipyrine in mares, but only by 8%, although studies in humans with antipyrine and procainamide have not led to comparable conclusions.

9.9.4 *Subcutaneous and intramuscular injections*

Subcutaneous and i.m injections are greatly affected by blood flow to the injection site (Section 2.3.5). Thus injections into the legs are released into the blood more rapidly during ambulation and/or acute exercise. It is thought that this is more likely to be the result of a 'massage' effect of the muscles rubbing against each other and against the skin, rather than the blood flow effect *per se*, although it would be counter-intuitive to think that there would be no 'mixing' effect resulting from increased delivery of blood to the absorption site. This massage effect is analogous to the hand rubbing of the injection site widely practised as an effective means of helping absorption of injections, although this practice may have more to do with psychological soothing than drug absorption. Data in this context has come almost entirely from insulin (subcutaneous injections) and atropine (intramuscular injections). It is important for diabetic patients self-injecting with insulin to know and understand the significance, to their own particular needs, of the site of injection in relation to their level of physical activity.

9.9.5 *The liver*

Hepatic blood flow is reduced during acute exercise. This has been demonstrated with indocyanine green clearance studies and has obvious implications for drugs with blood flow-dependent clearance, such as lidocaine, the plasma concentrations of which are higher during exercise. However, verapamil and propranolol showed no change in clearance, and there is some evidence that the lidocaine observation is more related to posture than to hepatic blood flow controlled by blood flow shunting. Theophylline clearance is reduced in exercise, but there is evidence from antipyrine, diazepam, and sulfadimidine acetylation plasma studies that there are basically no effects on enzyme activity. Chronic exercise apparently does not cause a long-term increase in liver blood flow, and, generally speaking, few positive effects have been shown on drug metabolism. In one study, a relatively short half-life was observed with aminopyrine in trained athletes, and an analogous observation has been made with antipyrine. Prednisolone has also been shown to have a faster elimination rate in trained athletes.

9.9.6 *The kidneys*

Renal blood flow is reduced as the result of exercise, as is the glomerular filtration rate, shown by creatinine clearance. Plasma protein concentrations can change either way with exercise, because of fluid shunting, and urinary pH being reduced. As a consequence, atenolol renal clearance is reduced, as is that of digoxin, along with active tubular secretion of procainamide.

9.9.7 *Body temperature*

Acute exercise increases body temperature by 1–2 °C. It has been suggested that this could increase the kinetic energy of drug molecules, increasing the rate of diffusion across membranes in the gastrointestinal tract and kidney, and increasing the activity of the drug metabolizing enzymes. However, available data do not seem to support this.

References and further reading

Baraldo M. The influence of circadian rhythms on the kinetics of drugs in humans. *Expert Opin Drug Metab Toxicol* 2008; 4: 175–92.

Benowitz NL. Clinical pharmacology of nicotine: implications for understanding, preventing, and treating tobacco addiction. *Clin Pharmacol Ther* 2008; 83: 531–41.

Benowitz NL. Pharmacology of nicotine: addiction, smoking-induced disease, and therapeutics. *Annu Rev Pharmacol Toxicol* 2009; 49: 57–71.

Bruguerolle B. Chronopharmacokinetics. Current status. *Clin Pharmacokinet* 1998; 35: 83–94.

Campbell TC, Hayes JR. Role of nutrition in the drug-metabolizing enzyme system. *Pharmacol Rev* 1974; 26: 171–97.

Cheymol G. Effects of obesity on pharmacokinetics implications for drug therapy. *Clin Pharmacokinet* 2000; 39: 215–31.

Conney AH, Gilman AG. Puromycin Inhibition of enzyme induction by 3-methylcholanthrene and phenobarbital. *J Biol Chem* 1963; 238: 3682–5.

Curry SH. *Drug Disposition and Pharmacokinetics*, 3rd edn. Oxford: Blackwell Scientific, 1980.

Curry SH, Kwon HR. Influence of posture on plasma nitroglycerin. *Br J Clin Pharmacol* 1985; 19: 403–4.

Datz FL, Thorne DA. Cause and significance of cold bone defects on indium-111-labeled leukocyte imaging. *J Nucl Med* 1987; 28: 820–3.

Franconi F, Brunelleschi S, Steardo L, Cuomo V. Gender differences in drug responses. *Pharmacol Res* 2007; 55: 81–95.

Froy O. Cytochrome P450 and the biological clock in mammals. *Curr Drug Metab* 2009; 10: 104–15.

Green B, Duffull SB. What is the best size descriptor to use for pharmacokinetic studies in the obese? *Br J Clin Pharmacol* 2004; 58: 119–33.

Harris RZ, Jang GR, Tsunoda S. Dietary effects on drug metabolism and transport. *Clin Pharmacokinet* 2003; 42: 1071–88.

Hodge LS, Tracy TS. Alterations in drug disposition during pregnancy: implications for drug therapy. *Expert Opin Drug Metab Toxicol* 2007; 3: 557–71.

Kroon LA. Drug interactions with smoking. *Am J Health Syst Pharm* 2007; 64: 1917–21.

Persky AM, Eddington ND, Derendorf H. A review of the effects of chronic exercise and physical fitness level on resting pharmacokinetics. *Int J Clin Pharmacol Ther* 2003; 41: 504–16.

Schein JR. Cigarette smoking and clinically significant drug interactions. *Ann Pharmacother* 1995; 29: 1139–48.

Smith RG. An appraisal of potential drug interactions in cigarette smokers and alcohol drinkers. *J Am Podiatr Med Assoc* 2009; 99: 81–8.

van Baak MA. Influence of exercise on the pharmacokinetics of drugs. *Clin Pharmacokinet* 1990; 19: 32–43.

Walter-Sack I, Klotz U. Influence of diet and nutritional status on drug metabolism. *Clin Pharmacokinet* 1996; 31: 47–64.

Zevin S, Benowitz NL. Drug interactions with tobacco smoking. An update. *Clin Pharmacokinet* 1999; 36: 425–38.

10

Pharmacogenetics and Pharmacogenomics

10.1 Introduction

Genetic differences in drug response may be due to differences in pharmacodynamics (different receptor populations) or in drug disposition (differences in drug metabolizing enzymes and transporters). When genetic differences are due to a single gene mutation and the incidence is $> 1\%$ then such differences may be detectable in population studies as a bi- or trimodal distribution. There is interest in such polymorphisms because metabolism inactivates or activates not only drugs, but also carcinogens and procarcinogens, and much of the recent literature is devoted to assessing the role of genetics as a risk factor in cancer. The term pharmacogenetics is usually applied the study of drug interactions with a relatively restricted number of genes, whereas pharmacogenomics aims to study the effect of the entire complement of genes (i.e. the genome) on drug action. As this is a rapidly developing field, anything written one year is likely to be superseded within a few years. Thus, this chapter will use selected examples to illustrate the principles involved.

10.1.1 Terminology

In mammalian cells, *chromosomes* are thread-like structures in the nucleus, comprised of DNA and associated proteins. Typically there are two sets of chromosomes, arranged in pairs (*diploid*), one set being inherited from each parent. *Genes* are sequences of nucleic acids located on regions (*loci*) of chromosomes that define the characteristics or traits of the organism. They can be considered as the basic units of heredity. Different forms of a gene are known as *alleles* and it is possible to inherit the same alleles, in which case the individual is referred to as a *homozygote* and said to be *homozygous*, or when the alleles are different, a *heterozygote*. The genetic make up of an organism is known as the *genotype*. *Phenotype* refers to the physical characteristics exhibited and these can be influenced by inherited and environmental factors. The phenotype in heterozygotes will be largely determined by the interaction of the different alleles which can often be referred to as *dominant* or *recessive*. In the simplest case, a dominant allele will produce the same phenotype as that when both dominant alleles are present. Dominant alleles may be denoted R, and recessive ones r, so that heterozygotes are Rr whilst homozygotes are either RR or rr. The different phenotypes are referred to as being *polymorphic* (having different forms). Clinically it is the phenotype that is important but knowledge of the genotype may help to explain the phenomenon.

By convention genes (e.g. *CYP2C19*) and alleles (e.g. *CYP2C19*1*) are written in italics, and in capitals when referring to human genes, whilst the gene products (enzymes, transporters, etc.) are written in the standard font (e.g. CYP2C19*1 or CYP2C19.1). The term *wild-type* may be encountered. It was introduced to describe the form of allele found in nature, that is it was considered to be the 'standard' or 'normal' allele,

Drug Disposition and Pharmacokinetics,   By Stephen H. Curry and Robin Whelpton
© 2011 John Wiley & Sons, Ltd

others being mutant alleles. However, most genes exist in a variety of forms, the frequency of which varies depending upon the geographic range of the species and, in the case of humans, the extent to which populations have migrated and interbred.

Often alleles occur because of a *single-nucleotide polymorphism* (SNP), which can give rise to a protein in which one amino acid is substituted for another. In the case of enzymes this may result in reduced activity or no activity. The site(s) of the SNPs may be identified, for example in CYP1B1*3 cytosine is replaced by guanine, C4326G; this produces an enzyme where leucine at 432 is replaced by valine, Leu432Val. Occasionally SNPs lead to enzymes with increased activity. A *null allele* is one that either produces no protein or the protein lacks any function.

10.2 Methods for the study of pharmacogenetics

10.2.1 Studies in twins

An obvious way to investigate whether a phenomenon is genetically related is to investigate it in twins. The major influence of genetic control on drug disposition has been demonstrated by studying the half-lives of several drugs including antipyrine, dicoumarol, phenylbutazone and nortriptyline in identical (monovular) and fraternal (biovular) twins. The similarity in values for identical twins can be striking as with the study by Vesell and Page (1968) who investigated the elimination half-life of phenylbutazone after oral doses (Table 10.1).

Table 10.1   Paired plasma half-life values (days) for the decline in phenylbutazone in seven pairs of identical and non-identical twins (Vesell and Page, 1968)

	1	2	3	4	5	6	7
Identical twins	2.8	2.6	2.8	4.0	3.9	1.9	3.2
	2.8	2.6	2.9	4.0	4.1	2.1	2.9
Non-identical twins	7.3	2.9	2.6	1.9	2.1	2.3	2.8
	3.6	3.0	2.3	2.1	1.2	3.3	3.5

10.2.2 Phenotyping and genotyping

Early observations of the influence of genetics on drug disposition arose because subjects could be classified according to their phenotype, for example they were either 'slow' or 'fast' acetylators of isoniazid (see below). Today phenotyping may be carried out systematically, using drugs as 'probes' to determine a subject's metabolizer status. Mixtures of drugs have been developed and some bear the name of institutions in which they were developed, for example, the Pittsburgh, Indiana and Karolinska Cocktails (Table 10.2) These drugs may also be used when investigating the effects of age, sex and drug interactions on enzyme activity (see Section 17.10.2). The tests may simply determine the concentration of the test drug in plasma or urine at a defined time after it has been administered, or specific metabolites may be measured. Alternatively, serial samples may be collected so that *AUC* and oral clearance values can be calculated. The erythromycin breath test involves administering [^{14}C-methyl]-erythromycin and collecting breath (in a balloon) for measurement of $^{14}CO_2$, which reflects the degree of desmethylation. Intravenous and oral administration of midazolam has been used to differentiate between hepatic and gut wall (+ hepatic) metabolism by CYP3A4. Additionally, inhibitors may be given in an attempt to confirm the identity of the enzyme involved. A limitation to this approach is the lack of specificity of some substrates and inhibitors for some enzymes and transporters. This is a particular problem with CYP3A4, CYP3A5 and P-glycoprotein (P-gp), which have similar substrate specificities.

Table 10.2 Examples of substances used to phenotype individuals for drug metabolizing activity

Enzyme	Probe	Measurement	Sample/time
CYP1A2	Caffeine[a]	Caffeine/1,7-dimethylxanthine (paraxanthine)	Plasma – 8 h
	Caffeine[b]	Paraxanthine/Caffeine	Serum – 6 h
	Caffeine[c]		Plasma – 4 h
CYP2B6	Bupropion[d]	Hydroxylation	
CYP2C8	Amodiaquine[d]	Desethylation	
CYP2C9	Losartan[c]	5-Carboxylic acid metabolite (E-3174)	Urine – 8 h
	Tolbutamide[b]		Serum – serial samples
CYP2C19	Mephenytoin[a]	4-Hydroxymephenytoin	Urine – 8 h
	Omeprazole[c]	5-Hydroxyomeprazole	Plasma – 3.5 h
CYP2D6	Debrisoquine[a]	4'-Hydroxydebrisoquine/ (4'-Hydroxydebrisoquine + debrisoquine)	Urine – 8 h
	Dextromethorpran[b]	Dextrorphan	Urine – serial samples
	Debrisoquine[c]	4'-Hydroxydebrisoquine/ (4'-Hydroxydebrisoquine + debrisoquine)	Urine – 8 h
CYP2E1	Chlorzoxazone[a]	6-Hydroxychlorzoxazone/ chlorzoxazone	Plasma – 4 h
CYP3A4	Dapsone[a]	Dapsone hydroxylamine/ (Hydroxylamine + dapsone)	Urine – 8 h
(Hepatic)	Midazolam (i.v.)[b]		Serum – serial samples
(Hepatic + intestinal wall)	Midazolam (p.o)[b]	1'-Hydroxymidazolam	
	Cortisol	6-β-Hydroxycortisol/cortisol	Urine
	[14C]-Erythromycin	[14C]-CO$_2$	Breath – 20 min
CYP3A4/5	Quinine[c]	3-Hydroxyquinine	Plasma – 16 h
	Dextromethorphan	3-Methoxymorphinan	
CYP3A5	Midazolam[d]	1'-Hydroxymidazolam	
NAT2	Dapsone[a]	Monacetyldapsone	Plasma – 8 h
	Sulfadimidine	Acetylsulfadimidine	Plasma – 8 h

[a]Pittsburg cocktail (Frye *et al.*, 1997); [b]Indiana cocktail (Wang *et al.*, 2001); [c]Karolinska cocktail (Christensen *et al.*, 2003); [d]O'Donnell *et al.*, 2007.

Polymerase chain reactions (PCR), originally with restriction fragment length polymorphism (RFLP), are used to sequence genes and so identify particular genotypes and variant alleles. Furthermore, transfection of cDNA into organisms (e.g. *Escherichia coli*) or cell lines (e.g. COS) allows production of recombinant enzymes which can be sequenced to identify changes in amino acid composition and enzyme activity, expressed as Km and V_{max}, values for the substrates of interest. This can lead to confusion as a 'mutant' enzyme may be more active per weight of protein than the wild-type when tested *in vitro*, but if the variant allele results in much less enzyme being expressed, then the *in vivo* activity may be reduced.

10.3 *N*-acetyltransferase

There are two major forms of arylamine *N*-acetyltransferase. Substrates of type 1 include *p*-aminobenzoic acid, *p*-aminosalicylic acid and endogenous *p*-aminobenzylglutamate, whereas a number of primary aromatic amine and hydrazine drugs are acetylated by *N*-acetyltransferase type 2 (NAT2), including the examples of Figure 3.14. Over 30 variants of the *NAT2* gene have been identified. *NAT2*4* is considered the 'wild-type' allele.

10.3.1   Isoniazid

The bimodal nature of metabolism of the anti-tuberculosis drug, isoniazid, was known in the early 1950s, but it was the classic experiments of Evans *et al.* (1960) that demonstrated the genetic nature of the polymorphism. Plasma isoniazid concentrations were shown to be bimodally distributed when 483 subjects were given identical doses. A subset of results from 267 members of 53 families confirmed the hereditary nature of the phenomenon. Subjects with plasma concentrations $<2 \, mg \, L^{-1}$ were referred to as 'rapid inactivators' whilst those with lower concentrations were classed as 'slow-inactivators' (Figure 10.1).

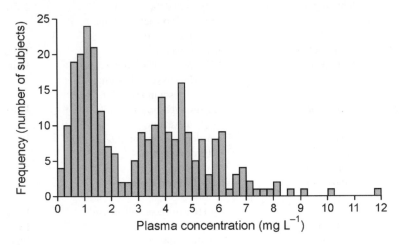

Figure 10.1   Frequency distribution for plasma isoniazid concentrations 6 hours after oral doses of $9.7 \, mg \, kg^{-1}$ in 267 members of 53 families. (Redrawn from Evans *et al.*, 1960.)

The rapid allele (R) is dominant, and so only homozygotes (rr) are slow acetylators. The distribution of fast to slow acetylators is approximately 50 : 50 in Caucasian and African Americans, so the gene frequency of the slow gene(s) must be ∼75%. Inuits and Japanese are primarily fast acetylators (95%) but some Mediterranean Jews are mainly slow (20% fast). Fast acetylators may require higher doses. Slow acetylators may develop a peripheral neuropathy due to the imine (Schiff's base) formed between isoniazid and pyridoxal, which depletes the vitamin, whereas rapid acetylators are prone to hepatotoxicity which is probably caused by *N*-acetylhydrazine that is released from the acetyl metabolite (Section 18.6.1).

It is sometimes possible to identify heterozygotes, but not from histograms of the type shown in Figure 10.1. For example, systemic clearance, rather than plasma concentration, was shown to correlate with the number of *NAT2*4 alleles (Kinzig-Schippers *et al.*, 2005).

10.3.2   Sulfonamides

The acetylation of the primary amine groups of several sulfonamides, including sulfadimidine (sulfamethazine) shows the same polymorphism as that of isoniazid. Because sulfadimidine concentrations could be measured using a relatively simple colorimetric assay, this drug has been used to test for acetylator status.

Following a test dose of 0.5 g, the proportion of N^4-acetylsulfadimidine in plasma was bimodally distributed, with an antinode at $\sim$40% (Figure 10.2). Thus, those with $>$40% acetyl metabolite in plasma are fast acetylators. With test doses of 2 g the antinode was $\sim$25%, presumably showing that at the higher dose the metabolism is becoming saturated. As with isoniazid, it has been shown that sulfadimidine elimination half-lives can be assigned to three distinct groups, reflecting the three phenotypes.

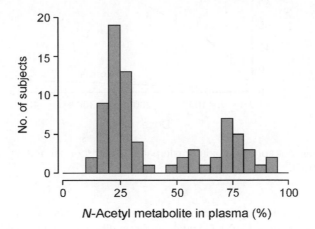

Figure 10.2 Distribution for percent *N*-acetyl metabolite in plasma 6 hours after a test dose sulfadimidine (0.5 g, p.o.). (Redrawn from Whelpton *et al.*, 1981.)

10.3.3 Other drugs

Other substrates of NAT2 include hydralazine, phenelzine and dapsone. The acetyl metabolites of these compounds are considered inactive in comparison to the parent drug. Generally, fast acetylators may not respond adequately to treatment whilst slow acetylators are more prone to adverse effects. Procainamide is unusual in that the acetyl metabolite has similar pharmacological properties to the parent drug and is marketed as acecainide.

10.4 Plasma cholinesterase

Several genotypes for plasma cholinesterase (pseudocholinesterase) have been discovered. Approximately 94% of the population are homozygous for the 'normal' allele and are designated EuEu. Of the atypical alleles, the one coding for a dibucaine-resistant form of the enzyme (Ea) is probably the most important clinically. EaEa homozygotes show prolonged paralysis when given the muscle relaxants suxamethonium (succinylcholine) and miracurium and may be sensitive to other drugs including, procaine, cocaine, pilocarpine, huperazine A and donepezil. Fluoride resistant (Ef) and silent (Es) alleles have been identified.

10.4.1 Suxamethonium

In normal subjects the duration of action of suxamethonium is approximately 5–10 minutes. Most of the injected dose is hydrolysed so that only some 5–10% reaches the acetylcholine receptors of the motor endplate. Drug that diffuses from the receptors is hydrolysed by the normally functioning enzyme. Approximately 1 in 3,000 people remain paralysed for an unusually long period following the drug (Figure 10.3). Should this occur during an operation then mechanical ventilation must be continued until the patient can breathe normally.

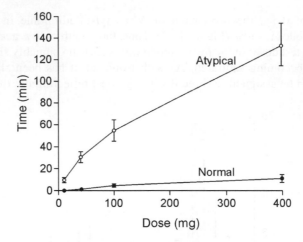

Figure 10.3   Duration of apnoea in adult male patients with normal and atypical cholinesterase (Kalow & Gunn, 1957).

As well as taking a family history, patients can be tested using standard cholinesterase assays in the presence of a standard concentration of the local anaesthetic dibucaine, which inhibits the normal enzyme activity by 80%. The enzyme from EaEa homozygotes is only inhibited by 20%. These values are known as the dibucaine number. Heterozygotes (EuEa) have dibucaine numbers of ~60, and although these individuals may have a longer duration of apnoea, it rarely lasts for more than 1 hour, and is not considered clinically important. A more serious situation arises in homozgyotes carrying the Es gene who have no pseudocholinesterase activity and the duration of apnoea may be over 8 hours. The frequency of this polymorphism is 1 : 100,000.

Demonstrating polymorphic hydrolysis of other drugs is complicated by the fact that they may also be substrates for the many other esterases that exist in plasma.

10.5   Cytochrome P450 polymorphisms

It is probable that polymorphisms exist for all the drug metabolizing cytochromes and several important clinical differences have been demonstrated for CYP2D6, CYP2C9 and CYP2C19. Individuals with two functioning genes are referred to as extensive metabolizers (EMs) and those with no, or only one, functioning gene, are classed as poor (PMs) and intermediate (IMs) metabolizers, respectively. People with more than two functioning genes are ultrarapid metabolizers (UMs).

10.5.1   Cytochrome 2D6

One of the first observations of a polymorphism in microsomal drug metabolizing enzymes was the exaggerated response to the obsolete antihypertensive, debrisoquine. The metabolism to the 4-hydroxy metabolite is catalysed by CYP2D6 and has been used to phenotype poor metabolizers (Table 10.2). Urine is collected for 8 hours following a test dose of debrisoquine (10 mg, p.o.) and the metabolic ratio (*MR*) calculated from:

$$MR = \frac{\% \text{ of dose as debrisoquine}}{\% \text{ of dose as 4-hydroxydebrisoquine}} \qquad (10.1)$$

Poor metabolizers are defined as those having a log(*MR*) value > 1.1 (i.e. a ratio > 12.5). At the same time as polymorphisms in debrisoquine metabolism were being investigated similar patterns were demonstrated for sparteine and nortriptyline. Subsequently, it was shown that the differences were due to different *CYP2D6* alleles. A large number of drugs are substrates for CYP2D6, including several β-blockers, neuroleptics and SSRIs (Table 3.1). It has been claimed that PMs obtain no pain relief from codeine as they are unable to metabolize it to morphine, while UMs show an exaggerated response to codeine. Tamoxifen is another drug that relies on CYP2D6 for its activation and PMs do not respond as well as EMs do to this drug.

The distribution of PMs varies amongst different ethnic groups. In Europeans it is ~7%, which means that some 20–30 million Europeans have no CYP2D6 enzymes. However it has also been estimated that 15–20 million Europeans have multiple copies of the gene. Approximately 2% of Swedish and 7% of Spanish people are UMs. The figure may be as high as 29% for Ethiopians. Many of these patients fail to respond to standard doses of CYP2D6 substrates and may be classed as non-responders. Figure 10.4 shows plasma nortriptyline concentrations in subjects carrying different numbers of functional genes (0–13) after a single dose. The differences in concentrations will be even more marked after multiple dosing to steady-state concentrations.

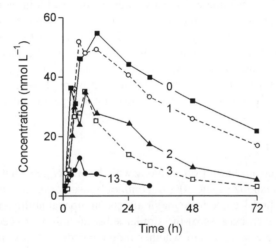

Figure 10.4   Mean plasma concentration of nortriptyline after a single oral dose in subjects with varying numbers (0–13) of *CYP2D6* genes as indicated on the lines. There were five subjects in each group apart from *n* = 1 for the subject with 13 genes (redrawn from Dalén *et al*, 1998).

10.5.2   Cytochrome 2C9

It has been estimated that CYP2C9 catalyses approximately 10% of P-450-mediated drug metabolism. *CYP2C9*2* and *CYP2C9*3* alleles arise from SNPs and the enzymes have been estimated to confer 70% and 10% of the intrinsic clearance of the wild-type enzyme (CYP2C9*1), respectively. Approximately, 35% of Caucasians have at least one *2 or *3 allele. Although rare, *CYP2C9*6* is a null allele, conferring no enzyme activity.

Substrates of this enzyme include phenytoin, tolbutamide, valproate, and warfarin (Table 10.3). In the past there have been reports of unusually long half-lives of some of these drugs in a small number of patients. Adverse drug reactions to phenytoin have be ascribed to patients having defective *CYP2C9* alleles, particularly in *3/*3 diplotypes, who comprise ~0.4% of Caucasians.

Table 10.3 Examples of CYP2C9 substrates

Analgesics	Anticonvulsants	Oral hypoglycaemics	NSAIDs
Paracetamol	Phenytoin	Tolbutamide	S-Naproxen
Phenacetin	Phenobarbital	Glibenclamide	Diclofenac
Aminopyrine	Valproate	Glipizide	Celecoxib
Oral anticoagulants	**Angiotensin II antagonists**	**SSRIs**	**Others**
Warfarin	Losartan	Sertraline	Fluvastatin
Dicoumerol	Candesartan	Venlafaxine	Sildenafil

NSAIDs, non-steroidal anti-inflammatory drugs.

10.5.3 Cytochrome 2C19

Polymorphic 4'-hydroxylation of S-mephenytoin was reported in the 1980s and it has since been shown that the enzyme responsible is CYP2C19. This enzyme catalyses the metabolism of several frequently prescribed drugs. It catalyses hydroxylation of the proton pump inhibitor, omeprazole and the desmethylation of diazepam, imipramine and citalopram. The incidence of poor metabolizers is higher in East Asian populations (13–23%) compared with Caucasians (2–5%). CYP2C19*1 is the wild-type allele whilst CYP2C19*2 or CYP2C19*3 are considered defective mutants that result in reduced enzyme activity. Thus, following a single dose of omeprazole the mean *AUC* value for *2/*2 homozygotes was nearly 10 times that measured for *1/*1. The *AUC* for heterozygotes was only twice that of the EM subjects.

The degree to which the pH of gastric contents was increased and the success of ulcer treatment with omeprazole was highly dependant on phenotype, with 100% success in PMs, but only 25% in EMs. It has been suggested that CYP2C19 genotyping is cost effective in predicting response to omeprazole and amoxicillin in the treatment of *Helicobacter pylori* infection and peptic ulcer.

10.5.4 Cytochromes 3A4/5

The *CYP3* alleles are clustered on chromosome 7 along with the *MDR1* gene that encodes P-gp. As CYP3A4/5 have been estimated to metabolize some 50% of commonly-used drugs and have important roles in first-pass metabolism, polymorphisms could have major affects on bioavailability, and hence pharmacological/toxicological activity. However, because of the similar substrate/inhibiter specificities of CYP3A4/5 and P-gp it is not always possible to ascertain whether individual differences in bioavailability are due to polymorphisms in the enzymes or the transporter. The situation is further complicated by the fact that high levels of these enzymes are expressed in enterocytes and hepatocytes. The enzymes can accommodate both small drugs, such as midazolam, and larger substrates, including ciclosporin. *CYP3A4* alleles have been identified but their frequencies are low.

Higher levels of CYP3A5 are expressed in Africans and saquinavir *AUC*s were 34% lower in 'CYP3A5 producers'. Two alleles arising from SNPs, *CYP3A5*3 and *CYP3A5*6 have relatively high frequencies and some subjects may not have any functioning CYP3A5. Dosage adjustments of immunosupressants, ciclosporin and tacrolimus may be required for some individuals, but the effect of polymorphism in *MDR1* may be a contributory factor (Section 10.6).

10.5.5 Other cytochrome P450 polymorphisms

Polymorphisms in CYP2B6 have been identified of which the variant *CYP2B6*6 appears to be the most important. The frequency of homozygotes for this allele is 3% in whites and 20% in blacks, and these individuals have higher plasma concentrations and increased adverse reactions with the anti-HIV drug, efavirenz. Cyclophosphamide is a prodrug with complicated activation and inactivation pathways, some of which are non-enzymatic. The first step, oxidation to 4-hydroxycyclophosphamide, is catalysed by CYP2B6 and individuals with the variant *CYP2B6*6* produce a protein that metabolizes the drug at a faster rate. However, the amount of enzyme expressed by *6/*6 homozygotes was considerably less than those with the wild-type allele. Polymorphism in nicotine oxidation has been attributed to an inactive variant allele of *CYP2A6*. An alternative explanation is that in some Asian subjects a *CYP2A6* gene may be deleted and may be responsible for reduced nicotine metabolism. Some individuals may have multiple copies of *CYP2A6*.

10.6   Alcohol dehydrogenase and acetaldehyde dehydrogenase

Alcohol dehydrogenase (ADH) is a dimeric enzyme made up of six separate subunits, encoded by three genes, *ADH₁*, *ADH₂* and *ADH₃*. Many combinations of isoenzymes exist, leading to different rates of metabolism amongst white, black African, and Asian populations.

   Aldehyde dehydrogenase (ALDH) is a mitochondrial enzyme. Some Asians have ALDH different from that of Caucasians and about 50% of Asians (principally Chinese) have inactive ALDH, leading to flushing and other unpleasant effects when these individuals consume ethanol. These effects are similar to those seen with disulfiram.

10.7   Thiopurine methyltransferase

Phenotyping or genotyping of thiopurine methyltransferase (TPMT) is used to guide treatment with azathioprine to avoid life-threatening acute toxicity. The incidence of very low TPMT activities is relatively high (1 : 300), whilst 11% of subjects have intermediate activity. 6-Mercaptopurine, derived from azathioprine is normally metabolized via one of three pathways: (i) methylation by TPMT, (ii) oxidation by xanthine oxidase or (iii) by hypoxanthine phosphoribosyltransferase to active thiopurine metabolites including 6- thioguanine nucleotides. Patients with low TPMT activity have unusually high levels of 6-thioguanidine incorporated into DNA, which is, in part, responsible for azathioprine toxicity.

10.8   Phase 2 enzymes

Because of the historical importance of *NAT2* polymorphism and the fact that it provides a simple, but clear, example of the issues involved, this phase 2 enzyme was discussed earlier. Other phase 2 enzymes, the UDP-glucuronosyltransferases (UGT), sulfotransferases (SULT) and glutathione transferases (GST) are super-families, much as the cytochrome P450s are a superfamily.

10.8.1   *UDP-glucuronosyltransferases*

The two major classes of *UGT* genes are *UGT*1 and *UGT*2 which produce at least 18 enzymes. UGT1A1 catalyses the glucuronidation of bilirubin. The wild-type allele is *UGT1A1*1*, but a common mutant, *UGT1A1*28* leads to a mild form of hyperbilirubinaemia, known as Gilbert's syndrome, which occurs in 5–10% of the population. As a consequence sufferers may be prone to the adverse effects of drugs or metabolites metabolized by UGT1A1. This has been shown to be the case with the anticancer drug irinotecan, which is metabolized to the active, and potentially toxic, metabolite known as SN-38. Glucuronidation and inactivation of this compound is catalysed by UGT1A1 and those with Gilbert's syndrome are more prone to neuropenia and diarrhoea. In 94% of people with Gilbert's syndrome two other UGTs are affected, one of which is thought to catalyse the glucuronidation of paracetamol.

3-Glucuronidation of morphine is catalysed by several UGTs, but only UGT2B7 has been shown to catalyse both 3- and 6-glucuronidation. Some studies of the allelic variant, *UGT2B7**2* have shown that the rate of glucuronidation is greater in carriers of this allele, however other studies have failed to substantiate this claim, and whether this genotype has any clinical significance is equivocal. Of course the situation is complicated by the fact that the 3-glucuronide is inactive whereas the 6-glucuronide is analgesic.

10.8.2   Sulfotransferases

The most widely studied sulfotransferase is SULT1A1, also known as 'thermostable' or phenol SULT, as it catalyses sulfation of a large number of endogenous and exogenous phenols including paracetamol and the 4-hydroxy active metabolite of tamoxifen. SULT1A1 and 1A3 are highly expressed in platelets thereby facilitating study. A common variant is SULT1A1.2, in which G638A substitution results in Arg213His in the allozyme. This enzyme is less thermally stable, has reduced activity compared to SULT1A1.1 and a shorter half-life. SULT1A1.3 arises from a A667G SNP leading to a Met223Val substitution in the enzyme.

Ethnic variations have been described. The frequency of the *1A1**1* allele is 0.914 in Chinese but only 0.656 and 0.477 in Caucasians and African Americans, respectively. The frequency of the *1A1**2* allele is 0.332 and 0.294 in the latter groups. The incidence of the *1A1**3* allele is 0.229 in African Americans. Much of the research into the functional effects these alleles has concerned sulfation of flavanoids, which may protect against cancer, 17-β-oestradiol, which has been implicated in breast cancer, and 4-hydroxytamoxifen. It has been suggested that women carrying the *1A1**2* or *1A1**3* alleles sulfate the active metabolite of tamoxifen to a lesser extent and so have higher exposure to the drug. However they also have reduced sulfation of oestradiol. Despite the role of sulfation in the metabolism of paracetamol, there appears to be no definitive observation on the impact of SULT variants.

10.8.3   Glutathione S-transferases

Human GST families are designated by uppercase Greek letters, such as alpha (A), mu (M) pi (P) and theta (T). Polymorphisms are known in the *GSTP1* gene, and deletions occur frequently with *GSTM1* so that 50% of the population are homozygous for the null allele. The frequency of deletions with *GSTT1* is ethnically determined, ~20% of Europeans and ~60% of Orientals and Africans being *GSTT1**0/**0*. High levels of GSTT1 are found in erythrocytes. Thus, it would appear that there is potential for phenotypic polymorphism in GSH conjugation, however most studies have concentrated on the role of these alleles in the development of cancers.

With regard to the effects on drug disposition, the clearance of busulfan, an alkylating agent used for treating chronic myelogenous leukaemia, was significantly lower in *GSTA1**A/**B* heterozygotes than in those with homogenous *GSTA1**A/**A*. The plasma concentrations were correspondingly higher in the heterogeneous group. Ekhart *et al.* (2009) investigated the effects of pharmacogenetics on the oxidation and conjugation of the alkylating anticancer drug thiotepa, which is metabolized to tepa. The conclusion was that patients who were homozygous for a *GSTP1* variant allele had greater exposure to the two compounds.

10.9   Transporters

The widespread distribution of multifunctional transport proteins results in these being of prime importance in pharmacogenetics. Interest is primarily in those genes that encode P-gp, multi-drug resistance associated protein, organic anion transport polypeptides, and organic cationic transporters. Some of these transporters are located in the basolateral membranes whilst others are in the luminal membrane and may work in concert to eliminate drugs. The complexity of these combinations can be appreciated from the simplified diagram of Figure 10.5.

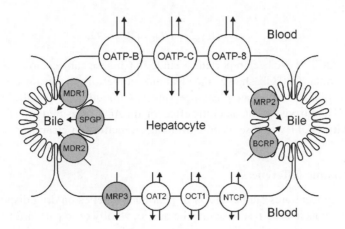

Figure 10.5 Simplified diagram of hepatic transporters. Organic anion transport proteins (OATP-B, OATP-C and OATP-8), organic anion transporter 2 (OAT2), organic cation transporter 1 (OCT1) and sodium-dependent taurocholate transporting polypeptide (NTCP) facilitate the uptake of solutes into hepatocytes. Multidrug resistance-associated protein 3 (MRP3) is an efflux pump. The hepatic canalicular efflux proteins include multidrug resistance proteins (MDR1 and MDR2), MRP2, sister of P-glycoprotein (SPGP) and breast cancer resistance protein (BCRP). The efflux proteins utilize ATP (not shown). Adapted from Tirona & Kim, 2002.

The *MDR1* gene which encodes for P-gp is highly polymorphic. One variant, C3435T, has been shown to result in half the normal P-gp expression in the duodenum and this was associated with higher digoxin concentrations in these subjects. Some 24% of the subjects were homozygous for this SNP. Similarly, the frequency in children being treated for HIV, was C/C (44%) C/T (46%) and T/T (10%). The children that were heterozygous, had higher plasma nelfinavir concentrations and lower oral clearance (although it is not clear whether the clearance values were corrected for differences in bioavailability). Children in this C/T group responded more quickly to treatment than the others. On the other hand, Kim *et al.* (2001) found conflicting results with fexofenadine but pointed out that additional SNPs were possible – C1236T and G677T, and Chowbay *et al.* (2003) showed that haplotypes where all three positions were substituted with thiamine (T-T-T) had increased *AUC* and C_{max} values for ciclosporin.

Allelic variations have been described in the various OATP families. An example of a functional effect of mutations is the clearance of pravastatin, which is transported into hepatocytes by OATP-C and into rena tubular cells by OAT3 (Nishizato *et al.*, 2003). In subjects with the *OATP-C**15 variant the total and non-renal clearances were lower than those subjects without this allele. The one subject who was homozygous *15/*15, had the lowest *CL* value and the largest *AUC*.

An investigation of the role of mutations in organic anion transporter (OCT) genes, in the disposition of the loop diuretic, torasemide, revealed that mutations in OCT4 (located in the luminal membrane of proximal tubular cells) rather than OCT1 or OCT3 variants (found in the basolateral membrane) reduced the renal clearance of this drug.

OCT2 is the major organic cation transporter in the basolateral membrane of renal proximal tubular cells and metformin is a suitable probe as 98% of a dose is eliminated via the kidneys, and the renal clearance (CL_R) ranges from $\sim$400–600 mL min^{-1}, indicating tubular secretion. There is little plasma protein binding so creatinine clearance, CL_{cr} has been used to estimate filtration, enabling the contribution from secretion, CL_{sec} to be estimated:

$$CL_{sec} = CL_R - CL_{cr} \qquad (10.2)$$

At least 28 variants of OCT2 have been reported. A study in Chinese subjects to investigate the functional effects of a common variant (G808T), showed the clearance of metformin was lowest in T/T homozygotes, while T/G heterzygotes had intermediate clearance values. Cimetidine, a specific inhibitor, reduced the metformin *AUC* in G/G homozygotes but had little effect on the *AUC* measured in T/T subjects. In a study in Caucasians, low activity OCT1 was associated with reduced metformin clearance.

10.10   Pharmacodynamic differences

Naturally this chapter concentrates on the effects of pharmacogenetics on drug disposition but it is worth remembering that variations in drug response may be due to, wholly or in part, differences in drug targets. Genetic variants in β-adrenoceptors have been long known. Differences in sensitivity to warfarin also arise because of mutations in the *VKORC1* gene which encodes for vitamin K epoxide reductase. African Americans are relatively resistant to warfarin. Recent interest has been into polymorphism in the gene that codes for angiotensin-converting enzyme (ACE).

References and further reading

Aoki M, Terada T, Ogasawara K, Katsura T, Hatano E, Ikai I, Inui K. Impact of regulatory polymorphisms in organic anion transporter genes in the human liver. *Pharmacogenet Genomics* 2009; 19: 647–56.

Chowbay B, Cumaraswamy S, Cheung YB, Zhou Q, Lee EJ. Genetic polymorphisms in MDR1 and CYP3A4 genes in Asians and the influence of MDR1 haplotypes on cyclosporin disposition in heart transplant recipients. *Pharmacogenetics* 2003; 13: 89–95.

Christensen M, Andersson K, Dalen P, Mirghani RA, Muirhead GJ, Nordmark A, *et al.* The Karolinska cocktail for phenotyping of five human cytochrome P450 enzymes. *Clin Pharmacol Ther* 2003; 73: 517–28.

Dalén P, Dahl ML, Bernal Ruiz ML, Nordin J, Bertilsson L. 10-Hydroxylation of nortriptyline in white persons with 0, 1, 2, 3, and 13 functional CYP2D6 genes. *Clin Pharmacol Ther* 1998; 63: 444–52.

Ekhart C, Doodeman VD, Rodenhuis S, Smits PH, Beijnen JH, Huitema AD. Polymorphisms of drug-metabolizing enzymes (GST, CYP2B6 and CYP3A) affect the pharmacokinetics of thiotepa and tepa. *Br J Clin Pharmacol* 2009; 67: 50–60.

Evans DAP, Manley KA, McKusick VA. Genetic control of isoniazid metabolism in man. *Br Med J* 1960; 2: 485–91.

Frye RF, Matzke GR, Adedoyin A, Porter JA, Branch RA. Validation of the five-drug 'Pittsburgh cocktail' approach for assessment of selective regulation of drug-metabolizing enzymes. *Clin Pharmacol Ther* 1997; 62: 365–76.

Guillemette C. Pharmacogenomics of human UDP-glucuronosyltransferase enzymes. *Pharmacogenomics J* 2003; 3: 136–58.

Ingelman-Sundberg M. Pharmacogenetics of cytochrome P450 and its applications in drug therapy: the past, present and future. *Trends Pharmacol Sci* 2004; 25: 193–200.

Kalow W, Gunn DR. The relation between dose of succinylcholine and duration of apnea in man. *J Pharmacol Exp Ther* 1957; 120: 203–14.

Kim RB, Leake BF, Choo EF, Dresser GK, Kubba SV, Schwarz UI, *et al.* Identification of functionally variant MDR1 alleles among European Americans and African Americans. *Clin Pharmacol Ther* 2001; 70: 189–99.

Kinzig-Schippers M, Tomalik-Scharte D, Jetter A, Scheidel B, Jakob V, Rodamer M, *et al.* Should we use *N*-acetyltransferase type 2 genotyping to personalize isoniazid doses? *Antimicrob Agents Chemother* 2005; 49: 1733–8.

Lennard L, Van Loon JA, Weinshilboum RM. Pharmacogenetics of acute azathioprine toxicity: relationship to thiopurine methyltransferase genetic polymorphism. *Clin Pharmacol Ther* 1989; 46: 149–54.

Nishizato Y, Ieiri I, Suzuki H, Kimura M, Kawabata K, Hirota T, *et al.* Polymorphisms of OATP-C (SLC21A6) and OAT3 (SLC22A8) genes: consequences for pravastatin pharmacokinetics. *Clin Pharmacol Ther* 2003; 73: 554–65.

O'Donnell CJ, Grime K, Courtney P, Slee D, Riley RJ. The development of a cocktail CYP2B6, CYP2C8, and CYP3A5 inhibition assay and a preliminary assessment of utility in a drug discovery setting. *Drug Metab Dispos* 2007; 35: 381–5.

Tirona RG, Kim RB. Pharmacogenomics of organic anion-transporting polypeptides (OATP). *Adv Drug Deliv Rev* 2002; 54: 1343–52.

Vesell ES, Page JG. Genetic control of drug levels in man: phenylbutazone. *Science* 1968; 159: 1479–80.

Wang Z, Gorski JC, Hamman MA, Huang SM, Lesko LJ, Hall SD. The effects of St John's wort (*Hypericum perforatum*) on human cytochrome P450 activity. *Clin Pharmacol Ther* 2001; 70: 317–26.

Whelpton R, Watkins G, Curry SH. Bratton-Marshall and liquid-chromatographic methods compared for determination of sulfamethazine acetylator status. *Clin Chem* 1981; 27: 1911–4.

11

Developmental Pharmacology and Age-related Phenomena

11.1 Introduction

There is a saying: 'Children are not just small adults.' However, and paradoxically, this statement is often followed by a discussion of dosing based on weight, such as in $mg\,kg^{-1}$ terms, as in classical pharmacology.

Interest in both the young and the old as special populations is now well established, especially in phenomena related to pharmacokinetics. Development of hepatic function and renal processes has received intensive study. Studies of drug responses, and of drug concentrations in body fluids, at different ages are commonplace. This reflects a dramatic change since the early days of clinical pharmacology. However, the study of extreme age groups remains hampered by severe limitations. For example, analytical methods suitable for the small blood samples obtainable from children have required specific development, and ethical problems persist for studies in both the very young and very old age groups.

A major problem has been dissociating effects of age itself from effects of age-related practices and phenomena, such as smoking, caffeine, alcohol and other drug consumption, and dietary and exercise habits. Also, age-related disease is an inevitable complication. There remains a great need for longitudinal studies in particular individuals, but the time required for such work is an obvious limitation. Any claim for an age-related observation must be interpreted in this light.

11.2 Scientific and regulatory environment in regard to younger and older patients

Awareness that there was a growing need to ensure that drug therapies for the paediatric population were studied with the same level of scientific and clinical rigour as adult therapeutic agents developed slowly between 1968 and 1998. In 1968, children were described by Shirkey as 'therapeutic orphans', but it was not until 1998 that the Food and Drug Administration (FDA) issued its 'Final Rule' on requirements for assessing safety and effectiveness in paediatric patients. Companies had been reluctant to study drugs in children because of the complexity, difficulty and expense of clinical trials in children, preferring to leave dosing recommendations to prescribers, who tended to assume that children merely required relatively low doses in order to respond similarly in the way that adults do. The result was empirical use of medicines without prior evidence-based efficacy and safety studies, with a high incidence of off-label or off-licence use of drugs (i.e. use in ways not approved by the relevant authorities). While often satisfactory for drugs for which there was a long history of experience, the outcome with newer drugs could be beneficial or harmful.

Legislative initiatives have catalysed a vast improvement in this situation, such that by 2009, the approved prescribing information on over 100 drugs had been brought up-to-date in regard to paediatric populations as

Drug Disposition and Pharmacokinetics, By Stephen H. Curry and Robin Whelpton
© 2011 John Wiley & Sons, Ltd

the result of specific clinical studies designed to make this possible. The scientific hurdles that first had to be overcome included:

- Parent, and ethics committee/institutional review board (IRB) attitudes had to be changed from a basically negative stance to realization that studies in children were needed for the benefit of children.
- Experimental techniques in regard to frequency of sampling and volume of blood removed in pharmacokinetic studies required refinement.
- Specialized research units in tune with paediatric issues, both physical and psychological, needed to be established.
- Population pharmacokinetic protocols had to be developed. In these, relatively small numbers of samples from each individual, such as those collected at clinic visits, but from relatively large populations of children, permitted the use of scatter-plot graphs for identification of pharmacokinetic profiles.
- Ethical approaches to the protection of children with regard to vulnerability to coercion, as well as use of placebo controls, had to be established.

Similar considerations have driven growing interest in the geriatric population, although legislative influence has not been as strong. Older patients tend to have multiple diseases, including those affecting the organs involved in drug disposition and fate, experience a disproportionate incidence of drug interactions, have poor compliance, are vulnerable to coercion when they are involved and have traditionally been excluded from research protocols, or have not had available to them appropriately equipped research units where high quality protocols are adopted. The epidemiological approach to collection of research data has again been very valuable, especially with such initiatives as the Framingham study (Levy and Brink, 2005) and, as with children, population pharmacokinetics has assisted in scientific rigour.

11.3   Terminology

One difficulty in age-related studies has been in defining when a neonate becomes a child, when a child becomes an adult, and when an adult becomes elderly. This terminology has evolved to agreement on the following basic divisions of the segments of life for most purposes:

- Premature   born before anticipated date of full-term development
- Neonate   birth – 1 month
- Infant   1 month – 2 years
- Child   2–12 years
- Adolescent   12–16 years
- Older adult   65 and over.

11.4   Physiological and pharmacokinetic processes

11.4.1   Absorption

The pH of the stomach varies from 6 to 8 at birth, dropping to pH 1 to 3 within a few hours. This is followed by 8 to 10 days of virtual achlorhydria. Gastric acid secretion is lower in premature newborns. Gastric pH is again lower by day 30. Adult values of gastric pH are reached after approximately 3–5 years, but some literature suggests that it can take as long as 12 years for the complete adult pattern to emerge. In adulthood, the pH increases with age so that approximately 35% of persons over 60 have achlorhydria. Gastric emptying is prolonged in neonates, reaching adult levels at 6–8 months. Growth of intestinal flora progresses with development. Biliary excretion and the splanchnic circulation show postnatal development and there is low bile acid secretion in newborns with implications for enterohepatic circulation involving conjugation and hydrolysis. In the elderly, there is a general delay in absorption of nutritional substances with age.

Intramuscular injections are subject to the changes in muscle blood flow, such as those which occur in the first days of life, and to temperature-dependent phenomena such as vasoconstriction. The high water content of the skeletal muscle mass may also be a factor.

The skin is more permeable to drugs in the newborn and the young, because of the thinner stratum corneum and greater hydration. This was shown to be very important when hexachlorophene was used when bathing babies, and remains an important reason for not using topical salicylic acid preparations, in particular in young children.

11.4.2 *Binding and tissue distribution*

Tissue distribution relates to body weight, and is obviously related to age as adults are larger than children. However, it is probably the proportion of particular constituents which is most important. The proportion of water in premature babies is as much as 85% of body weight. This drops to 55–75% in full-term babies, and 50–70% in adults. The extracellular fluid is 40% of body weight in neonates compared to 20% in adults. Adult values are reached at 10–15 years of age. The proportion of intracellular fluid (evident by calculation of differences) is relatively low at birth , but quickly reaches adult values which remain unchanged throughout the remainder of life – these differences are of no significance in dosing. The fat proportion increases with age and shows wide fluctuation at various stages in life. All of this will affect tissue-to-plasma concentrations ratios, for both lipophilic and hydrophilic drugs. Additionally, the relative acidosis of the newborn will affect the tissue distribution of weak electrolytes.

Protein binding can vary with age, although albumin concentrations are normal, total protein concentrations are lower in newborns. This implies that there are lower concentrations of proteins other than albumin in newborns. Adult values are reached at 10–12 months. This difference in binding, combined with a risk of high blood bilirubin concentrations consequent on immaturity of the conjugating systems, compounded by an immature blood-brain barrier, leads to the risk of kernicterus in the newborn being aggravated by drugs such as sulfonamides, which displace bilirubin from those binding sites that are available. Also in regard to binding, plasma free fatty acid concentrations, and also cholesterol, vary over a wide range and show age-related trends. These measurements are variously affected by diet, activity, obesity, caffeine and ethanol amongst other things. In fact, although some links between these measures and binding of drugs to albumin have been noted, no relation between, in particular, plasma free fatty acids and drug response connected with this appear to have been shown.

11.4.3 *The blood–brain barrier*

Whilst most of the basic structure of the human brain is formed before birth, neuron proliferation and migration continue into the postnatal period. The blood–brain barrier is immature at birth, reaching full development by about 6 months. Synaptic connections between neurons continue to develop in later years. At the other extreme of life, ageing of the microcirculation can result in significant alteration in the blood–brain barrier with function remaining intact, but with the susceptibility to disruption by external factors, such as hypertension and drugs, increasing. There is little evidence that overall transport functions change with age, but changes in select carrier-mediated transport systems in the blood–brain barrier definitely occur, such as with choline, glucose, butyrate and triiodothyronine. These age-related changes are the result of either alteration in the carrier molecules or the physicochemical properties of the cerebral microvessels. In regard to drugs, the specialized 'tight-junctions' and specific efflux mechanisms of the blood–brain barrier exclude most xenobiotics. In a small number of cases, drugs penetrate the brain via active transport. However, most useful centrally acting drugs diffuse into and out of the brain, and this diffusion is unlikely to be much affected by age. The exact sum of these influences on any particular drug example may be complex, but the overwhelming evidence that both newborns and elderly people are especially sensitive to centrally acting drugs suggests that, on balance, the two extremes of age result in the reduced ability of the brain to exclude xenobiotics, as well as, perhaps, relatively high pharmacological sensitivity.

11.4.4   Liver function

The ratio of weight of liver to total body weight in children is 30–35% of that in adults, but then the liver becomes relatively smaller as people grow older. The important drug metabolic reactions that occur in adults are to be found at the extremes of age. However, there are some important differences. *N*-Methylation, the reverse of *N*-desmethylation occurs in children, but is virtually absent later in life. This results in the conversion of theophylline, once a popular treatment for childhood asthma, into caffeine, which is basically undesirable in children. Also, sulfation is especially well developed in children, resulting in relatively rapid metabolism of paracetamol (acetaminophen). Acetylation is slow in children.

The rat liver shows postnatal development reaching the adult substructure at 5–7 days. In contrast, the human foetal liver shows complete differentiation at the third month of gestation. The blood supply changes at birth. Concentrations of factors such as NADP, and other essential constituents of the P-450 system show complicated changes in early life. In one study, undifferentiated total P-450 concentrations were as high in human foetal liver as in adults. A major complication is the degree of enzyme induction that affects particular samples, depending on exposure of mother, foetus and baby to environmental chemicals, and to drugs. This is difficult to measure. Both pre- and postnatal enzyme induction occurs, and this has led to the use of phenobarbital as a treatment for problems caused by low levels of bilirubin elimination, which lead to neonatal jaundice resulting from low hepatic ability to form glucuronides, mentioned earlier, at birth. In contrast, the activity of β-glucuronidase may be sevenfold the adult level in neonates, with implications for enterohepatic circulation (Section 3.3.8.1).

In one study attempting to characterize this, the newborn of mothers treated with phenytoin showed an ability to metabolize the drug comparable with that of the mothers, and carbamazepine metabolism was faster than that in the mothers. In this regard, the need to compare neonatal metabolism with that of the mother, preferably by the use of cord blood, has been stressed. This is infinitely preferable to comparison of neonates with unrelated adults.

In humans, alcohol dehydrogenase is detectable at 2 months of gestation and reaches adult values at 5 years of age. In contrast, foetal liver has high steroid hydroxylating ability. Experimental animals vary in their patterns. For example, rats show high *N*-hydroxylating activity at 2–3 days and 27 days, but lower activity in between. Rabbits show increasing activity from birth onwards. In humans, hydrolytic activity (e.g. plasma esterases) and acetylation develop postnatally taking a few weeks to reach adult values. Reduction shows immaturity at birth, as does conjugation (glucuronidation, mentioned earlier, plus, paracetamol apart, sulfate and glutathione conjugation).

Elderly people show reduced liver activity as evidenced by lower clearance and longer half-life values for various drugs. It is presumed that reduction in protein synthesis as well as other factors, as discussed below, plays a part in this.

11.4.5   Changes in CYP-450 isoforms during the life cycle

While there is an abundance of information on drug clearance and half-life as a function of age, there is relatively little on the specific isoforms of the P-450 system in this context. The experimental approach to this topic includes *in vitro* work with liver samples from aborted foetuses, and/or postmortem dissections, biopsies from adult donors undergoing abdominal surgery for unrelated reasons, and work-up of microsomes and RNA for measurement of activity and the study of mechanisms, plus *in vivo* work with specific probe substrates including 'cocktails' (Section 10.2.2.).

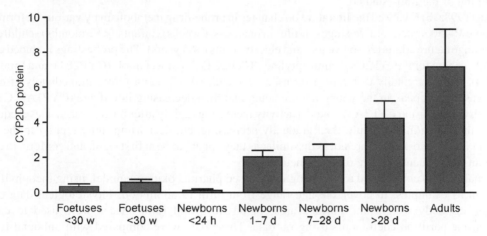

Figure 11.1   Age-related variation of CYP2D6 protein in human liver. Results are the mean ± s.e. mean of individual values. (After Treluyer *et al.*, 1991.)

One of the earliest studies relevant to this question was performed by Treluyer *et al.* (1991). Studying CYP2D6 and using dextromethorphan as a substrate, it was shown that activity rises in the first few weeks of life, regardless of gestational age. Figures 11.1 and 11.2 show the age-related variations in CYP2D6 protein in human liver in seven age groups, and also the age-related variations in CYP2D6 RNA in the same samples. Dextromethorphan metabolism correlated well with the data. It is striking that the protein concentrations increased across the life-span, but the RNA levels fell in the adult. This can be taken as indicating that regulation varies with age at the transcriptional level. Complementing this work, Agundez *et al.* (1997) studied DNA from blood cells, using standard PCR techniques and restriction fragment polymorphism methods. Their work particularly focussed on nonagenarians, and on three enzymatic polymorphisms that affect *CYP2D6*, *NAT2* and *CYP2E1* genes and the activity of their enzymatic gene products. The three

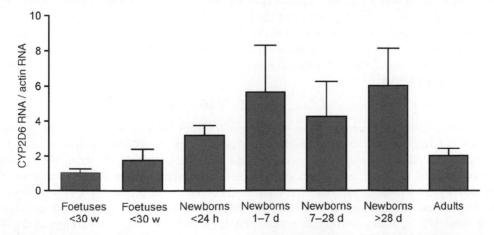

Figure 11.2   Age-related variation of CYP2D6 RNA in human liver (see Figure 11.1 for analogous information). (After Treluyer *et al.*, 1991.)

genotypes were compared in 41 nonagenarians and over 100 younger volunteers (mean age in their 30s). No qualitative or quantitative differences were found in the mutations underlying the three polymorphisms studied, nor in the expected enzymatic phenotypes. Thus longevity did not seem to be related to any special configuration of the traits studied.

Tanaka (1998a, b) reviewed the literature on changes in probe-drug metabolism by various isoforms of the cytochrome P-450 system at five stages in life: neonates (<4 weeks), infants (<12 months), children (<19 years), young/mature adults (20–64 years), and elderly adults (>65 years). The probe drugs included caffeine (CYP1A2), phenytoin (CYP2C9), amitriptyline (CYP2C19), paracetamol (CYP2E1) and midazolam (CYP3A4). Two marginally different patterns were identified: (i) activity low immediately after birth, then increasing to a peak in the young/mature adult, and then decreasing in old age (CYP1A2, CYP2C9, CYP2C19, CYP2D6 and CYP3A4); and (ii) activity increasing rapidly after birth to reach a level equivalent to that of the young/mature adult, then gradually decreasing, but decreasing more rapidly in the elderly (CYP2E1). These two patterns are not as dissimilar as they appear to be at first sight, and conform to existing ideas about the influence of age on drug metabolism.

Bjorkman (2005) constructed a physiologically based pharmacokinetic model, using theophylline and midazolam as examples, to cover the age range from birth to adulthood. Physiological data on such parameters as body weight and surface area, blood flow, and fluid spaces were used to calculate tissue : plasma partition coefficients using rat data. These data were compared with unbound fractions of the drugs in human plasma, and unbound intrinsic clearance was estimated for various age groups for CYP1A2, CYP2E1 and CYP3A4. Volume of distribution, renal clearance and elimination half-life data were estimated and compared with literature data. Predictions were generally good and the relationships between pharmacokinetic parameters and age reported in other literature were supported. Other more recent studies generally support the conclusions reached by Treluyer, Agundez, Tanaka and Bjorkman, including the existence of examples, such as metoprolol (Goryachkina *et al.*, 2008) where there are apparently virtually no age-related changes. Thus observations of *in vivo* patterns of age-related pharmacokinetic parameters, protein concentrations and transcription factors, and physiological factors, such as hepatic blood flow, should permit rational future consideration of any particular age-related phenomenon in this context.

11.4.6   Renal function

A relatively small proportion of cardiac output reaches the kidneys in the very young. Also, there is relatively slow glomerular and tubular development, and the loop of Henle is incomplete. Glomerular filtration in human beings is immature at birth, but postnatal development is rapid, shown by inulin clearance usually reaching adult values at 2–3 days, although occasionally needing 15–20 days to reach maturity. Between one month and 20 years of age, glomerular function remains stable. For the age group 20–90 years, inulin clearance follows the formula:

$$CL_{\text{inulin}} = 153.2 - (0.96 \times age) \tag{11.1}$$

indicating a slow decline. However, it is more usual to assess kidney function using creatinine clearance (see Section 6.3.1.1). Creatinine distributes throughout total body water, shows little binding or tissue localization, and is eliminated almost entirely by renal excretion. However, it exhibits renal tubular movement equally in both directions, so while its clearance calculates glomerular filtration, it can be subject to changes in tubular activity, if for instance, the balance of movement in the two directions were to become disturbed.

Active tubular reabsorption and tubular secretion develop postnatally; in general transport maximum values are reduced in neonates. *p*-Aminohippurate (PAH) clearance reaches adult values at 25–30 days, but the range is 3–35 days. The mass of functioning tubular cells is reduced at birth, as are tubule length, blood flow to the peritubular area, and energy supply processes. In contrast, no age-dependent differences in non-ionic diffusion have been noted, although the pH of the urine is somewhat reduced in neonates.

Creatinine clearance, normalized for body weight, is highly depressed in premature newborns, and is depressed in the neonate. However, it develops rapidly, reaching a maximum at 6 months, when it is twice the adult rate. Thereafter it diminishes at approximately 1% a year. From age 20 onwards creatinine clearance ($mL\,min^{-1}$) is given by the following formulae:
Males:

$$CL_{\text{creatinine}} = \frac{(140-age) \times W}{70} \tag{11.2}$$

Females:

$$CL_{\text{creatinine}} = \frac{(140-age) \times W}{85} \tag{11.3}$$

where *age* is in years and *W* in kg.

In the practice of medicine it is not uncommon to use serum creatinine measurements to estimate creatinine clearance, avoiding the need for urine collection. These estimates are prone to error, actually leading to underestimates of renal clearance, as serum creatinine reflects both production and elimination of creatinine, but they serve many diagnostic needs. Clearly, in the relatively simple case of a drug, such as gabapentin, that is excreted unchanged, it is thus possible to calculate an appropriate dose that allows for the decline in renal function with age using serum creatinine as a guide. This appears to be rarely done in such a systematic way, although empirically, these scientific facts affect the thinking behind prescribing practices. There are other physiological functions that decline with age at about the same rate as renal function, affecting measurements of cardiac and respiratory function in particular. Clearly, these are functions that can affect drug disposition.

The development of the renal elimination processes is presumed to underlie the changes in the half-life of ampicillin in neonates and infants shown in Table 11.1, which would dictate the use of relatively small doses of ampicillin in neonates, and the relatively high doses of tobramycin sometimes proposed for use in older children (see Section 11.5).

Table 11.1 Half-life of ampicillin in four groups of neonates and infants aged between 2 and 68 days following intramuscular doses of $10\,mg\,kg^{-1}$ (Axline *et al.*, 1967)

Age range (d)	Mean half-life in serum (h)
2–7	4.0
8–14	2.8
15–30	1.7
31–68	1.6

NB Ampicillin is excreted unmetabolized.

11.4.7 Metabolic and pharmacodynamic phenomena

It is to be expected that drug responses might vary with age as the result of pharmacodynamic sensitivity or metabolic differences. However, compared with the abundance of pharmacokinetic data, examples are few. Among those available, salicylates cause metabolic disturbances in all age-groups, but the disturbance can be especially great in children under one year of age. Disturbances in acid-base balances are easily provoked in young children, and this has implications for the use of electrolytes in pharmacological procedures. Many diuretics act selectively by reaching concentrations in the kidney higher than those elsewhere, and because filtration and tubular phenomena are not fully developed in neonates, this selective action can be less easily obtained for both pharmacokinetic reasons (see earlier) and also because the maximum effect obtainable on Na^+ transport, which is the basis for action of many diuretics, is less in the neonate. This can, in turn, lead to the need for higher diuretic doses for a standard effect in young patients.

11.5   Body surface area versus weight

Small objects have greater surface area in relation to volume than do large objects. This includes small people. The use of body surface area for normalizing doses across the life span is frequently mentioned. However, it is used relatively rarely. The concept can actually be traced back for over 150 years. The relationship between surface area, weight and height is given by the equation (amongst others):

$$S = W^{0.425} \times H^{0.725} \times 71.84 \qquad (11.4)$$

where S is body surface area in square metres, W is body weight in kilograms, and H is height in centimetres. This can be logarithmically transformed to:

$$\log S = 0.425 \log W \times 0.725 H \times 1.8564 \qquad (11.5)$$

to permit straight line plotting. This relationship was determined by Dubois and Dubois in 1916.

Nomograms exist permitting this information to be used to calculate surface area from height and weight for any individual. These nomograms have been reprinted in multiple publications, such as textbooks of paediatric clinical pharmacology, and the Documenta Geigy Scientific Tables. Clearly, surface area increases with increases in weight and height. In fact, the increase in surface area with age is linear, and there are good correlations between surface area and cardiac output, renal blood flow, and glomerular filtration rate. These correlations are better than those with body weight. The ratio of surface area to weight decreases with height and weight, and the ratio of surface area to height increases with height and weight. The concept is that therapeutic precision would improve if doses were related to surface area rather than weight.

As a practical example, while ciclosporin is usually dosed on the basis of weight in adults, starting with an i.v. dose of approximately $3 \, mg \, kg^{-1}$, dosing in children starts with either $2 \, mg \, kg^{-1}$ or $60 \, mg \, m^{-2}$ i.v. In a 27.2 kg (60 lb) child, $2 \, mg \, kg^{-1}$ is 54.4 mg, and Figure 11.3 compares this with doses calculated on the basis of $60 \, mg \, m^{-2}$ in five theoretical individuals weighing 27.2 kg (60 lb.), but with heights varied from 76 to 137 cm (30 to 54 inches). The highest of these doses is over 50% more than the lowest. The highest of these doses is over 50% more than the lowest. In practice, dosing on the basis of body surface area is limited almost entirely to the field of cancer chemotherapy, and then to children, where i.v. administration is widely used. A variation on this is incorporation of creatinine clearance into the calculations, to allow for the variations in renal function. Beyond this, perhaps with the possible exception of the prescribing of potentially toxic i.v. antibiotics, the use of body surface area in calculating doses appears to be positively discouraged by authorities such as the American Academy of Pediatrics, at least in part because it can lead to prescribing of drugs in ways different from those approved by the FDA. There seem to have been few, if any, systematic studies of human pharmacokinetic parameters as a function of body surface area, because, by definition, the new drug discovery system seeks new chemical entities that show linearity with weight in their drug dispositional properties.

However, some nursing literature still recommends the use of weight for premature and full-term neonates, and surface area for children.

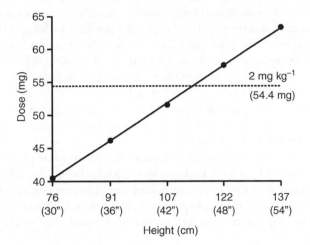

Figure 11.3 Ciclosporin dose (mg) calculated on the basis of surface area for five individuals weighing 27.2 kg (60 lb) and ranging in height from 76 cm (30 inches) to 137 cm (54 inches); the dashed line is the dose in mg that would have been given on the basis of weight alone (2 mg kg^{-1}).

Only for one of the drugs, sotalol, studied in the Paediatric Initiative (Section 11.2) was body surface area shown to be an important covariate. Smaller children (body surface area <0.33 m^2) showed a tendency for a larger change in the QT interval (QTc) and an increased frequency of prolongation of the QTc interval, as well as greater β-receptor blocking effects, so that it is considered that the best therapeutic approach with this drug is individualized dosing based on body surface area.

11.6   Age groups

11.6.1   Neonates and children

Summarizing from the foregoing, very young children show changes principally in metabolism, excretion and the blood–brain barrier, plus pharmacodynamic sensitivity. As an example, interesting data have been presented for ampicillin. Table 11.1 shows mean half-times in four groups of infants of various ages. The drug was excreted slowly by the youngest infants. As this drug is dependent for its removal from the body on excretion by the kidney, it is immediately obvious that this is an example of an immature kidney at birth. Both glomerular filtration and active tubular secretion are immature in the newborn and excretion of ampicillin is dependent on both of these processes. In addition to the data quoted in the table, two other observations are important. First, the half-lives for the drug in serum were studied in the same patients at different ages – that is serially. The serum half-lives always declined, reaching a steady value around 15 days, independent of the weight or gestational age of the infants. This suggested that the maturation is purely a postnatal phenomenon and is possibly a substrate-dependent enzyme induction phenomenon, of either an enzyme or of a tubular transport protein.

As discussed earlier, children, as opposed to newborn infants, have mature systems for handling drugs, but children are of course smaller than adults. Over the years a number of different systems have been devised for calculating the doses of drugs to be given to children. These doses are usually a fraction of the adult dose, dependent on the age and weight of the child, with the full adult dose being introduced at puberty.

11.6.2   Elderly

There is a general presumption that the aged are unusually sensitive to drugs, and that this may be caused by subclinical impairment of cardiac output and hepatic and/or renal function associated with old age. Certainly, clinical evidence seems to show that the elderly are especially sensitive to drugs, so that the idea of increased sensitivity is now unquestioned. The principal factors seem to be liver and kidney impairment, and also a pharmacodynamic contribution, such as increased central nervous system (CNS) sensitivity. The difficulty arises when there is no clinical evidence of impairment of elimination. Other factors such as loss of weight and changes in perfusion of tissues perhaps caused by atherosclerosis, as well as target organ sensitivity outside the context of CNS sensitivity probably play a part. Changes in neurotransmitter concentrations with ageing will lead to differences in drug response both centrally and peripherally. The drug examples below illustrate key points to consider in this population.

Warfarin has been vigorously investigated in relation to age. At the same plasma warfarin concentration there was a greater inhibition of vitamin K dependent clotting factors in elderly people, but this was accompanied in at least one study by no important differences in pharmacokinetics, even though there was reduced plasma albumin and reduced binding to plasma proteins. In an analogous study, the rat was used as an animal model to show that the major difference on ageing is sensitivity of the target organ (Shepherd *et al.*, 1977).

Nitrazepam, although now virtually obsolete in modern therapeutics, was so fully investigated, that it remains of interest as a model compound, as it was at one time a compound of considerable concern in regard to its safety in elderly people. The aged experienced more unwanted CNS depression from this drug than did other people. In one particular well-controlled study the peak concentration was lower, the volume of distribution was higher, and the terminal-phase half-life was unexpectedly shorter in elderly people who had a wide variety of pathology including heart failure, leukaemia and diabetes. In contrast, in another study, no differences were seen between healthy young and old persons, and in yet other studies a prolonged half-life has been recorded in older people – this latter observation has become the most prevalent one. Protein binding and clearance studies have shown equivocal results. There was a difference in response in both of the initial studies, indicating that: (a) that the elderly are more sensitive pharmacodynamically; and (b) pathology is more likely than age to be the reason for the pharmacokinetic changes (in general, the apparent volume of distribution increased, for reasons that are not entirely clear, leading to delayed elimination) (Greenblatt and Allen, 1978).

Lithium has been thoroughly investigated. There were significant correlations between daily lithium dose and age over a wide range, and also between measured lithium concentration and age, probably because of age-related reductions in renal clearance. However, the dose dropped by only 45% over a 60-year scale, and levels dropped by only 23%. On average, the 80-year-old required only two-thirds the dose required by a 20-year-old to obtain the same concentration. The correlation for this relationship was *statistically* significant, ($r = -0.2354$, $p < 0.05$; $n = 82$); however, as $r^2 = 0.055$, the effect of age accounts for only 5.5% of the variation in concentration. Thus, the clinical significance is slight (Hewick *et al.*, 1977).

Phenytoin was similarly studied. In a carefully evaluated program, serum phenytoin was correlated positively with age according to the equation:

$$\log(serum\ concentration) = 1.471 + 0.0046\ age + residual\ error$$
$$r = 0.31,\ t = 4.1,\ p<0.001$$

(11.6)

The increase in concentration was a result of a reduction in V_{max} with age. The study which included measurement of weight, sex and height, as well as age, in epileptic patients, concluded that adjustment of the dosage according solely to age of a patient would achieve only a marginal improvement in therapy. In any case, phenytoin serum concentrations are routinely monitored during therapy (Houghton *et al.*, 1975).

11.7 Further examples

Data on variations in pharmacokinetics with age have now been collected on hundreds of drugs and published in thousands of papers. Data compilations such as those in *Goodman and Gilman's The Pharmacological Basis of Therapeutics* tend to present data on age effects on 50% of the drugs listed, with multiple observations on each drug (e.g. on bioavailability, clearance, volume of distribution, half-life, etc.). Often, where no effect of age is listed, it is because no studies have been conducted, not necessarily because there is no effect. Two observations stand out as illustrating trends:

- For both the young and the old, data are often equivocal, with both decreases and increases in pharmacokinetic parameters documented, along with greater variance than occurs in the reference adult population.
- The single most common observation in these various special populations is reduced clearance in both the young and the elderly, with consequent increases in half-life.

Further reading and references

Agúndez JA, Rodríguez I, Olivera M, Ladero JM, García MA, Ribera JM, Benítez J. CYP2D6, NAT2 and CYP2E1 genetic polymorphisms in nonagenarians. *Age Ageing* 1997; 26: 147–51.

Allegaert K, van den Anker JN, Naulaers G, de Hoon J. Determinants of drug metabolism in early neonatal life. *Curr Clin Pharmacol* 2007; 2: 23–9.

Allegaert K, Verbesselt R, Naulaers G, van den Anker JN, Rayyan M, Debeer A, de Hoon J. Developmental pharmacology: neonates are not just small adults. *Acta Clin Belg* 2008; 63: 16–24.

Allegaert K, Van den Anker JN, Debeer A, Cossey V, Verbesselt R, Tibboel D, Devlieger H, de Hoon J. Maturational changes in the *in vivo* activity of CYP3A4 in the first months of life. *Int J Clin Pharmacol Ther* 2006; 44: 303–8.

Almeida L, Minciu I, Nunes T, Butoianu N, Falcao A, Magureanu SA, Soares-da-Silva P. Pharmacokinetics, efficacy, and tolerability of eslicarbazepine acetate in children and adolescents with epilepsy. *J Clin Pharmacol* 2008; 48: 966–77.

Axline SG, Yaffe SJ, Simon HJ. Clinical pharmacology of antimicrobials in premature infants. II. Ampicillin, methicillin, oxacillin, neomycin, and colistin. *Pediatrics* 1967; 39: 97–107.

Bebia Z, Buch SC, Wilson JW, Frye RF, Romkes M, Cecchetti A, Chaves-Gnecco D, Branch RA. Bioequivalence revisited: influence of age and sex on CYP enzymes. *Clin Pharmacol Ther* 2004; 76: 618–27.

Benedetti MS, Whomsley R, Canning M. Drug metabolism in the paediatric population and in the elderly. *Drug Discov Today* 2007; 12: 599–610.

Bjorkman S. Prediction of drug disposition in infants and children by means of physiologically based pharmacokinetic (PBPK) modelling: theophylline and midazolam as model drugs. *Br J Clin Pharmacol* 2005; 59: 691–704.

Blake MJ, Gaedigk A, Pearce RE, Bomgaars LR, Christensen ML, Stowe C, et al. Ontogeny of dextromethorphan *O*- and *N*-demethylation in the first year of life. *Clin Pharmacol Ther* 2007; 81: 510–6.

Blanco JG, Harrison PL, Evans WE, Relling MV. Human cytochrome P450 maximal activities in pediatric versus adult liver. *Drug Metab Dispos* 2000; 28: 379–82.

Dallimore D, Anderson BJ, Short TG, Herd DW. Ketamine anesthesia in children – exploring infusion regimens. *Paediatr Anaesth* 2008; 18: 708–14.

de Wildt SN, Ito S, Koren G. Challenges for drug studies in children: CYP3A phenotyping as example. *Drug Discov Today* 2009; 14: 6–15.

DeWoskin RS, Thompson CM. Renal clearance parameters for PBPK model analysis of early lifestage differences in the disposition of environmental toxicants. *Regul Toxicol Pharmacol* 2008; 51: 66–86.

Gibson DM, Bron NJ, Richens A, Hounslow NJ, Sedman AJ, Whitfield LR. Effect of age and gender on pharmacokinetics of atorvastatin in humans. *J Clin Pharmacol* 1996; 36: 242–6.

Goryachkina K, Burbello A, Boldueva S, Babak S, Bergman U, Bertilsson L. CYP2D6 is a major determinant of metoprolol disposition and effects in hospitalized Russian patients treated for acute myocardial infarction. *Eur J Clin Pharmacol* 2008; 64: 1163–73.

Greenblatt DJ, Allen MD. Toxicity of nitrazepam in the elderly: a report from the Boston Collaborative Drug Surveillance Program. *Br J Clin Pharmacol* 1978; 5: 407–13.

Grochow LB, Colvin M. Clinical pharmacokinetics of cyclophosphamide. *Clin Pharmacokinet* 1979; 4: 380–94.

Hewick DS, Newbury P, Hopwood S, Naylor G, Moody J. Age as a factor affecting lithium therapy. *Br J Clin Pharmacol* 1977; 4: 201–5.

Houghton GW, Richens A, Leighton M. Effect of age, height, weight and sex on serum phenytoin concentration in epileptic patients. *Br J Clin Pharmacol* 1975; 2: 251–6.

Ismail Z, Triggs EJ, Smithurst BA, Parke W. The pharmacokinetics of amiloride–hydrochlorothiazide combination in the young and elderly. *Eur J Clin Pharmacol* 1989; 37: 167–71.

Johnson TN, Thomson M. Intestinal metabolism and transport of drugs in children: the effects of age and disease. *J Pediatr Gastroenterol Nutr* 2008; 47: 3–10.

Juma FD, Rogers HJ, Trounce JR. The pharmacokinetics of cyclophosphamide, phosphoramide mustard and nor-nitrogen mustard studied by gas chromatography in patients receiving cyclophosphamide therapy. *Br J Clin Pharmacol* 1980; 10: 327–35.

Kohler E, Sollich V, Schuster-Wonka R, Jorch G. Lung deposition after electronically breath-controlled inhalation and manually triggered conventional inhalation in cystic fibrosis patients. *J Aerosol Med* 2005; 18: 386–95.

Levy D, Brink S. *A Change of Heart: How the Framingham Heart Study Helped Unravel the Mysteries of Cardiovascular Disease*. New York: Alfred A. Knopf, 2005.

Li J, Zhao J, Hamer-Maansson JE, Andersson T, Fulmer R, Illueca M, Lundborg P. Pharmacokinetic properties of esomeprazole in adolescent patients aged 12 to 17 years with symptoms of gastroesophageal reflux disease: A randomized, open-label study. *Clin Ther* 2006; 28: 419–27.

MacLoed SM, Radde IC, *Textbook of Pediatric Clinical Pharmacology*. Littleton: PSG Publishing Company, 1985.

Marier JF, Dubuc MC, Drouin E, Alvarez F, Ducharme MP, Brazier JL. Pharmacokinetics of omeprazole in healthy adults and in children with gastroesophageal reflux disease. *Ther Drug Monit* 2004; 26: 3–8.

McInerny TK,(Editor in Chief) *American Academy of Pediatrics Textbook of Pediatric Care*. Elk Grove Village: American Academy of Pediatrics, 2009.

Rodier PM. Developing brain as a target of toxicity. *Environ Health Perspect* 1995; 103 (Suppl 6): 73–6.

Rodriguez W, Selen A, Avant D, Chaurasia C, Crescenzi T, Gieser G, *et al.* Improving pediatric dosing through pediatric initiatives: what we have learned. *Pediatrics* 2008; 121: 530–9.

Shah GN, Mooradian AD. Age-related changes in the blood–brain barrier. *Exp Gerontol* 1997; 32: 501–19.

Shepherd AM, Hewick DS, Moreland TA, Stevenson IH. Age as a determinant of sensitivity to warfarin. *Br J Clin Pharmacol* 1977; 4: 315–20.

Shirkey H. Therapeutic orphans. *J Pediatr* 1968; 72: 119–20.

Skowno JJ, Broadhead M. Cardiac output measurement in pediatric anesthesia. *Paediatr Anaesth* 2008; 18: 1019–28.

Stevens JC, Marsh SA, Zaya MJ, Regina KJ, Divakaran K, Le M, Hines RN. Developmental changes in human liver CYP2D6 expression. *Drug Metab Dispos* 2008; 36: 1587–93.

Tallian KB, Nahata MC, Lo W, Tsao CY. Pharmacokinetics of gabapentin in paediatric patients with uncontrolled seizures. *J Clin Pharm Ther* 2004; 29: 511–5.

Tanaka E. *In vivo* age-related changes in hepatic drug-oxidizing capacity in humans. *J Clin Pharm Ther* 1998; 23: 247–55.

Tanaka E. Clinically important pharmacokinetic drug–drug interactions: role of cytochrome P450 enzymes. *J Clin Pharm Ther* 1998; 23: 403–16.

Treluyer JM, Jacqz-Aigrain E, Alvarez F, Cresteil T. Expression of CYP2D6 in developing human liver. *Eur J Biochem* 1991; 202: 583–8.

Vachharajani NN, Shyu WC, Smith RA, Greene DS. The effects of age and gender on the pharmacokinetics of irbesartan. *Br J Clin Pharmacol* 1998; 46: 611–3.

Wigal SB, Gupta S, Greenhill L, Posner K, Lerner M, Steinhoff K, *et al.* Pharmacokinetics of methylphenidate in preschoolers with attention-deficit/hyperactivity disorder. *J Child Adolesc Psychopharmacol* 2007; 17: 153–64.

Yaddanapudi S, Batra YK, Balagopal A, Nagdeve NG. Sedation in patients above 60 years of age undergoing urological surgery under spinal anesthesia: comparison of propofol and midazolam infusions. *J Postgrad Med* 2007; 53: 171–5.

Zhao J, Li J, Hamer-Maansson JE, Andersson T, Fulmer R, Illueca M, Lundborg P. Pharmacokinetic properties of esomeprazole in children aged 1 to 11 years with symptoms of gastroesophageal reflux disease: a randomized, open-label study. *Clin Ther* 2006; 28: 1868–76.

Zuppa AF, Barrett JS. Pharmacokinetics and pharmacodynamics in the critically ill child. *Pediatr Clin North Am* 2008; 55: 735–55. xii.

Zuppa AF, Mondick JT, Davis L, Cohen D. Population pharmacokinetics of ketorolac in neonates and young infants. *Am J Ther* 2009; 16: 143–6.

12

Effects of Disease on Drug Disposition

12.1 Introduction

Drug responses are affected by disease states because of changes in both pharmacokinetics and pharmaco-dynamics. This is especially apparent with diseases that affect the processes of drug disposition and pharmacokinetics – absorption, protein binding, metabolism, and excretion. However, our ability to reach conclusions concerning general principles has historically been impaired by specific difficulties:

- Diseases rarely occur in isolation, and categorization of patients as being in one particular disease group is simplistic in approach. For example, liver disease can lead to compensation by renal activity, and vice versa, or both can be impaired in parallel. Experimental difficulties arise with studies of one excretory organ when another one is involved as a complicating factor but not studied.
- It may be difficult to analyse for single factors. Subdivision of liver disease is a case in point – in many of the early studies it was rare to find a study restricted to one of: (i) acute viral hepatitis; (ii) cirrhosis; (iii) drug induced hepatic disease; or (iv) other problems.
- Diseases commonly occur at particular stages in life and therefore it can be difficult to separate the effects of disease on pharmacokinetic properties of drugs from such factors as age.
- Patients in disease groups are commonly treated with drugs that affect each other, such as enzyme-inducing agents, as well as several drugs at any one time. For example, patients with liver disease are likely to have been treated with a large number of drugs including sedative drugs that complicate objective assessment of any central nervous system (CNS) impairment caused by the disease.
- Measurements can be more difficult to make in patients compared with healthy controls because of the complexity of the scientific problem involved and the extent to which intervention in the life of a patient is possible.
- Longitudinal studies are needed, using the pre-disease subject as his or her own control. Also, studies are often at different stages of what is sometimes referred to as 'decompensation' and this has not, in the past, been adequately recorded in the literature.
- Studies with diseased patients can raise ethical questions that do not apply to control subjects.
- Some drugs are designed to utilize healthy organs to relieve influences on diseased organs, for example the use of diuretics to reduce the fluid load on the heart, while others are designed to reverse pathology at its site, for example the use of oral hypoglycaemic drugs to modify insulin utilization.

Effects of disease should be assessed in terms of the mechanisms of drug disposition and assessments of changes in drug plasma concentrations, as outlined in other chapters concerned with special populations and situations, such as age (Chapter 11), and drug interactions (Chapter 17).

Drug Disposition and Pharmacokinetics, By Stephen H. Curry and Robin Whelpton
© 2011 John Wiley & Sons, Ltd

12.2 Gastrointestinal disorders and drug absorption

12.2.1 General considerations

It is to be expected that gastrointestinal pathology will affect drug response by changing drug absorption. However, the pattern for any condition is complex. For example, changes in pH do not necessarily affect absorption because of the relation between pH, site of absorption, gastric emptying, and such. Similarly, in coeliac disease multiple changes occur, including increased rate of gastric emptying, increased gastric acid secretion and prolonged reduction in pH after eating, reduced surface area for absorption, increased permeability of the gut wall, and deceased local enzyme concentrations. These factors interact to accelerate or decelerate absorption, and it is not surprising that a mixed pattern of changes has been reported.

12.2.2 Inflammatory conditions of the intestines and coeliac disease

Inflammatory bowel conditions include Crohn's disease and ulcerative colitis. Their cause is basically unknown. They are characterized by abdominal cramps and diarrhoea. Crohn's disease affects the full thickness of the intestinal wall, most commonly occurring in the lower part of the small intestine, and in the large intestine. In contrast, ulcerative colitis affects only the large intestine, and does not affect the full thickness of the bowel wall. Crohn's disease is especially common in young people, and is associated with inflammatory conditions of organs other than the gastrointestinal tract, such as the eyes and joints. Inflammatory diseases have the potential to change the surface area available for absorption, the thickness of the intestinal wall and therefore the distance over which diffusion takes place, intestinal pH, mucosal enzymes that metabolize drugs, intestinal microflora, gastric emptying and peristalsis, and transporters that control inward and outward movement of nutrients and drugs. It is not surprising therefore that a variety of different observations has been made with various drugs in these conditions. For example, in Crohn's disease, the absorption of clindamycin and propranolol has been shown to be increased in extent, while that of many other drugs is decreased. The production of α_1-acid glycoprotein (AAG) is increased. The expression of CYP3A4 and P-glycoprotein (P-gp) levels were significantly higher in biopsy samples from a group of children with Crohn's disease compared with controls. These differences could account for decreased bioavailabilities. An *in vivo* study with radiolabelled prednisolone in adults showed reduced bioavailability in Crohn's patients (on the basis of urinary excretion and *AUC* measurements). Faecal excretion was greater in Crohn's patients. In the case of paracetamol (acetaminophen), the mean rate constant of absorption was not reduced in Crohn's disease, the conclusion being that any pharmacokinetic differences that occurred were related to slower gastric emptying. Similarly, absorption of trimethoprim, methyldopa, and lincomycin has been shown to be reduced while that of sulfamethoxazole was increased. Enhanced absorption of macromolecules has been observed in Crohn's disease (e.g. horseradish peroxidase) using biopsy samples and *in vitro* methods of study.

Thus it can be anticipated that a drug dependent on passive diffusion is likely to be relatively slowly absorbed in these patients, while a drug heavily affected by mucosal CYP3A4 or P-gp might show highly variable absorption. A drug affected by gastric emptying might be expected to show delayed absorption, for example methyldopa, which showed reduced plasma levels of both the parent drug and one of its metabolites in Crohn's patients, which was attributed to slower gastric emptying (Figure 12.1). In contrast, in coeliac disease, the parent drug concentrations were also reduced, but metabolite concentrations were increased, suggesting stimulation of intestinal mucosal drug metabolizing enzymes. Also, as mentioned earlier, in coeliac disease there is an increase in the rate of gastric emptying. Table 12.1 summarizes a number of observations regarding gastrointestinal pathology and drug absorption.

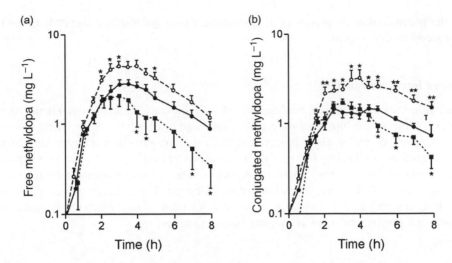

Figure 12.1 (a) Plasma concentration of methyldopa and (b) conjugated methyldopa after a single oral dose in normal subjects (•), patients with coeliac (○) and Crohn's disease (■). Values are mean ± SEM, $^*p < 0.05$, $^{**}p < 0.001$. (Redrawn from Renwick *et al.*, 1983.)

Table 12.1 Some examples of influences of gastrointestinal pathology on drug absorption

Condition	Drug(s) affected	Nature of effect
Prolongation of gastric emptying	L-dopa	More than usual destruction by gastric acid
Achlorhydria	Aspirin and cefalexin	Impaired absorption because of pH effects
Gastric stasis/pyloric stenosis	Paracetamol	Impaired absorption
Shigella gastroenteritis/fever	Ampicillin and iron	Impaired/reduced absorption
Malabsorption syndromes	Tetracycline and digoxin	Absorption reduced
Biliary tract disease	Cefalexin	Reduced bile secretion affects solubilization prior to absorption
Coeliac disease	Many drugs	Examples exist of both decreased and increased absorption (see text)
Mucoviscidosis	Cefalexin	Decreased absorption
Crohn's disease	Many drugs	Examples exist of both decrease and increased absorption (see text)
Diarrhoea in children	Ampicillin/co-trimoxazole	Impaired absorption/no change
Gastroenteritis in children	Sulfonamides	Decreased renal clearance resulting from relative acidosis

12.3   Congestive heart failure

In congestive heart failure (CHF) the cardiac output (the volume of blood pumped per minute) is reduced so that insufficient quantities of oxygen and nutrients are delivered to the tissues for their normal functioning. Associated with CHF are atrial fibrillation and flutter, which are disorders of the electrical discharge patterns of the heart, causing relatively fast atrial contraction, and also causing the ventricles to contract more rapidly and less efficiently than normal. There are obviously many other disorders of the heart that affect patient health – for pharmacokinetic purposes the primary need is to focus on reduced perfusion of the organs that are involved in the pharmacokinetic processes of absorption, tissue distribution, metabolism, and excretion, caused by reduced cardiac output.

12.3.1   Altered intestinal function

Chronic heart failure patients showed increased bowel wall thickness in the terminal ileum, the ascending colon, the descending colon, the transverse colon, and sigmoid (pelvic) colon. Increases in intestinal permeability to carbohydrates have been observed on the basis of observations of the lactulose/mannitol ratio and sucralose excretion. A relatively low level of D-xylose absorption has been taken to indicate occurrence of bowel ischaemia in CHF patients There are higher concentrations of adherent bacteria in the bowel mucous of CHF patients. CHF can cause inflammation in the gut. There are obvious implications for these changes on drug absorption, but they appear not to have been tested. Methods available for this are transcutaneous sonography, chemical assays, and biopsy samples.

12.3.2   Altered liver blood flow

There is a general supposition that CHF must reduce the rate and possibly the extent of tissue penetration by drugs. This has been investigated, for example, with lidocaine. The apparent volume of distribution of the drug was reduced, but no changes in the terminal-phase half-life were seen. Relatively high plasma levels were therefore recorded. A similar drop in the volume of distribution of procainamide has been noted. Another interesting observation with lidocaine was that the active metabolite, desmonomethyl lidocaine accumulates in the plasma in CHF, leading to CNS toxicity. Because lidocaine metabolism is limited by hepatic blood flow, which is lowered in heart failure, the explanation for this has to lie in lessened metabolism. However, at first sight the accumulation of the metabolite as a consequence of this is difficult to understand, but it may be that the pathological effect also leads to reduced further metabolism, and indeed, reduced renal excretion, of the metabolite.

12.3.3   Altered rate of metabolism – aminopyrine

In one of the earliest studies of CHF and pharmacokinetics, aminopyrine (amidopyrine) was used as a test substance with its metabolically-labile methyl group labelled with ^{14}C. (Hepner *et al.*, 1978). Desmethylation leads to the radioactivity appearing in exhaled air, as ^{14}C-carbon dioxide and/or ^{14}C-formaldehyde. In the patients, 2.6% of the dose was recovered in this way, compared with 5.6% in healthy volunteers. Also, the total body clearance of the aminopyrine was 29.7 mL min^{-1} in CHF compared with 125.1 mL min^{-1} in the controls. The apparent volume of distribution was greater (63.6 L) in the patients compared with the controls (43.6 L). The breath test values also correlated with clinical improvement in response to treatment. This test, as a general method of investigation of drug metabolism, is further discussed in the liver disorders (Section 12.4).

12.3.4   Altered route of metabolism – glyceryl trinitrate

CHF patients often require higher doses of glyceryl trinitrate (GTN, nitroglyerin) for a useful effect. This drug has two pathways of metabolism, a high-affinity pathway operating in the nanomolar concentration range, in which 1,2-glyceryl dinitrate (1,2-GDN) is formed, and a low-affinity pathway operating in the micromolar range, in which 1,3-glyceryl dinitrate (1,3-GDN) is formed. The hypothesis that at a given GTN-induced blood pressure reduction, the CHF group would present with higher GTN concentrations, and decreased 1,2-GDN/GTN and 1,2-GDN/1,3-GDN ratios in comparison with healthy subjects, was validated. Patients with CHF have attenuated GTN responsiveness (i.e. they require more GTN for a given clinical effect) and decreased relative formation of 1,2-GDN compared with controls, showing altered biotransformation in CHF.

12.3.5 Altered clearance – mexilitine

The apparent oral clearance (CL/F) was calculated from serial serum mexilitine assays in a large number of patients, 210 with CHF and 374 non-CHF controls. The ratio was reduced in the CHF group ($0.264 \pm 0.093 \, \text{L h}^{-1} \text{kg}^{-1}$ compared with control values of $0.393 \pm 0.082 \, \text{L h}^{-1} \text{kg}^{-1}$; data are mean $\pm$ SD, $p < 0.05$), indicating a difference in hepatic and/or renal activity towards the drug, resulting from blood flow and/or enzyme activity differences. There could also have been differences in bioavailability. The patients were sorted by New York Heart Association (NYHA) class into four groups. There was a relationship between the ratio and the NYHA class as an objective measure of severity of CHF, with CL_{Oral} being lower in patients in the class representing the higher level of CHF (Kobayashi *et al.*, 2006).

12.3.6 Congestive heart failure plus renal problems – toborinone

The pharmacokinetics of toborinone were studied in patients with CHF and concomitant renal and/or hepatic disease. Glomerular filtration rate was measured using sodium iothalamate (amidotrizoate, diatrizoate) clearance and hepatic function was measured using a caffeine metabolism test. Indocyanine green clearance was also measured (Section 12.4.3). There were four test groups: CHF alone, CHF plus renal disease, CHF and hepatic disease, and controls. No significant differences were observed among the four groups in pharmacokinetic parameters. However, systemic clearance correlated with various measures of hepatic and renal impairment. For example, CL was reduced in parallel with reduced creatinine clearance, glomerular filtration rate, indocyanine green clearance, desmethylation activity, and liver blood flow.

12.3.7 Decompensated and treated CHF – torasemide

Members of a group of 12 CHF patients were given test doses of torasemide (torsemide) before and after haemodynamic parameters and clinical signs and symptoms of decompensated CHF were resolved. Plasma drug and urinary sodium concentrations were measured. Before resolution of the pathology, urinary sodium levels and urine output were relatively high. A significant increase in torasemide AUC occurred as the result of the treatment. However, there were no significant increases in the maximal excretion rate of the drug between the two test doses. Thus, CHF status had an impact on the pharmacokinetics of the drug, and on the relationship between plasma levels and effect, but did not affect the maximal excretion rate of the drug. It is of interest that there is an analogy in this with liver disease (Bleske *et al.*, 1998).

12.3.8 Oedema

It is particularly difficult to determine whether any pharmacokinetic changes in oedema are caused by the underlying disease or by the oedema itself. For example, frusemide (furosemide) has been well studied in this context, as it is the drug of choice for rapid reversal of oedema in both CHF and renal failure. The rate of absorption of oral doses has been shown to be decreased in oedema of CHF although with no change in bioavailability – C_{max} values were generally decreased although paradoxically the time to achieve C_{max} was described as 'reduced or unchanged' (Vrhovac *et al.*, 1995). Plasma protein binding of this drug is decreased in CHF, decompensated liver disease, and nephrotic syndrome, probably because of hypoalbuminaemia. The elimination half-life can be unchanged, as in CHF and in cirrhosis, or prolonged, as in the oedema of chronic renal failure. Thus the pharmacokinetic changes are thought to relate to the diseases, rather than the oedema.

Digoxin is also of interest in relation to oedema, although, apart from expressions of much-needed caution in its dosing in oedematous patients, because of its narrow therapeutic window, no clear picture is available. However, it is reasonable to conclude that the large interindividual differences that occur in the pharmacokinetic properties of this drug in oedematous CHF patients relate in some way to variations in its volume of distribution, and to the reduction in oedema consequent on its use affecting this apparent volume of distribution. When evaluating digoxin kinetics it is important to assess whether the plasma concentrations could have been inflated by cross-reaction with digoxin-like immunoreactive substances, DLIS (Section 19.1.4).

Digoxin, and also frusemide, as treatments for oedema, provide opportunities for longitudinal studies with 'before and after' pharmacokinetic investigations. In contrast, this has not been the case with theophylline, which has shown a higher bioavailability accompanied by reduced clearance and a longer half-life in oedema associated with CHF, probably connected with liver blood flow changes. Study of this drug at the same time that an oedema-reducing drug was given would obviously be possible.

12.4 Liver disease

12.4.1 *Pathophysiology*

Cirrhosis is destruction of normal liver tissue leaving non-functioning scar tissue surrounding areas of functioning liver tissue. Many conditions can lead to cirrhosis, including alcoholic liver damage and chronic hepatitis. Basically, this is a progressing problem. *Hepatitis* is inflammation of the liver. It can be caused by viruses and/or chemicals, and can be acute or chronic. *Jaundice* causes a yellow discoloration of the skin and whites of the eyes, resulting from an increase in bilirubin in blood. One source of jaundice is *cholestasis*, or a reduction or stoppage of bile flow (obstructive jaundice). *Ascites* is fluid in the abdominal cavity, associated with liver disease, especially cirrhosis. There is also *liver encephalopathy*, or impairment of CNS function associated with liver disease, and, as a result of any kind of liver disease, there may be changes in liver blood flow as caused by '*shunts*' opening up to functionally replace blood vessels serving damaged liver tissue. The word 'decompensation' is used to indicate the gradual loss of liver function. There is an 'intact' liver theory, which states that chronic liver disease is associated with a reduced number of hepatocytes that function normally, and are normally perfused, as well as with the development of intrahepatic portosystemic shunting.

12.4.2 *Liver blood flow, binding to plasma proteins, and intrinsic hepatic clearance*

Liver disease can affect liver blood flow, creating the potential for effects on drugs with high extraction ratios such as propranolol and lidocaine, or hepatic intrinsic clearance, creating the potential for effects on drugs such as metoprolol, and (as the binding proteins are synthesized in the liver) on plasma protein binding, creating the potential for effects on warfarin and naproxen. Pharmacokinetic assessment should employ the equations presented in Chapter 7, with particular emphasis on the following two equations that apply to situations in which the percent protein bound is above or below 90%:

$$CL_H = Q_H E \tag{7.8}$$

$$CL_H = Q_H \left(\frac{f_u CL'_{int}}{Q_H + f_u CL'_{int}} \right) \tag{7.17}$$

and the simplifications that can be made with this equation when the extraction ratio is below 0.3 or above 0.7 (Section 7.4.1).

Another equation can be useful in this context if it can be shown that all of an orally administered dose is absorbed into the intestinal epithelium cells and hence passes into the portal circulation:

$$CL_{Oral} = \frac{CL}{F} = \frac{D}{AUC} = f_u CL'_{int} \qquad (12.1)$$

i.e. when $F = 1$, see Equation 7.18.

Verbeeck (2008) has emphasized the importance of *unbound clearance* for drugs with protein binding greater than 90%, and has also expressed the opinion that there is no great need for consideration of hepatic models of any complexity greater than that of the homogeneous single pool model in this context.

12.4.3   *Methods of investigation*

- The standard alanine transaminase and aspartate transaminase (ALT/AST) clinical laboratory tests are included in the standard chemical screen used by doctors to test the general health of a broad range of patients, even in routine physical examinations. They test liver function, and to a lesser extent heart, muscle and brain function, but no relationship with drug metabolism is apparent.
- The Child–Pugh score utilizes information on serum bilirubin and albumin, prothrombin time, encephalopathy, and ascites, each classified into three groups based on the levels of severity, to generate a score that assesses decompensation as it occurs over time while liver disease progresses. The score rises from ≤4 (effectively normal) to 10–15 (severe disease). It does not take into account any contribution from glomerular filtration. The clearances of several compounds have been correlated with Child–Pugh values. These include caffeine and erythromycin, but there is no obvious general pattern. There is another similar test (MELD) that takes serum creatinine into consideration, and is effective in predicting three-month mortality in patients awaiting liver transplants.
- Indocyanine green (ICG) is almost entirely eliminated through the bile. It has a relatively high extraction ratio (0.5–0.8) and its clearance changes with hepatic blood flow. It is given intravenously, and there is a correlation with severity of liver disease and also indicators of liver function such as bilirubin concentrations and prothrombin time. However, ICG clearance has not been shown to be superior to the Child–Pugh score.
- Galactose is also a convenient compound with a high extraction ratio. It has been known to show impaired clearance when given intravenously to liver patients and so to provide an indication of the severity of the disease. More recently, a single point method, based on the plasma concentration at the end of an infusion was shown to correlate with a wide variety of indicators of liver function, including AST and ALT, serum bilirubin, albumin, and prothrombin time. The galactose concentration was relatively high in patients with chronic hepatitis, cirrhosis, and hepatocellular carcinoma, compared with that in healthy volunteers. Several drug studies have shown a correlation between the galactose single-point (GSP) method of assessment and pharmacokinetic parameters, such as promazine clearance (Figure 12.2). The broader applicability of this method remains to be determined.
- Metabolic cocktails are used for assessing the various isoforms of CYP450 ('metabolic liver cell dysfunction tests'). At one time, antipyrine was widely used to non-specifically assess P-450 *in vivo*. Antipyrine is well absorbed, heavily dependent on microsomal metabolism for its clearance, has very low plasma protein binding, and has a volume of distribution equal to that of of body water. It has a low extraction ratio. Antipyrine has been superseded by the use of various more specific substrates for the various isoforms of P-450 (Table 10.2). The technique is applicable in liver disease as in age and other special population studies.
- Lorazepam is used in much the same way as the metabolic cocktail but as an assessment of glucuronide formation.

- Biopsy samples have been used in the study of glucuronyl transferase, sulfotransferase, acetyltransferase and glutathione transferase and correlations observed between conjugation of model substrates such as 2-naphthol, *p*-aminobenzoic acid, and benzo(a)pyrene-4,5-oxide, and severity of liver disease (Figure 12.3). There are obvious implications for drugs eliminated primarily by conjugation, and for metabolites of drugs that undergo phase I drug metabolic reactions.
- Ultrasound and electromagnetic methods for measuring liver blood flow also exist.

It should be obvious that there is no easy laboratory test that is sufficiently non-invasive and general in its application to fulfil the role that creatinine clearance fulfils in the case of renal disease (Section 12.5.2).

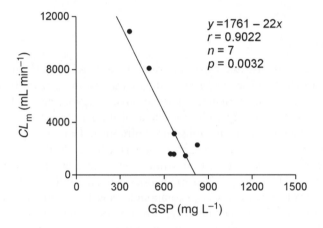

Figure 12.2   Correlation of unbound metabolic clearance of promazine with galactose single point determination (GSP).

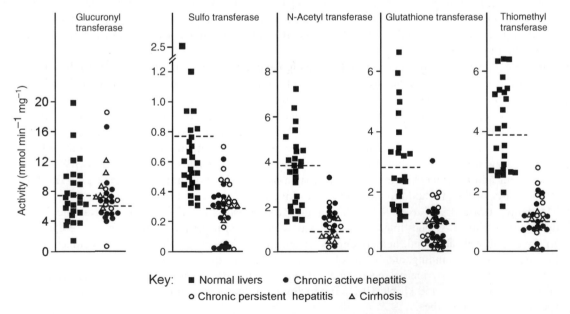

Figure 12.3   Effects of liver disease on conjugation reactions. Broken lines represent the mean value for each group. (Redrawn from Pacifici *et al.,* 1990.)

12.4.4   Examples

Table 12.2 tracks the evolution of ideas, during the early days of the study of drug metabolism in liver diseases, about how cirrhosis affects the ability of the liver to metabolize drugs. It summarizes the conclusions of a number of authors, only some of whom were able to demonstrate disease-induced changes in drug metabolism. In other studies, acute viral hepatitis caused a reduction in the rate of metabolism of hexobarbital, diazepam, pethidine and antipyrine. No change was seen in the metabolism of phenobarbital or

Table 12.2   Metabolism of a selection of drugs in hepatic cirrhosis (Curry, 1980)

Drug	Experimental observations	Comments
Pentobarbital	Overall metabolism essentially the same in cirrhotics and controls; no difference in pharmacological sensitivity	Study mainly a historical curiosity. Non-specific spectrophotometric assay; studies in first hour after i.v. dose only
Phenylbutazone	Plasma levels of drugs declined *slightly faster* in patients with advance alcoholic cirrhosis	Example of counter-intuitive result in early research; later studies showed that liver patients receive large quantities of enzyme inducing drugs that could partially offset any cirrhosis-induced decline in enzyme activity
Phenylbutazone, aminopyrine, salicylic acid, dicoumarol and antipyrine	Phenylbutazone and antipyrine had *shorter* half-life values in cirrhotics; others longer	Authors concluded that cirrhosis has little or no effect, although dicoumarol difference was substantial ($t_{1/2}$ increased by approximately 65% in cirrhosis)
Chloramphenicol	Overall metabolism essentially normal in cirrhosis	Further example of counter-intuitive result
Ethanol	Rates of metabolism: controls 98 mg kg^{-1} h^{-1} and cirrhotics 102 mg kg^{-1} h^{-1}	Only in cirrhotics with the most advance liver disease and severe jaundice was the rate reduced
Tolbutamide	Rate of overall metabolism unchanged; rate of hydroxylation impaired	First observation of different outcome dependent on specificity of research
Digoxin	No difference in metabolism between cirrhotics and controls	Research based on urinary and faecal half-life measurements with tritiated digoxin; no excretion differences
Phenytoin	Slower metabolism in cirrhotics	Phenytoin shows saturation of the drug metabolizing enzymes within the clinical range of concentrations, thereby enhancing the probability of an effect of cirrhosis.
Thiopental	Duration of effect longer in cirrhotics	Thiopental effect controlled by redistribution between tissues; could indicate volume of distribution changes or increased brain sensitivity.
Morphine	First (?) demonstration of enhanced brain sensitivity to a centrally-acting drug	No pharmacokinetic changes
Chlorpromazine	Plasma concentrations in cirrhotics and controls similar, regardless of previous drug treatment. Electroencephalogram (EEG) different in controls and affected differently by chlorpromazine	Supports the idea that patients with liver disease are especially sensitive to centrally-acting drugs, regardless of their capacity to metabolize such compounds (see text)

(continued)

Table 12.2   (*Continued*)

Drug	Experimental observations	Comments
Amylobarbital	Serum concentrations in patients and controls similar, but protein binding different, and rate constants of metabolism from plasma water reduced in liver patients with low plasma albumin – no correlation with effect which was measured in early stages after i.v. injection	Supports idea that patients with certain types of liver disease do show impaired metabolism of drugs, but that compensatory changes occur in distribution compartments, leading to similar concentrations in blood
Paracetamol	Half-life in controls 2.9 ± 0.3 hours; in patients with liver damage 7.6 ± 0.8 hours (both mean $\pm$ SEM)	In paracetamol overdose. May be the result of the drug damaging the liver
Suxamethonium	Close correlation between: (a) severity of liver disease and rate of hydrolysis by plasma cholinesterase; and (b) duration of apnoea	Suxamethonium is unusual among the compounds in this table, in that it is metabolized by plasma esterase and not microsomal oxidation

phenytoin, but the binding of phenytoin was reduced. The half-life of tolbutamide was shortened, and there was also an increase in the unbound concentration; it has been suggested that tolbutamide is displaced from protein by bilirubin. In cirrhosis, the half-life of phenobarbital was prolonged, with a reduction in metabolite formation, and an increase in urinary excretion of the unchanged drug. The albumin concentration was negatively correlated with the phenobarbital half-life, showing that the drug had a longer half-life when albumin concentrations were lower – probably two parallel effects. It has been suggested that the plasma albumin concentration might indicate the degree of parenchymal damage. Also in cirrhosis, lidocaine, diazepam and pethidine (meperidine) showed reduced hepatic clearance, and this was not attributable to perfusion changes, one of the proposed mechanisms.

Interestingly, in cirrhosis, phenytoin and warfarin (which has its action in the liver) show inconsistent changes. In regard to drug-induced hepatic disease, there is a significant reduction in the rate of metabolism of paracetamol, phenytoin and phenobarbital following paracetamol damage. In regard to other problems, azotemia has been shown to be associated with reduced thiopental binding. Thiopental anaesthesia is prolonged in patients with hypoalbuminaemia caused by chronic liver disease.

It is of especial interest that oral bioavailability is substantially increased in cirrhosis for drugs that have a moderate to high hepatic extraction ratio. Examples range from verapamil, with a 1.6-fold increase, to chlormethiazole, with an 11.6-fold increase, and include pethidine, morphine, nifedipine, midazolam, and most of the β-blocking drugs. It seems likely that this results from porto-caval shunting and reduced exposure to CYP3A4.

Some further examples include: (i) protein binding of sulfonamides is reduced in liver disease; (ii) (+)-tubocurarine is normally bound to plasma globulins, and resistance to this drug in hypergamma-globulinaemia due to liver disease has been attributed to increased protein binding; (iii) in liver biopsy samples, the P-450 content and *N*-demethylase activity has been shown to be normal in 'mild to moderate' hepatitis, dropping by 50% in 'severe' hepatitis and cirrhosis, with no changes in reduced nicotinamide adenine dinucleotide phosphate (NADPH) cytochrome C reductase; (iv) the hepatotoxicity of paracetamol is greater in patients treated with enzyme inducing agents; (v) there is decreased conjugation of salicylate with glycine in liver disease; (vi) hydrolysis of procaine and aspirin is slowed in patients with liver disease; (vii) propranolol has shown decreased systemic clearance, with a fall in serum albumin and rise in bilirubin, and (interestingly) prolongation of prothrombin index, but the reduction in protein binding which also occurred supposedly did not relate to changes in proteins – there was an increase in volume of distribution, and changes in liver blood flow and plasma protein pool size seemed critical in the overall changes; (viii) alcohol dehydrogenase, and catalase are lower in patients with liver disease, but NADPH-dependent ethanol oxidizing systems are higher.

Many of the observations discussed in this section above were made before the discovery of the different isoforms of P-450, and their differential ability to metabolize their various substrates, including the significance of CYP3A4 in presystemic elimination. One of the mystery observations of the time was the fact that in many cases the ability of the liver to metabolize drugs seemed to be conserved until the very last stages of decompensation. More recently it has been shown that the different isoforms are conserved for different times (Figure 12.4). CYP2C19 activity is lost early in the process of decay of liver function, while CYP2E1 is conserved until the hepatorenal syndrome is established. CYP2D6, which is responsible for a relatively high proportion of drug metabolism, is conserved for a relatively long time, and CYP1A2 is in-between CYP2D6 and CYP2E1. Unfortunately, these data are not comprehensive, and, in particular, do not include CYP3A4. However, substrates for the CYP3A family of isoforms, such as nifedipine and midazolam, show impaired metabolism in cirrhosis.

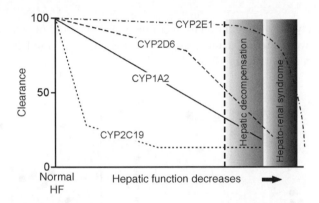

Figure 12.4   Sequential progressive model of hepatic dysfunction: implications for clearance of drugs predominantly metabolized by individual CYP pathways in liver. (Redrawn from Frye *et al.*, 2006.)

Suxamethonium (succinylcholine) is of especial interest. It is metabolized by pseudocholinesterase enzymes, which are synthesized in the liver. However, the metabolism of suxamethonium occurs in plasma, to which the enzyme migrates after it is synthesized. Furthermore, this hydrolysis reaction can be reproduced *in vitro*. Figure 12.5 shows the metabolism of suxamethonium *in vitro* by pseudocholinesterase enzymes from the plasma of patients with impaired hepatic function, including impaired synthesis of plasma pseudo-cholinesterase enzymes.

12.4.5   Drug effects

Studies of pharmacodynamics in relation to liver disorders have been mostly limited to β-blocking drugs, diuretics, and drugs that depress the CNS. However, topotecan is an example of an experimental anticancer drug that shows a lack of a difference between percent decrease in absolute neutrophil count (ANC) and percent decrease in platelet count as a function of the *AUC* of total topotecan during an initial course of therapy in cancer (malignant solid tumours) patients with and without impaired hepatic function. Quite apart

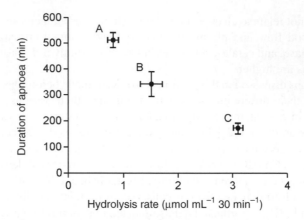

Figure 12.5 Relationship between enzymatic hydrolysis of suxamethonium and duration of apnoea after 0.6 mg kg^{-1} in patients with severe (A), moderate (B) liver disease and normal livers (C). Data are mean ± SEM. (After Birch *et al.*, 1956, and Foldes *et al.*, 1956.)

from the fact that this study showed attention to liver function in the field of cancer, this study has special interest in that the data were studied using scatter diagrams and E_{max} models of the pharmacokinetic/ pharmacodynamic relationships (Figure 12.6).

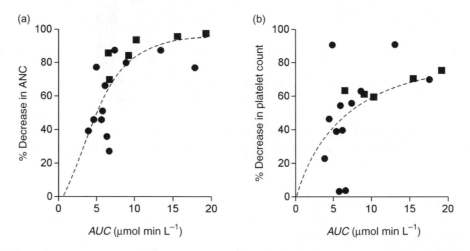

Figure 12.6 The effect of topotecan on (a) neutrophil count (ANC) and (b) platelet count as a function of *AUC*. Patients with and without liver dysfunction are represented by solid circles and solid squares respectively. (After O'Reilly *et al.*, 1996.)

In regard to β-blocking drugs it has been proposed that a decreased therapeutic effect should be expected because there is a decrease in the sensitivity to the chronotropic effects of isoprenaline in liver disease, as well as a reduction in β-adrenoceptor density in mononuclear cells. Such a decrease in effect has been observed with propranolol.

Diuretics such as frusemide (furosemide), and also triamterene, torasemide and bumetamide, also show decreased effects in liver patients, characterized by a need for a relatively high concentration of the diuretic in the renal tubules in order to induce an effect. This could be caused by a reduction in the number of nephrons and by the maximum response per nephron, in the kidney of course, but consequent on the liver disease. However, this is likely to be offset by reduced hepatic clearance leading to higher concentrations of the drug in the kidneys, so that the effect is maintained but with a different pharmacodynamic/pharmacokinetic relationship.

As regards CNS drugs, at one time it was rare that drug responses were adequately measured in patients with hepatic problems, even though such patients are particularly sensitive to CNS drugs. A study with chlorpromazine highlighted this. The interest in chlorpromazine lay in the fact that, in spite of the similarity in the plasma levels, significant differences between patients and controls, in the encephalogram (EEG) and in the drug effects on the EEG, were noted. This confirmed the widespread suspicion that patients with liver disease are unusually sensitive to behavioural effects of drugs with central actions, quite apart from any differences in drug metabolism.

An interesting study of the effect of liver disease on plasma levels of another centrally-acting drug, amylobarbital, has been reported (Mawer *et al.*, 1972). Two groups of liver patients and a control group of healthy subjects were given single doses of amylobarbital by intravenous injection. Plasma levels of the drug were similar in all three groups, but the degree of protein binding of amylobarbital was reduced in one of the two groups of patients. The members of this group were characterized by their abnormally low concentrations of albumin in serum. So with lower protein binding and the same total drug concentrations in serum, concentrations in serum water were obviously higher this group, and, furthermore the half-life of the terminal phase was longer. Both groups of patients showed an increase in the rate constant of the faster (redistribution) phase of the double exponential graph of concentration against time. The slower phase is an indication of the combined rates of metabolism and excretion, probably influenced, at least in part, by the rate of return of the drug from tissues. The faster phase is an indication of tissue penetration by the drug.

It thus appears that there may in fact be a reduction in the capacity of the diseased liver to metabolize amylobarbital, but only in those patients who have reduced concentrations of albumin, which is considered to be a parallel but independent consequence of reduced hepatic function. However, it appears that the body at least partially compensates this impaired function, by modifications in the rate and extent of drug penetration into tissues. In the end, similar total concentrations in plasma result. Finally, in this amylobarbital study, the clinical response to amylobarbital was similar in the three groups. However, it was assessed at the early time points when there were no major drug concentration differences between the two groups. These differences were most dramatic at later time points, when the effect had largely worn off. Thus the extreme complexity of the relationship between liver disease and drug metabolism is emphasized.

The examples of Table 12.2 demonstrate the development of knowledge in this context from counterintuitive early results to detailed understanding.

12.5   Renal impairment

12.5.1   *General considerations*

Kidney failure can be acute, with rapid decline in function but equally rapid recovery if the pathology is reversed, or chronic with slow progression and little chance of recovery. In either case there is accumulation of metabolic waste products in the blood (azotaemia or uraemia) especially urea and creatinine, plus the potential for anaemia, acidosis, decreases in blood calcium and vitamin D, and increases in blood phosphate, parathyroid hormone, and potassium. Acute kidney failure can be caused by impairment of the blood supply to the kidney (e.g. congestive heart failure, bleeding, dehydration, shock, or liver failure – hepatorenal syndrome) or obstruction of urine flow (e.g. prostate enlargement or other tumours) or injury (e.g. allergic reactions within the kidney, toxic chemicals, conditions affecting glomerular filtration, blocked blood vessels or crystal deposits). Chronic kidney failure can result from high blood pressure, urinary tract obstruction, glomerulonephritis, structural abnormalities, diabetes, or autoimmune disorders. Kidney function is typically evaluated by means of creatinine clearance, which is considered elsewhere (Section 3.2.1.4). Creatinine is freely diffusible at the glomerulus, and there is approximately equal transfer in each direction across the renal tubule, so creatinine clearance effectively measures glomerular filtration rate (GFR). Shortcuts based on measurement of serum creatinine tend to underestimate creatinine clearance, although they do provide a useful clinical tool.

The claim that kidney disease reduces the rate of elimination of drugs cannot be challenged. However, it should be realized that drugs may be excreted unchanged to any extent, varying from 0 to 100%. Clearly, a drug that is 100% dependent on the kidney for its removal from the body might be greatly affected by kidney disease. In contrast, a drug that is not excreted unmetabolized is less likely to be affected. However, the excretion of the metabolites of this type of drug, as opposed to the excretion of the unchanged drug, will almost certainly be affected. The metabolites will accumulate in plasma, leading to an exaggerated response if the metabolites contribute to the pharmacological effect or, possibly toxicity, that is not seen when the metabolites are excreted normally. If the presence of large quantities of metabolites reduce the rate of metabolism of the unchanged drug, by metabolite inhibition, then there is the theoretical possibility of accumulation of unmetabolized drug. Additionally, renal impairment is likely to lead to varying degrees of water loading and this may lead to modification of the concentrations of the drug in the fluid compartments of the body, including plasma.

12.5.2   Mathematical approach

The pharmacokinetic significance of changes in GFR is conveniently considered from a worked example for digoxin. The rate constant of elimination (k) is given by:

$$k = k_R + k_{NR} \tag{12.2}$$

where k_R is the rate constant of renal elimination and k_{NR} is the rate constant of non-renal elimination (hepatic and other metabolism and miscellaneous non-renal processes), and k_R is proportional to creatinine clearance, CL_{cr}.

The normal half-life of digoxin is 1.7 days, and digoxin is eliminated ~65% unchanged in the urine, so Equation 12.2 becomes:

$$k = 0.65k + 0.35k \tag{12.3}$$

and because $t_{1/2} = 0.693/k$ (Equation 1.10), then $k = 0.408 \, \text{d}^{-1}$, from which k_R and k_{NR} are 0.265 and 0.143 d^{-1}, respectively. The value of k_R applies when GFR is 'normal', i.e. 125 mL min^{-1}. We can calculate the half-life for partial renal failure represented by, say a CL_{cr} of 50 mL min^{-1}. Thus, as k_R is proportional to CL_{cr}, the 'new' k_R is:

$$k_R = \frac{50}{125} 0.265 = 0.106 \, \text{d}^{-1}$$

As there is no reason, in this example, to suppose that k_{NR} has changed, the value of the elimination rate constant when $CL_{cr} = 50$ mL min^{-1}, is $0.106 + 0.143 = 0.249 \, \text{d}^{-1}$. Solving for half-life in this state, $t_{1/2} = 2.78$ d. Thus, the half-life is not proportional to creatinine clearance. In this example, a 60% reduction in GFR will lead to a 65% increase in half-life. For a shortcut, nomograms exist for clinical dose adjustment based on serum creatinine.

In older literature, a more general approach to this is to be found, using the equation:

$$k = k_{NR} + \alpha CL_{cr} \tag{12.4}$$

where α is a factor of proportionality expressing the relation between the renal elimination rate of the drug and CL_{cr}. According to this equation, for any drug, a graph of k versus CL_{cr} will be one of:

- A straight line through the origin if the drug is entirely eliminated by the kidneys ($k_{NR} = 0$; $\alpha > 0$).
- A line parallel with the x-axis and intercepting the y-axis if the drug is entirely eliminated extrarenally ($k_{NR} > 0$; $\alpha = 0$).
- A straight line intercepting the y-axis if both routes of elimination are involved ($k_{NR} > 0$; $\alpha > 0$).

Gabapentin is an example of a drug that would show the characteristics of the first of these cases. A limited number of highly lipophilic drugs, such as the tricyclic antidepressants virtually meet the specification for the second case. Digoxin is in the third category.

More modern literature favours a direct analogue of the equations above for the situation in which drug clearance data are available, rather than rate constants or half-life values. Thus:

$$CL_{ur} = CL_{NR} + CL_R \frac{CL_{cr(ur)}}{CL_{cr(N)}} \tag{12.5}$$

Where CL_{ur} is the systemic clearance of the drug in the uraemic patient (subscript ur) and N stands for 'normal.' This follows from the additivity of clearance, the use of creatinine clearance as the indicator of renal function, and the fact that renal clearance of the drug is proportional to creatinine clearance. It also assumes that there is no effect of renal impairment on non-renal elimination of the drug (see later). Thus, if the ratio of the two creatinine clearances is known, along with the drug clearances for normal subjects, the clearance in the uraemic patient can be calculated. Alternatively, if the normal total body clearance and f_e, the fraction excreted unchanged in the normal case, are known, then:

$$CL_{ur} = CL(1 - f_e) + f_e CL \frac{CL_{cr(ur)}}{CL_{cr(N)}} \tag{12.6}$$

Equation 12.6 is for systemic drug clearance in the patient with partial renal impairment, in terms of systemic clearance (CL) in a 'normal' person (no impairment) adjusted for the fraction excreted by non-renal processes plus CL adjusted for the fraction excreted unchanged by the kidney multiplied by the ratio of the relevant creatinine clearances. A further useful relationship is then possible, by dividing Equation 12.6 by CL:

$$\frac{CL_{ur}}{CL} = (1 - f_e) + f_e \frac{CL_{cr(ur)}}{CL_{cr(N)}} \tag{12.7}$$

12.5.3   Examples

In practice, the presence of renal impairment is a reason for care with any drug, but plasma level studies have been largely concerned with drugs excreted mostly unmetabolized. The majority of the work has concerned antibiotics, and a selection of values of the rate constants of elimination in anuric patients and in patients with normal kidney function is shown in Table 12.3. The first four columns show the parameters for normal subjects (N) while the last two columns are the elimination rate constant, $k_{(anuric)}$ and half-life, $t_{1/2(anuric)}$, respectively, in anuric patients. Of course, there is no equivalent column for apparent first-order rate constants for renal excretion, $k_{R(anuric)}$, as these patients were not producing urine. These data were mostly compiled from databases such as that in *Goodman and Gilman's the Pharmacological Basis of Therapeutics*, but data in anuric patients is from Dettli *et al.* (1974).

Table 12.3   Pharmacokinetic data relevant to drug elimination in anuric patients and patients with normal function

Drug	f_e	$t_{1/2(N)}$ (h)	$k_{(N)}$ (h^{-1})	$k_{R(N)}$ (h^{-1})	$k_{(anuric)}$ (h^{-1})	$t_{1/2(anuric)}$ (h)
Desipramine	0.02	22.0	0.032	0.0006	ND[a]	ND
Cefaloridine	0.059	1.36	0.51	0.03	0.03	23.1
Erythromycin	0.13	1.6	0.43	0.056	0.13	5.33
Sulfamethoxazole	0.14	9.8	0.071	0.0099	0.07	9.9
Chloramphenicol	0.15	3.3	0.21	0.032	0.2	3.47
Tetracycline	0.58	10.6	0.065	0.038	0.008	86.6
Digoxin	0.65	41.0	0.017	0.011	0.008	86.6
Gabapentin	1.0	6.5	0.11	0.11	0.0053	132.0

[a] Not determined. Note that in the symbols in this table, as in the text, N indicates normal, and R indicates renal.

Desipramine is included as an example with virtually no excretion of the parent drug, and for which it can be presumed that renal failure would have no effect on the half-life, while gabapentin is at the other extreme, and would presumably not be eliminated at all in an anuric patient. Digoxin is the outstanding example of a clinical need to adjust dosage in renal failure. Five of the compounds are antimicrobial drugs. One of them, chloramphenicol, is partly dependent on metabolism and excretion in bile for its removal from the body. Sulfamethoxazole is a sulfonamide; it also undergoes metabolism. These two compounds showed quite small changes in their overall kinetics, even though about one-sixth excreted in the unchanged form by the kidney – they were apparently the least affected of the seven compounds. Erythromycin, with a percentage excretion unchanged similar to that of sulfamethoxazole and chloramphenicol showed a three-fold increase in half-life, but cefaloridine, with a very small percentage excretion unchanged, showed a dramatic increase in half-life. It should be noted that the data in Table 12.3 were taken from a vast amount of data available on the relationship between renal failure and drug elimination for individual examples, for the purpose of illustrating possibilities. Data of this type are available for approximately 150 drugs, with heavy emphasis on antiinfective agents. Tabulations of this data are to be found in texts concerned principally with patient care. Generally speaking this database shows a greater impact of renal failure on drugs with the higher levels of elimination through the kidney. It is now standard clinical practice to base drug dosage with some of these compounds on creatinine clearance.

It has been suggested that various other differences arise as the result of renal failure. For example, there is apparently decreased absorption of oral iron, a carrier-mediated process, and it has also been suggested that nitrofurantoin shows impaired absorption, as there is no build-up in plasma although urinary excretion is reduced. It is probable that a compensating increase in hepatic elimination explains this. This is of considerable interest, as renal and hepatic impairment can occur together, or can compensate for each other.

There is evidence of protein binding changes in renal failure. Binding of sulfonamides is reduced in nephrectomized patients. Phenytoin binding is also reduced, but binding of basic drugs, such as desipramine, quinidine, dapsone and diazoxide is unchanged. In the case of diazoxide, this causes an increased hypotensive effect. Metabolism may be affected by renal failure. This has been shown by studies with isoniazid, sulfisoxazole, hydralazine (all acetylated), procainamide (the hydrolysis reaction, although both acetylation and hydrolysis occur) and hydrocortisone (reduction). Renal failure can also change receptor sensitivity because of its influence on electrolyte and acid-base balance.

Increased hepatic elimination has been presumed for amylobarbital, phenytoin and propranolol in patients in renal failure, on the basis of reduced renal clearance without corresponding changes in overall pharmacokinetic properties. Pindolol, normally 40% excreted unchanged shows decreased renal excretion but no half-life change (and no protein binding change). Something similar occurs with cefaloridine. The hydroxylated metabolites of barbiturates, normally considered inactive, achieve importance in uraemia, probably because of changes in end organ sensitivity plus their accumulation to high concentrations. The half-life of sotalol is 5 h in normal subjects, 42 h in renal patients, and 7 h in the same patients on dialysis. It is not bound to plasma protein.

Tissue distribution may be affected. For example, the distribution of ampicillin within the kidneys of patients with chronic glomerulonephritis is different from that of control subjects, and there will be clearly be difficulty in achieving therapeutic levels of drugs when the kidney is the target organ. Interestingly, it has been shown that the distribution of kanamycin is unaffected by renal failure, and this has been attributed to lack of protein binding of this drug.

There are some methodological problems especially relevant to renal failure and hepatic problems. First, drugs are eliminated at rates and to extents. One can change or both can change. Rates of renal and hepatic elimination are interrelated only by biological factors, but the two extents, summed, can only equal the dose. Studies which fail to take this into account can be quite misleading. Second, half-life is directly proportional to volume of distribution and inversely proportional to whole body clearance (Equation 4.7) All three parameters are derived from one set of data, a series of plasma concentration measurements. Researchers who think that a change in one necessarily means no change in the others are in danger of missing the vital implications of data.

12.6 Thyroid disease

12.6.1 General considerations

The thyroid gland is driven by the pituitary, which secretes thyroid stimulating hormone (TSH). The thyroid gland can be overactive (hyperthyroidism, or thyrotoxicosis, including Grave's disease), or underactive (hypothyroidism), or its control by TSH can be abnormal. The normal gland is described as euthyroid. The thyroid is particularly susceptible to cancer, inflammation (thyroiditis), and lack of, or overload with, dietary iodine (e.g. Derbyshire neck) or drug-derived iodine (e.g. from amiodarone). The gland secretes tetraio-dothyronine (L-thyroxine, T4) which is deiodinated to triiodothyronine (T3). Administration of L-thyroxine as a therapeutic agent can be to restore missing normal hormone (restoring the euthyroid condition in hypothyroidism), or it can be to suppress endogenous thyroxine production (restoring the euthyroid condition in hyperthyroidism), by means of a feed-back mechanism that suppresses TSH production.

Hyperthyroidism can result in an increased rate of gastric emptying, reduction in intestinal transit time, increased cardiac output and blood flow to tissues such as the liver and kidney, increased activity of hepatic microsomal enzymes and therefore hepatic intrinsic clearance, an increase in glomerular filtration rate, and reduced concentrations of albumin and α_1-acid glycoprotein in plasma. Hypothyroidism, generally speaking, does the opposite. All of these changes have the potential to alter drug disposition. Also, their reversal by means of effective therapy makes possible the longitudinal study of patients in both impaired and euthyroid states.

12.6.2 Examples

There are numerous examples of compounds and their pharmacokinetic properties that illustrate the general principles implicit in the previous two paragraphs. For example, in hyperthyroidism, the absorption rate of riboflavin and paracetamol is increased, and the rate of oxidative metabolism of tolbutamide, theophylline, antipyrine and metoprolol is increased. The rate of glucuronidation of oxazepam is increased, the renal clearance of digoxin is increased, and the plasma protein binding of propranolol and warfarin is decreased. Conversely, in hypothyroidism, the rate of absorption of paracetamol is decreased, the rate of oxidative metabolism of antipyrine is decreased, the renal clearance of digoxin is decreased, and the plasma protein binding of propranolol is increased. However, in hypothyroidism, opposite effects of hyperthyroidism do not *necessarily* occur, as with riboflavin absorption, oxazepam glucuronidation, and warfarin binding, and, in some cases (e.g. phenytoin and diazepam metabolism) there are no effects in either direction, and/or data are conflicting, or just not available. Three well-investigated cases, of digoxin, propranolol and warfarin, are worthy of further attention.

Digoxin bioavailability is especially affected by intestinal P-gp expression. This has been measured by means of reverse transcriptase-polymerase chain reaction of MDR1 messenger ribonucleic acid (mRNA) and immunohistochemical examination, in healthy volunteers before and after suppression of TSH with

L-thyroxine. To complement this, the pharmacokinetics of the P-gp substrate talinolol were assessed after i.v. and oral administration. L-thyroxine increased duodenal MDRI mRNA expression, and P-gp, 1.4-fold and 3.8-fold respectively – there were minor changes in talinolol half-life and *AUC*, consistent with reduced systemic availability of the oral doses under the influence of the L-thyroxine.

This work is relevant to the case of digoxin, with which relatively high doses of the drug are needed to control the heart rate of patients with hyperthyroidism and atrial fibrillation, not particularly because of any changes in absorption *per se*, volume of distribution, biliary clearance, renal clearance, or non-renal clearance (although some changes have been observed), but because of the induction of P-gp leading to increased efflux of digoxin from the duodenal tissue against the flow of absorption, reducing the proportion of the dose reaching the general circulation.

Warfarin requirements are increased in hypothyroidism but reduced in thyrotoxicosis. In part, the effect of warfarin on the regeneration of reduced vitamin K, an essential cofactor in the activation of clotting factors II, VII, IX and X by post-translational carboxylation of glutamate residues leads to diminished synthesis of active coagulation factors and accumulation of their non-functional precursors, and thus thyrotoxic patients exhibit an increase in prothrombin time. Although there are some reports of changes in the half-life of warfarin and in its binding, in total it is obvious that there are no dramatic changes in pharmacokinetic parameters of warfarin of sufficient magnitude to be associated with this, so the warfarin potentiation is thought to reflect an enhanced pharmacodynamic effect. Steady-state plasma concentrations of the coagulation factors are decreased, because of an enhanced degradation rate in thyrotoxicosis, and shorter plasma half-life values of the coagulation factors. The opposite is observed in hypothyroid states.

Propranolol clearance after intravenous injections is increased by approximately 50% in the hyperthyroid state, an effect attributed to increased liver blood flow. However, there is also increased clearance after oral dosing that is attributed to increased hepatic microsomal intrinsic clearance. There is also a slight reduction in protein binding. These changes result in a reduction in serum triiodothyronine concentrations, leading to the idea that propranolol is a useful treatment for certain hyperthyroid conditions, such as short-lived excessive secretion of thyroid hormone, which occurs in 'thyroid storm'. This occurs by virtue of propranolol inhibiting the sympathetic drive caused by thyroid hormone. Therefore it inhibits its own enhanced clearance. Simultaneous determination of intravenously and orally administered propranolol in hyperthyroid and euthyroid patients has shown markedly higher plasma concentrations of the drug following the oral doses, with smaller increases following the i.v. doses, and no differences in half-life, indicating increased clearance in the hyperthyroid state, but to an added degree with oral dosing, raising the question of whether CYP3A4 is particularly affected in the hyperthyroid state (Figure 12.7).

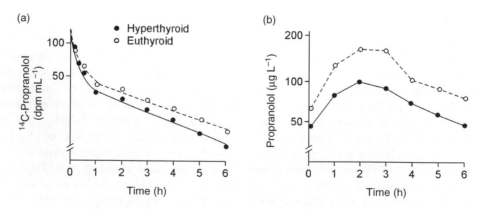

Figure 12.7   Simultaneous determination of i.v. (a) and oral (b) kinetics of propranolol in patients when hyperthyroid and euthyroid.

Specific information regarding the effect of thyroid function on individual drug-metabolizing isoforms of CYP 450 seems to be unavailable. However, in one particular study, the addition of triiodothyronine to primary hepatocytes significantly *reduced* CYP3A4 protein and mRNA, and, conversely, experimental hypothyroidism in rats led to CYP3A4 induction and CYP2C11 suppression. Thyroidectomized rats exhibited a marked increase in CYP8B1 protein and mRNA levels, whereas treatment of normal rats with L-thyroxine significantly reduced CYP8B1 activity and mRNA. No human studies appear to have been conducted as yet with substrates for specific isoforms, so the propranolol data remains the only basis at present for an expectation of enhanced presystemic metabolism of drugs by CYP3A4 in the hyperthyroid state (Sarlis and Gourgiotis, 2005).

12.7   Summary

A vast amount of knowledge exists on the effect of diseases on the pharmacokinetic and pharmacodynamic properties of drugs. Much of this knowledge consists of isolated pieces of information about single examples, and the search for themes of broader application can be arduous. With respect to diseases that effect drug disposition, the following general principles are well established:

- Experimental difficulties involved in separating disease effects from other factors pose difficult problems when seeking general principles.
- Gastrointestinal disorders can modify both the rate and extent of absorption, with increases or decreases dependent on the interplay of pH, gastric motility, surface area available for absorption, and other pathological changes, and on the chemical and biochemical properties of the particular drug, permitting a logical prediction of the effect on the basis of general principles.
- CHF leads to changes in drug absorption, hepatic and renal blood flow, hepatic intrinsic clearance, and presystemic elimination, mostly reductions in rate, and the effect on any particular drug will depend on which processes are predominant. It is not easy to separate the effects of oedema from those of the disease.
- Hepatic disorders reduce blood flow, hepatic intrinsic clearance, and plasma protein binding, but the consequences of these effects are modified in many cases by late onset, for multiple reasons including compensatory enzyme induction. Clearance calculations provide useful approaches to quantifying these effects. The significance for any particular drug will depend on the interplay and predominance of the various factors in its disposition. Centrally acting drugs show enhanced pharmacodynamic effects.
- Renal disease can cause reduced glomerular filtration and modified renal tubular transport. These changes mainly affect drugs dependent on urinary excretion of the unmetabolized drug, and this can be especially important in the dosing of digoxin, with its narrow margin if safety, and a whole range of antibiotics. Clearance calculations facilitate safe dosing of these drugs. There is a considerable incidence of concomitant occurrence of hepatic disease in renal disease patients.
- With exceptions, hyperthyroidism leads to enhanced rates of the drug disposition processes and hypothyroidism to the opposite.

References and further reading

Albers S, Meibohm B, Mir TS, Laer S. Population pharmacokinetics and dose simulation of carvedilol in paediatric patients with congestive heart failure. *Br J Clin Pharmacol* 2008; 65: 511–22.

Bleske BE, Welage LS, Kramer WG, Nicklas JM. Pharmacokinetics of torsemide in patients with decompensated and compensated congestive heart failure. *J Clin Pharmacol* 1998; 38: 708–14.

Birch JH, Foldes FF, Rendell-Baker L. Causes and prevention of prolonged apnea with succinylcholine. *Curr Res Anesth Analg* 1956; 35: 609–33.

Bouvier N, Millart H. Pharmacokinetics of beta blockers in hyperthyroidism. Therapeutic relevance. *Therapie* 1998; 53: 523–31.

Branch RA. Drugs in liver disease. *Clin Pharmacol Ther* 1998; 64: 462–5.

Busenbark LA, Cushnie SA. Effect of Graves' disease and methimazole on warfarin anticoagulation. *Ann Pharmacother* 2006; 40: 1200–3.

Curry SH. *Drug Disposition and Pharmacokinetics*, 3rd edn. Oxford: Blackwell Scientific, 1980.

Delco F, Tchambaz L, Schlienger R, Drewe J, Krahenbuhl S. Dose adjustment in patients with liver disease. *Drug Saf* 2005; 28: 529–45.

Dettli L, Ohnhaus EE, Spring P. Proceedings: Digoxin dosage in patients with impaired kidney function. *Br J Pharmacol* 1972; 44: 373P.

Dettli LC. Drug dosage in patients with renal disease. *Clin Pharmacol Ther* 1974; 16: 274–80.

Fakhoury M, Lecordier J, Medard Y, Peuchmaur M, Jacqz-Agrain E. Impact of inflammation on the duodenal mRNA expression of CYP3A and P-glycoprotein in children with Crohn's disease. *Inflamm Bowel Dis* 2006; 12: 745–9.

Feely J. Clinical pharmacokinetics of beta-adrenoceptor blocking drugs in thyroid disease. *Clin Pharmacokinet* 1983; 8: 1–16.

Foldes FF, Swerdlow M, Lipschitz E, Van Hees GR, Shanor SP. Comparison of the respiratory effects of suxamethonium and suxethonium in man. *Anesthesiology* 1956; 17: 559–68.

Frye RF, Zgheib NK, Matzke GR, Chaves-Gnecco D, Rabinovitz M, Shaikh OS, Branch RA. Liver disease selectively modulates cytochrome P450–mediated metabolism. *Clin Pharmacol Ther* 2006; 80: 235–45.

Hays MT. Thyroid hormone and the gut. *Endocr Res* 1988; 14: 203–24.

Hepner GW, Vesell ES, Tantum KR. Reduced drug elimination in congestive heart failure. Studies using aminopyrine as a model drug. *Am J Med* 1978; 65: 371–6.

Holt S, Heading RC, Clements JA, Tothill P, Prescott LF. Acetaminophen absorption and metabolism in celiac disease and Crohn's disease. *Clin Pharmacol Ther* 1981; 30: 232–8.

Hu OY, Tang HS, Sheeng TY, Chen TC, Curry SH. Pharmacokinetics of promazine in patients with hepatic cirrhosis–correlation with a novel galactose single point method. *J Pharm Sci* 1995; 84: 111–4.

Kobayashi M, Fukumoto K, Ueno K. Effect of congestive heart failure on mexiletine pharmacokinetics in a Japanese population. *Biol Pharm Bull* 2006; 29: 2267–9.

Lawrence JR, Sumner DJ, Kalk WJ, Ratcliffe WA, Whiting B, Gray K, Lindsay M. Digoxin kinetics in patients with thyroid dysfunction. *Clin Pharmacol Ther* 1977; 22: 7–13.

Liu Z, Fang S, Wang L, Zhu T, Yang H, Yu S. Clinical study on chronopharmacokinetics of digoxin in patients with congestive heart failure. *J Tongji Med Univ* 1998; 18: 21–4.

Malin M, Isolauri E, Pikkarainen P, Karikoski R, Isolauri J. Enhanced absorption of macromolecules. A secondary factor in Crohn's disease. *Dig Dis Sci* 1996; 41: 1423–8.

Mawer G, Miller NE, Turnberg LA. Metabolism of amylobarbitone in patients with chronic liver disease. *Br J Pharmacol* 1972; 44: 549–60.

Mikkaichi T, Suzuki T, Onogawa T, Tanemoto M, Mizutamari H, Okada M, *et al.* Isolation and characterization of a digoxin transporter and its rat homologue expressed in the kidney. *Proc Natl Acad Sci USA* 2004; 101: 3569–74.

Nishio N, Katsura T, Inui K. Thyroid hormone regulates the expression and function of P-glycoprotein in Caco-2 cells. *Pharm Res* 2008; 25: 1037–42.

O'Connor P, Feely J. Clinical pharmacokinetics and endocrine disorders. Therapeutic implications. *Clin Pharmacokinet* 1987; 13: 345–64.

O'Reilly S, Rowinsky E, Slichenmyer W, Donehower RC, Forastiere A, Ettinger D, *et al.* Phase I and pharmacologic studies of topotecan in patients with impaired hepatic function. *J Natl Cancer Inst* 1996; 88: 817–24.

Pacifici GM, Viani A, Franchi M, Santerini S, Temellini A, Giuliani L, Carrai M. Conjugation pathways in liver disease. *Br J Clin Pharmacol* 1990; 30: 427–35.

Petersson M, Rundqvist B, Bennett BM, Adams MA, Friberg P. Impaired nitroglycerin biotransformation in patients with chronic heart failure. *Clin Physiol Funct Imaging* 2008; 28: 229–34.

Renwick AG, Higgins V, Powers K, Smith CL, George CF. The absorption and conjugation of methyldopa in patients with coeliac and Crohn's diseases during treatment. *Br J Clin Pharmacol* 1983; 16: 77–83.

Rominger JM, Chey WY, Chang TM. Plasma secretin concentrations and gastric pH in healthy subjects and patients with digestive diseases. *Dig Dis Sci* 1981; 26: 591–7.

Sandek A, Bauditz J, Swidsinski A, Buhner S, Weber-Eibel J, von Haehling S, *et al*. Altered intestinal function in patients with chronic heart failure. *J Am Coll Cardiol* 2007; 50: 1561–9.

Sandell EP, Hayha M, Antila S, Heikkinen P, Ottoila P, Lehtonen LA, Pentikainen PJ. Pharmacokinetics of levosimendan in healthy volunteers and patients with congestive heart failure. *J Cardiovasc Pharmacol* 1995; 26Suppl 1: S57–62.

Sarlis NJ, Gourgiotis L. Hormonal effects on drug metabolism through the CYP system: perspectives on their potential significance in the era of pharmacogenomics. Curr Drug Targets Immune *Endocr Metabol Disord* 2005; 5: 439–48.

Shaffer JA, Williams SE, Turnberg LA, Houston JB, Rowland M. Absorption of prednisolone in patients with Crohn's disease. *Gut* 1983; 24: 182–6.

Shenfield GM. Influence of thyroid dysfunction on drug pharmacokinetics. *Clin Pharmacokinet* 1981; 6: 275–97.

Siegmund W, Altmannsberger S, Paneitz A, Hecker U, Zschiesche M, Franke G, *et al*. Effect of levothyroxine administration on intestinal P-glycoprotein expression: consequences for drug disposition. *Clin Pharmacol Ther* 2002; 72: 256–64.

Slaughter RL, Lanc RA. Theophylline clearance in obese patients in relation to smoking and congestive heart failure. *Drug Intell Clin Pharm* 1983; 17: 274–6.

Tammara B, Trang JM, Kitani M, Miyamoto G, Bramer SL. The pharmacokinetics of toborinone in subjects with congestive heart failure and concomitant renal impairment and/or concomitant hepatic impairment. *J Clin Pharmacol* 2002; 42: 1318–25.

Tang HS, Hu OY. Assessment of liver function using a novel galactose single point method. *Digestion* 1992; 52: 222–31.

Taylor TV. Non-invasive investigation of the upper gastrointestinal tract using technetium-99m. *Ann R Coll- Surg - Engl* 1971; 61: 37–44.

Verbeeck RK. Pharmacokinetics and dosage adjustment in patients with hepatic dysfunction. *Eur J Clin Pharmacol* 2008; 64: 1147–61.

Vrhovac B, Sarapa N, Bakran I, Huic M, Macolic-Sarinic V, Francetic I, Wolf-Coporda A, Plavsic F. Pharmacokinetic changes in patients with oedema. *Clin Pharmacokinet* 1995; 28: 405–18.

Wiersinga WM. Propranolol and thyroid hormone metabolism. *Thyroid* 1991; 1: 273–7.

Yonkers KA, Kando JC, Cole JO, Blumenthal S. Gender differences in pharmacokinetics and pharmacodynamics of psychotropic medication. *Am J Psychiatry* 1992; 149: 587–95.

Zini R, Riant P, Barre J, Tillement JP. Disease-induced variations in plasma protein levels. Implications for drug dosage regimens (Part II). *Clin Pharmacokinet* 1990; 19: 218–29.

13

Quantitative Pharmacological Relationships

13.1 Introduction

As stated in Chapter 1, it is quite common to read that there are two branches of pharmacology: pharmacodynamics and pharmacokinetics. At a more specific level, many authors prefer to distinguish between the description of overt effects of drugs and mechanistic studies (as subdivisions of pharmacology), limiting the use of the word pharmacodynamics to studies of the relationship between drug concentrations, and mechanisms, at sites of action. This chapter and the one that follows it are concerned with the relationship between drug concentrations in body fluids, principally plasma, and the intensity and duration of drug effects, sometimes called 'dose–response and time–action relationships'.

It is generally accepted that the effects of drugs are usually greater when exposure is greater, and that the effects of drugs that are rapidly removed from the body disappear quickly. Conversely, reduced exposure is thought to reduce effect, either by reduction in dose, or by use of a drug that is rapidly removed from the body. However, when these relationships are studied in detail relatively complex relationships are observed. During the last twenty-five years there has been explosive growth in our understanding of what has come to be known as the pharmacokinetic/pharmacodynamic (PK/PD) relationship. PK/PD modelling uses measurements of overt effects, and of drug concentrations in plasma and at other sites, to explore mechanisms. This is conveniently studied through:

- Fundamentals of dose–effect relationships (dose–response curves, recorded principally *in vitro*).
- Study of similarities and differences between *in vitro* and *in vivo* data.
- Consideration of the potential for variation.
- Specific examples of dose–effect and concentration–effect relationships *in vivo*.
- PK/PD modelling, which integrates pharmacokinetic concepts with concentration–effect relationships.
- Concepts of loading and maintenance doses for long-term treatment.

13.2 Concentration–effect relationships (dose–response curves)

Dose–effect relationships follow from the nature of the reversible interaction between a drug and its receptor. A drug may be an agonist and exert a stimulatory effect on the receptor, or it may be an antagonist and act by binding to the receptor to prevent the effects of an agonist (Section 1.1).

The effects of a drug do not increase indefinitely with increasing doses, as there is a finite number of receptors. The reversible binding of a drug, D, to its receptor, R, can be assessed by the Law of Mass Action:

$$D + R \rightleftharpoons DR$$

Drug Disposition and Pharmacokinetics,   By Stephen H. Curry and Robin Whelpton
© 2011 John Wiley & Sons, Ltd

where DR is the drug–receptor complex. The equilibrium constant (Section 1.5.1), K_D, is related to the molar concentrations of the species at equilibrium:

$$K_D = \frac{[D][R]}{[DR]} \tag{13.1}$$

The total concentration of receptors is the sum of the bound and non-bound receptors, $[DR] + [R]$, and the proportion, or fraction, of receptors bound, f_A, can be obtained by rearrangement of Equation 13.1:

$$f_A = \frac{[DR]}{[DR] + [R]} = \frac{[D]}{[D] + K_D} \tag{13.2}$$

This is the equation of a rectangular hyperbola [Figure 13.1(a)].

K_D has units of concentration and is numerically equal to the concentration at which 50% of the receptors are occupied. The lower the value of K_D the greater is the *affinity* (i.e. the tendency to bind) of the agonist for its receptor.

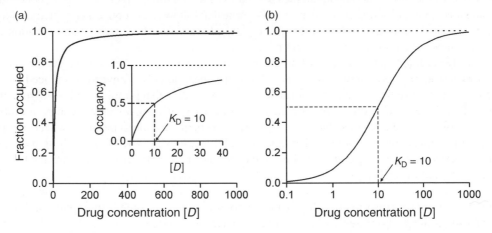

Figure 13.1   (a) Fraction of receptors occupied (occupancy) as a function of drug concentration. K_D is numerically equal to the concentration at which half the receptors are occupied (Inset). (b) Logarithmic transformation converts the hyperbola into a symmetrical sigmoid shape typical of log(dose)–response curves.

The measured response to a drug is related, but not necessarily directly proportional to receptor occupancy. Agonists that interact with their receptors to produce the maximum response possible are referred to as 'full' agonists whereas agonists that are incapable of producing the maximum response, even when all the receptors are occupied, are referred to as *partial agonists* (Figure 13.2). A partial agonist in the presence a full agonist may antagonize the effects of the latter and so partial agonists are sometimes called *agonist-antagonists*.

Figure 13.2 illustrates some important features of agonists and log(concentration)–response curves. First, it is usual to plot the response as the percentage of the maximum that a full agonist can produce. If the responses are from an *in vitro* experiment using isolated tissues, say guinea pig ileum, then different pieces of tissue will give different responses, but normalizing the results to the maximum obtainable with each tissue allows comparison between experiments and average responses can be plotted. Such dose–response curves, where the response is continuously variable between the minimum and maximum response are 'graded' response curves. For *in vivo* data 'quantal' response curves may be used. With these, some measurable endpoint is defined, for example pain relief, and then the numbers of subjects from a test population that attain the endpoint are recorded for each concentration or dose of drug. The y-axis of Figure 13.2 for a quantal dose–response curve would be 'percentage of subjects responding', the maximum (100%) being when all the subjects responded to the treatment.

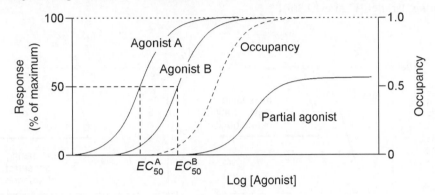

Figure 13.2 Log(agonist concentration)–response curves. Agonist A is more potent than B because it has a lower EC_{50}. The partial agonist cannot elicit the maximum response even when all the receptors are occupied.

In Figure 13.2, the curves for agonists A and B have been positioned to the left of the occupancy curve illustrating that not all the receptors have to be occupied for the response to be maximal. Stephenson (1956) introduced the term *efficacy* to describe the way in which agonists vary in the response they produce, even when they occupy the same number of receptors. The partial agonist is less efficacious than the full agonists because it cannot elicit as great a response. The full agonists are considered equi-efficacious as they both have maximal efficacy. The responses to agonist A occur at lower concentrations than those for agonist B. Agonist A is more *potent* than agonist B, that is less drug is required to obtain a defined effect. Potency is usually assessed from the concentration producing 50% of the maximum response, EC_{50}. For a quantal response, EC_{50} or ED_{50}, is the dose at which 50% of the *test* population respond. It may seem paradoxical, but partial agonists can be highly potent. This is true of the opiate agonist, buprenorphine, which has high affinity for the μ-opioid receptor. As a consequence, it can displace other opiate analgesics such as morphine, reducing the pain relief and possibly causing withdrawal symptoms in addicts.

A further consideration is that the slopes of dose–response curves may differ. Some drugs have steep curves whilst others have shallow curves. The response at any concentration, C, can be written in terms of the maximum response, E_{max} and EC_{50}:

$$\text{response} = \frac{E_{max}C^\gamma}{EC_{50}^\gamma + C^\gamma} \tag{13.3}$$

where γ is a factor that defines the slope, or shape, of the concentration–effect curve. Pharmacodynamic models that use Equation 13.3 are referred to as E_{max} models. When γ is large the slope of the curve is steep and the effect may appear to be almost all or nothing because the response may go from the minimal to the maximum effect over a very small range of concentrations.

13.2.1 Antagonism

Antagonists have affinity for receptors but not efficacy. Consequently, their effects can only be assessed in the presence of a suitable agonist. Antagonists can be competitive, that is they compete with the agonist for receptor binding sites, or non-competitive, these often bind covalently to the receptor. Competitive antagonism is reversible because the antagonist can be displaced by increasing the concentration of agonist, and, provided sufficient agonist is used the maximal tissue response can be obtained. This results in log(agonist concentration)–response plots constructed in the presence of differing concentrations of antagonist being parallel [Figure 13.3(a)].

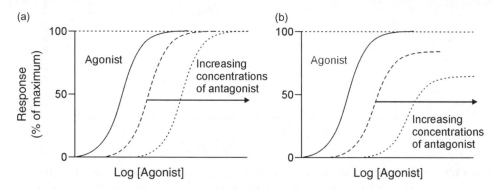

Figure 13.3 (a) Competitive antagonists cause a parallel shit in log(agonist concentration)–response curve to the right. (b) Non-competitive antagonists move the agonist curve to the right with a decreasing maximum response, leading to a non-parallel shift.

A non-competitive antagonist reduces the maximum tissue response. In effect, the numbers of viable receptors are progressively reduced with increasing concentrations of non-competitive antagonist so the maximum response declines. This leads to a non-parallel shift of the log(antagonist concentration)–response curve [Figure 13.3(b)].

The potency of antagonists can be assessed from curves of the type shown in Figure 13.3. Dose–response (*DR*) or concentration–response (*CR*) ratios are calculated:

$$CR = \frac{EC_{50} \text{ of agonist in presence of antagonist}}{EC_{50} \text{ of agonist alone}} \qquad (13.4)$$

for a range of antagonist concentrations. The more potent an antagonist the greater will be the value of *CR* for a given molar concentration of drug. The potencies of competitive antagonists are compared using the pA_2 values, where pA_2 is the negative logarithm of the molar concentration of antagonist when $CR = 2$. The notation is analogous to that used for pH, the p indicating a negative logarithm.

13.2.2 Variation

As stated, much of the foregoing referred to *in vitro* experiments, in which isolated tissue samples are exposed to fixed concentrations of drugs, and tissue responses are measured, as contractions of the tissue, release of biochemicals, chemical change of exogenous substrates, etc. In these experiments, variations in response occur to fixed concentrations, or a range of concentrations can induce a defined response. For any drug–effect combination, there will be a mean response to a fixed dose. Responses will be seen on either side of this mean and there will be a range of response. The distribution of responses can be assessed mathematically. The commonest distribution is the symmetrical 'normal' distribution. This involves the largest proportion of the responses being closest to the mean, with successively smaller proportions of the responses being at successively larger intervals from the mean. The usual measure of scatter in the normal distribution is the standard deviation, σ,

$$\sigma = \sqrt{\frac{\sum(x-\bar{x})}{n-1}} \tag{13.5}$$

which is the square root of the variance. The variance (σ^2) is obtained by summing the squares of the deviations of each result, x, from the mean result $\bar{x}$ and dividing by one less than the number (n) of observations ($n-1$), called the 'degrees of freedom':

$$\sigma^2 = \frac{\sum(x-\bar{x})}{n-1} \tag{13.6}$$

A large standard deviation indicates a wide scatter of results with a wide distribution curve. The number of degrees of freedom is equal to the number of data points that must be used to calculate the standard deviation. It is $n-1$ because the last deviation can be deduced from the fact that $\sum(x-\bar{x}) = 0$. For a normal distribution approximately 68% of the measurements lie within $\pm 1\sigma$ of the mean, $\sim$95% are within $\pm 2\sigma$ and $\sim$99.7% are within $\pm 3\sigma$ of the mean. Data should be tested for 'normality' before any parametric statistical tests are performed as otherwise the appropriate tests are non-parametric ones. Figure 13.4 shows that salicylate concentrations at 3 h in 100 subjects were normally distributed.

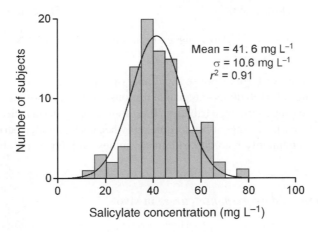

Figure 13.4 Serum salicylate concentrations in 100 subjects 3 h after 35 mg kg^{-1} sodium salicylate orally. Data were fitted to a normal distribution (solid line).

It should be noted that while the normal distribution is common in pharmacology, other distributions occur frequently. The commonest variation is skewing, with one side of the curve tailing much further from the mean than is the case with the other side. Bimodal distributions also occur. If skewed data can be transformed so that the results are normally distributed then parametric statistics can be applied to the transformed data. For example if the logarithms of the observations are normally distributed, then the statistics are performed on the logarithms of the observations.

One particular aspect of the distribution of pharmacological phenomena relates to whether responses *in vivo* are normally distributed with respect to dose or log(dose). The basic work was done with log (dose)–effect curves, for graded responses. When the response is measured as percent of the biological units affected, a so-called quantal response, the log(dose)–effect curve is an integrated frequency distribution diagram based on a log(dose). Thus original work implies that pharmacological responses are log normally distributed. However, *in vivo*, the exposure range is smaller, especially in human pharmacology, and data may appear to be normally distributed with respect to dose. Data collected in humans is sparse, consisting mainly of plasma concentration data, such as the salicylate data in Figure 13.4, which is not, strictly speaking, response data. These data, generally speaking, appear to indicate normal distribution with respect to dose in some cases and log(dose) in others. However, the data of Figure 13.5 are instructive. This figure shows the warfarin dose in a group of patients who achieved a prothrombin ratio of 1.7.

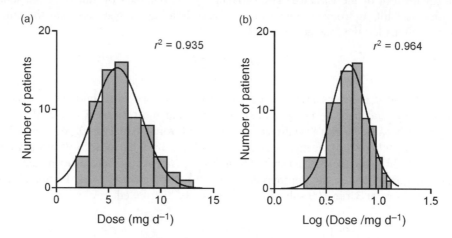

Figure 13.5 Warfarin dose (mg/day) to achieve a prothrombin ratio of 1.7 (INR = 3.4). (After White *et al.*, 1987, 1991).

Distributions are shown for warfarin response as number of patients in each dose category (thus relating incidence of effect to a range of fixed doses – a quantal response) related to dose, and also to log(dose). The statistics show a better fit to a log–normal distribution. It is often not easy in such circumstances to determine whether the distribution is normal or log–normal, principally because, as stated earlier, of the narrow range over which responses are commonly recorded in human studies. Similar results have been reported with suxamethonium.

13.2.3 *Some illustrations of dose–response curves* in vivo

Although it is not difficult to demonstrate a 'perfect' dose–response curve *in vitro*, demonstration of such perfect relationships in many other situations is much more difficult. Some examples are illustrated below.

- *In vivo*, the log(dose)–response curve may be a straight line from a restricted part of the available range. For example, in Figure 13.6, for sleeping times for phenobarbital in mice, the upper limit is fixed by a different drug response in that higher doses cause death. The lower limit is the recording of sedation rather than sleep.
- In human pharmacology it is often necessary to work at the lower end of the dose–response curve because of the prime importance of the safety and comfort of the subjects involved (Figure 13.7).
- In chemotherapy it is usual to work at the upper end of the curve, as antimicrobial drugs usually have a wide margin of safety and the maximum effect against the microbes is required (Chapter 14).
- In psychopharmacology it is commonly impossible to define or achieve the maximum possible effect, so the upper limit is less than 100% in a quantal curve. Additionally, there is a large subjective contribution (placebo effect) and some response is obtainable from a placebo preparation, so the graph does not necessarily go through the origin (Figure 13.8).
- Compensatory reflexes play a considerable part in modifying dose–response curves when the measurable effect is a change in blood pressure; this makes the observations highly time dependent (Section 14.2.2).

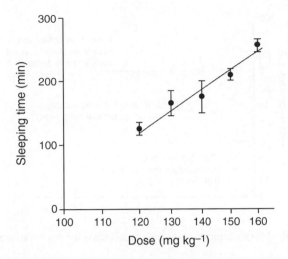

Figure 13.6 Dose–response relationship for phenobarbital concentrations (logarithmic scale) and sleeping time in mice. Each point is the mean of six to eight mice $\pm$ SEM. Below 120 mg kg^{-1} the drug had only a sedative effect. Above 160 mg kg^{-1}, some of the mice died.

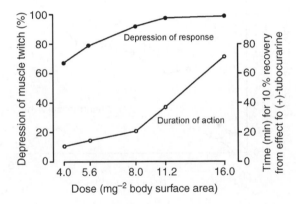

Figure 13.7 Dose–response curves for depression of standard muscle twitch response and for duration of action (time required for 10% of standard response to return) in human subjects given $(+)$-tubocurarine ($n \geq 20$ at each time point). (Redrawn from Walts & Dillon, 1968.)

- Latent pharmacological effects can show unusual dose–response relationships. For example, the effect of warfarin on clotting time occurs after a considerable delay and after the peak plasma level of the drug has passed. In such a situation, the perfect dose–response relationship is obscured (Chapter 14).
- In teratology, there is a dose range which leads to live births of malformed fetuses (Figure 13.9). There is a higher dose range that causes foetal death illustrated in ferrets by resorption, with no live births. Thus, while a dose–response curve for resorption is not particularly unusual, the dose–response curve for abnormal survivors has a peak, with a fall-off with both higher and lower doses.
- Preventive treatment with drugs is common in therapeutics. A relationship between degree of prevention (which desirably is 100%) and dose is very difficult to demonstrate, especially as many diseases requiring preventative treatment, notably mania and epilepsy, are episodic in their occurrence (Chapter 14).

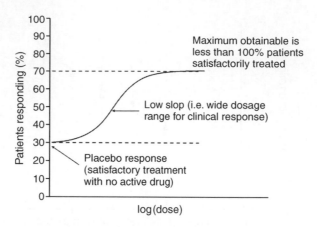

Figure 13.8   Model quantal dose–response curve for psychopharmacology.

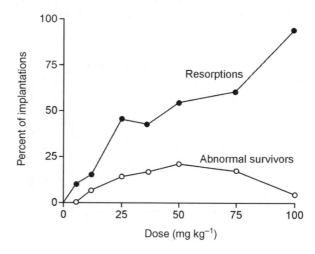

Figure 13.9   Dose–response curves for the teratogenic activity of trypan blue in ferrets. Foetuses that do not survive the pregnancy are reabsorbed (resorptions). The proportion of live abnormal foetuses declines as the dose increases. (Redrawn from Beck & Lloyd, 1964.)

13.3   The importance of relating dose–effect and time-action studies

Historically, while drug effects were related almost exclusively to dose, both *in vitro* and *in vivo*, by pharmacologists, the possibility that *time since dose* is important was known to the ancients from their experiences with alcohol and other naturally occurring pharmacological agents, and was even documented by Shakespeare in his play *Romeo and Juliet* (in which Friar Lawrence gave Juliet a drug designed to induce a death-like coma lasting exactly 42 hours, only to have Romeo turn up too early, misunderstand the situation, and commit suicide in grief just as Juliet was waking up). However, the recognition by scientists that time since dose is important in quantitative pharmacology is more recent. This fact was introduced in Chapter 4. It is clearly the basis for wear-off of effect, but, additionally, it actually changes the *relationship between effect and concentration in plasma*, and this has been known for a relatively short time. Pioneering work had been done by Levy and Nelson, and Wagner, in particular, in the 1960s on single dose-plasma level-effect relationships and on the duration of action of drugs as a function of dose (Levy, 1964a and b, and 1965; Levy and Nelson, 1965; Wagner, 1968a and b). At about the same time, Brodie and colleagues demonstrated how complicated the relationships are when drugs with multi-compartment distribution are studied in this context (e.g. Brodie, 1967). Earlier, Murphy and Lasagna (1961) using diuretics, had found that depending on whether a cumulative effect (24-h urine production) or an 'instant' effect (rate of urine flow at a particular time) was measured, different relationships of response were possible. Also in the 1960s, Levy and Nagashima (Levy and Nagashima 1969; Nagashima and Levy 1969) demonstrated the relative time courses of anticoagulant concentration and effect.

Thus, the relationship between effect and concentration of drug in plasma should not be expected to be constant or simple, and it can vary with time. More recently, the emphasis has been on combined modelling of effects and concentrations. As stated, the broadly-accepted dose–effect relationships of pharmacology were largely elucidated by *in vitro* experimentation, with organ bath drug concentrations as the reference point. The salicylate example in Figure 13.4 introduced, for our purposes, the concentration in plasma as the reference point for *in vivo* work. Time–action considerations require the presumption of a certain relationship between the concentrations of drugs in the fluids in the vicinity of the drug receptors *in vivo* and concentrations in plasma *in vivo*. Logic requires the following assumptions:

- Drug concentrations in the fluids in the vicinity of drug receptors *in vivo* and *in vitro* are analogous.
- The concentrations of drug–receptor complexes are a function of the concentration of drug molecules in the fluid in the vicinity of the receptors.
- Drug distribution through body fluids occurs by reversible diffusion and by binding processes.
- Drug concentrations in plasma water *in vivo* are analogous to the concentrations in the *in vitro* experiment; the distinction between plasma water and whole plasma should be particularly noted.

13.3.1   The fundamental single dose time-action relationship

With *in vitro* experimentation, the drug is added and removed by the experimenter, and the concentration changes little if at all between addition and removal. *In vivo*, the concentration of the drug in whole plasma and plasma water rises and falls with time, as shown in earlier chapters. *In vivo*, quantitative relationships between effect and drug concentrations are superimposed upon a changing drug concentration. The model presented as Figure 4.1, is the model for the relationship between effect and concentration for an oral or other non-intravenous dose. This model makes no presumptions regarding linearity of kinetics or response relationships. However, in some of the earliest studies of the relationship between concentration and effect, Gerhard Levy made two fundamental deductions:

- If the effect is measured in the middle of portion of an *in vivo* log dose–response relationship, and if the decline of plasma concentration is by first-order kinetics, then the effect wears off at a constant rate.
- If the same assumptions apply as in the previous bullet, doubling of the dose extends the duration of effect by one half-life.

However, the above only applies to a bolus intravenous injection and assuming a single compartment model applies. Even for a simple single compartment model, the duration of action will not necessary double for a doubling of an oral dose. The change in duration is dependant on where the minimum effective concentration is relative to $C_{max.}$ (Figure 13.10).

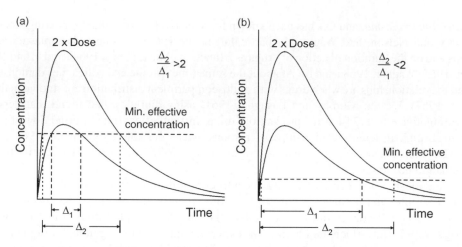

Figure 13.10   The proportionate change in the duration of effect (Δ) on doubling the dose depends on where the minimum effective concentration line is initially. (a) When the line is close to the maximum concentration, doubling the dose more than doubles the duration. (b) When the line is much lower, doubling the dose may have little effect on the duration of action.

Thus, when the simplest models for oral doses apply, an increase in dose causes a proportional increase in C_{max}, no difference in t_{max}, and a variable increase in duration of effect. Within this limitation, if the effect were proportional to dose, a doubling of dose would lead to a doubling of effect, and a less than doubling of duration of effect. Careful consideration of the dose–response relationships discussed earlier leads to the inevitable conclusion that the foregoing is a special case that is unlikely to occur. Thus, if the dose and concentrations are in the upper part of the dose–response curve then doubling of the dose and concentration will cause less than a doubling of effect. Conversely, if the dose and concentrations are in the lower part of the dose–response curve, then doubling of the dose and concentration will lead to more than a doubling of effect. Similarly, if the absorption of the drug is saturable, then there will also be less than the expected increase in effect when the dose is increased. By the same standard, a saturable first-pass effect can cause more than the expected increase in effect, unless the first-pass effect produces a metabolite through which the drug exerts its effect, then the opposite will occur. If metabolism is non-linear (e.g. phenytoin), then an increase in dose will cause more than the expected increase in duration of effect; there will also be more than the expected increase in C_{max}, and therefore in maximum effect, but worse still, a larger dose will have a later t_{max}. If metabolism occurs with zero-order kinetics (e.g. alcohol), then the duration of effect will be doubled if the dose is doubled, if all other simpler rules apply. If the drug shows saturation of protein binding in the therapeutic range, then doubling of the plasma concentration will cause more than a doubling of the concentration in plasma water, and more than a proportionate increase in effect, unless, of course, the receptors are saturated! If the drug exhibits two-compartment distribution there may be an even more complex relationship between dose, concentration and effect at any particular time, and, in all cases, the patterns shown during long-term multiple dosing may introduce extreme complexity, especially if all of these factors combine within a single drug example. And long term dosing (see next section) opens up the possibility of variations occurring during therapeutic regimens that do not occur with single doses, and disease opens up possibilities of variations caused by disease, and by the drug interactions that occur when multiple drugs are used in the individual patient. For all of these reasons, new drug discovery programmes are usually designed to produce new treatments that conform to the simplest models.

13.3.2   The fundamental multiple-dose concentration–effect relationship

This is represented in Figure 4.2. Again, no presumptions are made about particular kinetics applying, except that, in this case, elimination occurs by first-order processes. The effect is shown as first occurring when the plasma concentrations first rise above the threshold for effect, during, in this case the third dosage interval – continuous effect is established during the fifth dosage interval. The concentrations fluctuate between a maximum and a minimum, as described in earlier chapters, both of which are below the threshold for unwanted effects and above the threshold for desired effect.

The fundamental models shown in Figures 4.1 and 4.2, illustrated in part by the examples in this chapter, together with the observations of Gerhard Levy, provide an excellent starting point for our excursion into the world of PK/PD modelling, the principal topic of the next chapter.

References and further reading

Atkinson AJ, Abernethy DR, Daniels CE, Dedrick R. *Principles of Clinical Pharmacology*. New York: Academic Press, 2006.

Ariëns EJ. (Ed) *Molecular Pharmacology*. Vol 1. London: Academic Press, 1965.

Barlow RB. *Introduction to Chemical Pharmacology*. London: Methuen, 1964.

Beck F, Lloyd JB. Dosage–response curves for the teratogenic activity of trypan blue. *Nature* 1964; 201: 1136–7.

Brodie BB. Physicochemical and biochemical aspects of pharmacology. *JAMA* 1967; 202: 600–9.

Brunton Lazo J, and Parker K (Eds). *Goodman & Gilman's The Pharmacological Basis of Therapeutics*, 10th edn. New York, Macmillan, 2005.

Clark WG, Brater DC, Johnson AR, *Goth's Medical Pharmacology*. St.Louis: CV Mosby, 1992.

Curry SH. *Drug Disposition and Pharmacokinetics*. Oxford: Blackwell Scientific, 1980.

DiPiro JT, Talbert RL, Yee GC, Matzke GR, Wells BG, Posey LM. *Pharmacotherapy: A Pathophysiological Aprroach*, 7th edn. McGraw-Hill, 2008.

Katzung GB, Masters SB, Trevor AJ. *Basic and Clinical Pharmacology*, 11th edn. New York: Lange, 2009.

Levy G. Relationship between rate of elimination of tubocurarine and rate of decline of its pharmacological activity. *Br J Anaesth* 1964a; 36: 694–5.

Levy G. Relationship between elimination rate of drugs and rate of decline of their pharmacologic effects. *J Pharm Sci* 1964b; 53: 342–3.

Levy G. Apparent potentiating effect of a second dose of drug. *Nature* 1965; 206: 517–8.

Levy G, Nagashima R. Comparative pharmacokinetics of coumarin anticoagulants. VI. Effect of plasma protein binding on the distribution and elimination of bishydroxycoumarin by rats. *J Pharm Sci* 1969; 58: 1001–4.

Levy G, Nelson E. Theoretical relationship between dose, elimination rate, and duration of pharmacologic effect of drugs. *J Pharm Sci* 1965; 54: 812.

Murphy J, Casey W, Lasagna L. The effect of dosage regimen on the diuretic efficacy of chlorothiazide in human subjects. *J Pharmacol Exp Ther* 1961; 134: 286–90.

Nagashima R, Levy G. Comparative pharmacokinetics of coumarin anticoagulants. V. Kinetics of warfarin elimination in the rat, dog, and rhesus monkey compared to man. *J Pharm Sci* 1969; 58: 845–9.

Rang HP, Dale MM, Ritter JM, Flower R. *Rang and Dale's Pharmacology*. London: Churchill Livingstone, 2007.

Stephenson RP. A modification of receptor theory. *Br J Pharmacol Chemother* 1956; 11: 379–93.

Wagner JG. Kinetics of pharmacologic response. I. Proposed relationships between response and drug concentration in the intact animal and man. *J Theor Biol* 1968; 20: 173–201.

Wagner JG, Pharmacokinetics. *Annu Rev Pharmacol* 1968; 8: 67–94.

Walts LF, Dillon JB. Durations of action of d-tubocurarine and gallamine. *Anesthesiology* 1968; 29: 499–504.

White RH, Hong R, Venook AP, Daschbach MM, Murray W, Mungall DR, Coleman RW. Initiation of warfarin therapy: comparison of physician dosing with computer-assisted dosing. *J Gen Intern Med* 1987; 2: 141–8.

White RH, Mungall D. Outpatient management of warfarin therapy: comparison of computer-predicted dosage adjustment to skilled professional care. *Ther Drug Monit* 1991; 13: 46–50.

14

Pharmacokinetic/Pharmacodynamic Modelling: Simultaneous Measurement of Concentrations and Effect

14.1 Introduction

The objectives of modern analysis of drug action are to delineate the chemical or physical interactions between drug and target cell and to characterize the full sequence and scope of actions of each drug (Ross, 1996). This involves integrating the dose–response relationships considered in the previous chapter with time–action considerations.

In Chapter 4, and again in the previous chapter, we showed that, after single doses of drugs, while concentrations in plasma increase and decrease, so effects are presumed to increase and decrease. This is a fundamental expectation. No particular models, linear or non-linear need to be invoked in stating this principle, which can be considered descriptively. However, we can now examine in greater detail the ways in which this principle does and does not apply, and study the application of mathematical models to what has come to be known as the pharmacokinetic/pharmacodynamic (PK/PD) relationship.

A number of relevant scientific and medical observations date from before the formal organization of parallel PK and PD research. For example, it has long been known that:

- The effect may be closely related to the concentration in plasma so that the time of maximum concentration and effect coincide (e.g. i.v. thiopental, with which induction of effect is virtually instantaneous, and oral carebastine).
- The time of maximum effect can precede the time of maximum concentration (e.g. alcohol, glutethimide and isoprenaline).
- The time of maximum effect can occur later than the time of maximum concentration (e.g. LSD and warfarin).
- In some single dose cases there is an expectation of a permanent effect (e.g. treatment of a headache with an analgesic).
- The effect of a drug can be undetectable after a single dose, only appearing during multiple dosing therapy, but not obviously related to the growth of the plasma concentrations towards the pharmacokinetic steady-state (e.g. antibacterial drugs, and those used in epilepsy and psychiatry).

In this chapter we present model relationships of various different types between observed concentration and effect. Some of the examples chosen for this chapter are descriptive, with no mathematical models. Also,

Drug Disposition and Pharmacokinetics,   By Stephen H. Curry and Robin Whelpton
© 2011 John Wiley & Sons, Ltd

some of the mathematical models are presented with no applications. The objective is to illustrate the general principles applicable in key examples from what is now, a considerable volume of literature.

14.2   PK/PD modelling

14.2.1   *Objectives*

- To describe the time-courses of concentrations and effects in mathematical terms.
- To derive equations for concentration/response relationships at sites of action.
- To explain time-dependent relationships in mechanistic terms.
- To generate the ability to make predictions of dose/effect relationships, effect of changing dosage regimens etc.
- To facilitate comparisons between different drugs and selection of optimal compounds in structure/ activity work.
- To provide methods for the study of drugs for which there are no bioanalytical methods.

14.2.2   *Single-compartment, time-independent PK/PD models*

Although it is reasonable to expect a modified form of the $E_{\max}$ model (Equation 13.3) to apply, *in vivo* the range of concentrations may be limited and simpler equations can be appropriate, particularly if the concentrations are on the lower, almost linear, part of the dose–response curve [Figure 13.1(a)] The simplest model is where (i) the drug distributes into a single compartment, represented by plasma, and (ii) the effect is an instantaneous, direct function of the concentration in that compartment. In this situation, the relationship between drug concentration (C) and a pharmacological effect (E) can be simply described by the linear function:

$$E = SC \tag{14.1}$$

where S is a slope parameter. If the measured effect has some baseline value (E_0), when drug is absent (e.g. a physiological effect such as diastolic blood pressure or resting tension on the tissue in an organ bath), then the model may be expressed as:

$$E = E_0 + SC \tag{14.2}$$

The parameters of this model, S and E_0, may be estimated by linear regression. This model does not contain any information about efficacy and potency, cannot identify the maximum effect, and thus cannot be used to find the concentration for 50% effect or for a defined effect in 50% of patients (EC_{50}). At higher concentrations, plotting log(concentration) commonly makes the data linear within the approximate range 20–80% of maximal effect [Figure 13.1(b)]. This log transformation facilitates a graphical estimation of the slope (m) of the apparently linear segment of the curve:

$$E = m\ln(C_0 + C) \tag{14.3}$$

where C_0 is the baseline concentration (usually zero, but not for cases of add-on therapy or when administering molecules that are also present endogenously). In this equation, the pharmacological effect may be expressed, when C is zero, as:

$$E_0 = m\ln C_0 \tag{14.4}$$

As implied earlier, for larger ranges of concentration it may be necessary to use the E_{max} model (Equation 13.3). This allows the computation of an EC_{50}; and when two or more compounds are investigated, comparison of potencies. If there is a baseline response (E_0), then this may be added:

$$E = E_0 + \frac{E_{max} C^{\gamma}}{EC_{50}^{\gamma} + C^{\gamma}}$$

(14.5)

As before (Section 13.2) the sigmoidicity parameter or Hill coefficient, γ, accounts for situations when the slope of the curve is greater or less than 1. At the molecular level, $\gamma > 1$ indicates cooperativity of binding; indeed Hill introduced the coefficient to explain the binding of oxygen to haemoglobin ($\gamma \sim 2.8$). However, *in vivo* it is more difficult to explain why the value of the Hill coefficient is what it is. The slopes of dose-response curves vary between subjects, and some of the variation may reflect differences in plasma protein binding, as it is generally the non-bound drug that can diffuse to the site of action. From a practical point of view, the larger the value of the exponent, the steeper the dose–response curve. A very high exponent can be viewed as indicating an all-or-none effect (e.g. the development of an action potential in a nerve). Within a narrow concentration range, the observed effect goes from all to nothing or vice versa.

The equation equivalent to Equation 14.5 for the relationship between concentration and effect for an antagonist is:

$$E = E_0 + \frac{I_{max} C^{\gamma}}{IC_{50}^{\gamma} + C^{\gamma}}$$

(14.6)

where IC_{50} is the concentration causing 50% inhibition.

In vivo, these models, analogous to the classical dose or log dose–response curves of *in vitro* pharmacology, are limited to direct effects in single-compartment systems. These models make no allowance for time-dependent events in drug response.

14.2.3   Time-dependent models

The most common approach to *in vivo* PK/PD modelling involves sequential analysis of the concentration versus time and effect versus time data. The kinetic model provides an independent variable, such as concentration, *driving* the dynamics. Only in limited situations could it be anticipated that the effect influences the kinetics, for example, effects on blood flow or drug clearance itself.

Levy (1964), Jusko (1971) and Smolen (1971, 1976) described the analysis of dose–response time data. They developed a theoretical basis for the performance of this analysis from the data obtained from the observation of the time course of pharmacological response, after a single dose of drug, by any route of administration. Smolen (1976) extended the analysis to the application of dose–response time data for bioequivalence testing.

In dose–response–time models, the underlying assumption is that pharmacodynamic data give us information on the kinetics of drug in the *biophase* (i.e. the tissue or compartment precisely where the drug exhibits its effect). In other words, apparent half-life, bioavailability and potency can be obtained simultaneously from the dose–response–time data. Considering such a model, assuming (i) first-order input/output processes and (ii) extravascular dosing, the kinetic model then drives the effect function of the dynamic model. It is the *dynamic behaviour* which is described by the response model. A zero-order input and first-order output governs the *turnover* of the response. This permits consideration of situations where the plasma concentration represents delivery of the drug to an effect compartment; the time course of drug concentration and of effect (both in the biophase) is different from that simply observed in plasma concentrations.

14.2.3.1   Ebastine and carebastine

Ebastine is a novel H_1-antagonist. It is extensively metabolized to its carboxylic acid analogue carebastine. Investigations into the pharmacokinetics, the plasma concentrations, and the effects of carebastine, which appears to account for most, if not all, of the pharmacological potency, were conducted in healthy volunteers after oral doses (Vincent *et al.* 1983). The pharmacokinetics exhibited exponential growth and decay, probably reflecting a balance of formation from the ebastine and the elimination of the carebastine. The authors calculated an elimination half-life of 10–12 hours. The pharmacodynamic effect studied was percentage of the wheal area induced in a standardized trauma test, compared with placebo. The time to maximum wheal inhibition, which occurred between 2 and 6 hours after dosing with ebastine, corresponded with the time to peak plasma concentration of carebastine. The plasma concentrations were linearly and significantly correlated with the absolute percentage of the histamine-induced wheal area in some but not all of the subjects (Figure 14.1). Thus, if a hysteresis curve (Section 14.2.4) were to be drawn with the best of these data, it would fit the model B_1 of Figure 14.3. The most appropriate mathematical model would be one of the simplest of the equations, Equation 14.1. However, note that the wheal area was not 100% at the earliest time point, when the concentration was presumably zero, suggesting that Equation 14.2 would be more appropriate. This might have reflected experimental variation, or it could have reflected a small contribution to the effect by the unchanged drug, ebastine, at the earlier time points.

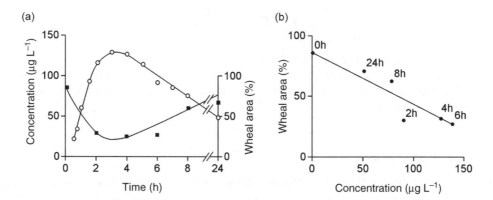

Figure 14.1   (a) Time course of the plasma concentrations of carebastine (open circles) and the histamine-induced wheal area (closed squares) in a representative subject after a single oral dose of 10 mg. of ebastine. (b) Relationship between plasma concentration and effect – the times of collection are shown. (Redrawn from Vincent *et al.*, 1988.)

14.2.3.2   Unequal distribution within plasma: ethanol and glutethimide

With regard to ethanol, it is now well-established that:

- The effect of alcohol is greater during the rising phase of the blood alcohol–time curve (i.e. when absorption is taking place). This is commonly called 'acute tolerance'.
- There is a concentration gradient during absorption with highest concentrations in the hepatic portal vein, intermediate concentrations in the circulation to the brain, and lowest concentrations in the antecubital vein, the usual site of blood sampling (all three are similar once equilibrium is reached).
- Alcohol crosses all membranes through pores, and its volume of distribution, including brain, is total body water, so a one-compartment pharmacokinetic model is appropriate for most purposes.

- The effects are unexpectedly high, per unit of blood concentration, during the early phase, because of the site of sampling (Table 14.1). This almost certainly applies to all orally administered centrally acting drugs, during the acute phase of oral dosing, including hypnotics in particular.

Table 14.1 Ethanol effects on the ascending and descending phases of the blood ethanol curve; Shipley–Hartford Abstraction Scale, errors of omission (Jones and Vega, 1972; Curry, 1980)

Group	Mean score $\pm$ SD
Ascending phase	2.8 ± 2.3
Descending phase	0.6 ± 0.8

F-ratio $= 7.56$ (p<0.01).

An appropriate PK/PD model for this case would be a model of the type in Equation 13.3 showing saturation (a maximum possible effect). However, many different measures are used to evaluate the effects of alcohol, including ratings of overt behaviour and psychomotor tests, some of which would have a baseline value. At low doses an apparent excitatory response may especially be seen. This also occurs in the initial phase following any dose. Additionally, different tests are needed for evaluation of mid-range moderate impairment and high dose stupor, so both Equations 14.5 and 14.6 will be applicable dependent on circumstances.

Analogous observations have been made with glutethimide (Figure 14.2). The ability of a subject to follow a moving light was studied with the subject's head in a fixed position (smooth tracking test). The inhibition of this ability reached its peak when the concentration of glutethimide was still rising. The peak concentration coincided with almost total disappearance of effect. It is probable that this, as with the alcohol observation, arose from inhomogeneity of drug concentrations in plasma. This phenomenon was extensively documented and reviewed by Win Chiou in 1989 (Section 7.9). A similar result is observed when the effect is rapidly neutralized by physiological reflex mechanisms, as with isoprenaline (see later). As already mentioned in relation to alcohol, this is sometimes termed acute tolerance.

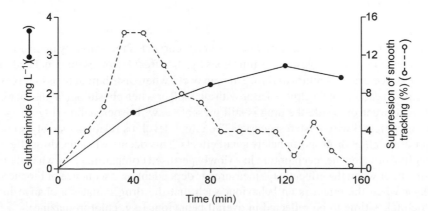

Figure 14.2 Plasma concentrations of glutethimide and percentage suppression of smooth tracking. (Redrawn from Curry & Norris, 1970.)

14.2.4   *Hysteresis*

The word hysteresis is derived from the Greek, meaning 'deficiency' or 'lagging behind'. It was coined by James Ewing to describe ferromagnetism – after removal of the magnetizing force the metal remains magnetized and only becomes non-magnetized very slowly. The crux of hysteresis is that the effects cannot be predicted from the input alone, it is necessary to know the current state of the system and for that, one needs to know the previous state. Thus, working from the standard growth and decay model for effect and for plasma concentrations, it is possible to plot effect (y-axis) against plasma concentrations (x-axis). If a line is drawn linking the data points *in the order in which they were recorded* a construction line of one of the four types in Figure 14.3. is derived.

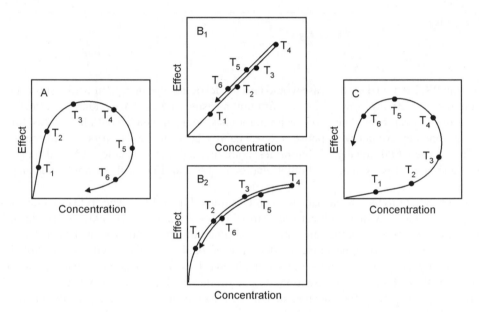

Figure 14.3   Model curves. A – clockwise hysteresis, B_1 – no hysteresis, straight line relationship, B_2 – no hysteresis, saturation evident, C – anticlockwise hysteresis. In the view of some experts, Case A should be termed "proteresis".

Hysteresis curve (A) is described as a clockwise hysteresis curve. It shows an early intense effect occurring at relatively low concentrations in plasma. As time goes by, the effect becomes maximal and then decreases before the plasma concentrations return to zero. This can occur when distribution to tissues plays a large part in the time course of the action of the drug, such as with alcohol, or when physiological mechanisms cause the effect of the drug to wear off while the drug is still present, as with isoprenaline (Figure 14.4).

Curve (C) is an anticlockwise hysteresis curve. The effect develops relatively slowly so that early high plasma concentrations elicit disproportionately small effects. This occurs when the drug effect is caused by an active metabolite, or when the receptors are in a slowly perfused compartment, or when there is a specific effect compartment, or when the effect that is measured is dependent on a sequence of biochemical changes (e.g. warfarin), or when the effect is on behaviour such that the drug initiates a change in for example perception, which takes time to be reflected in overall behaviour (e.g. chlorpromazine).

Curves (B_1) and (B_2) are the ones expected according to pharmacological principles, where hysteresis is not a confounding factor. They are rarely seen. Note that in this figure we have shown a graph of effect versus

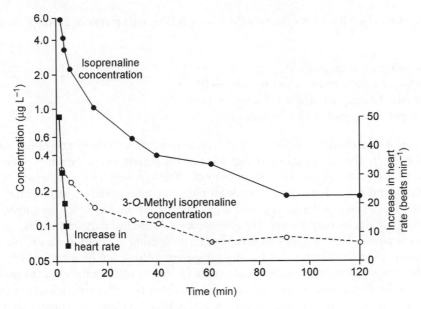

Figure 14.4 Concentration–effect relationship for isoprenaline in a dog given $0.64\,\mu g\,kg^{-1}$ intravenously. (Redrawn from Conolly *et al.*, 1971.)

concentration, not log(concentration). There is a near linear effect/concentration relationship in B_1, of the type that would be expected if the effect was measured at the lower left hand end of a conventional dose–effect curve, and B_2 exhibits the potential result of studies conducted at higher concentrations. As mentioned earlier, Figure 14.1. (carebastine) shows data on which a relationship of his kind (B_1) would be observed, restricted in the study quoted to the lower portion of the curve, where the effect is basically linearly related to concentration.

14.2.5   Pharmacokinetic distribution models

These models take into account the universal fact that even when a drug is delivered intravenously into a vein; it is not instantaneously at equilibrium between plasma and its site of action. There will be cases where the practicalities of experimental design dictate that there is no deviation from a close relationship between the time courses of effect and plasma concentration, such as with carebastine (above), but more commonly phenomena that are described in this section occur.

14.2.5.1   Thiopental and propofol

Thiopental is presented as the archetype intravenous anaesthetic, developed during the 1950s for use in military arena surgical hospitals (MASH) in particular. Thiopental shows redistribution after i.v. doses with a prominent α-phase occurring while the drug is distributing into one or more peripheral compartments. Clinically, a series of serially smaller doses every 20 minutes or so were needed to maintain anaesthesia with this drug, or it could be given as a continuous infusion with a gradually reducing rate.

Thiopental has been virtually superseded by more modern drugs, of which propofol is probably the most studied. The pharmacokinetic phenomena of the two drugs are similar, with the complication that propofol is formulated in excipients/solvents that must themselves be administered with great caution by slow intravenous infusion. There has been extensive PK/PD modelling with propofol; some of the characteristics being:

- Dosing by slow intravenous infusion.
- A central 'pool' or compartment with two sub-pools:
  - a rapidly equilibrating pool, which includes the brain
  - a slowly equilibrating pool, which includes fat.

Thus the situation with propofol is analogous to observations made with thiopental (see Figure 5.8), and it leads to a similar result. The relationship between effect and concentration in the rapidly-equilibrating pool would be describable with an equation relating intensity of effect to concentration in plasma. With thiopental, it is believed that no more than 2 minutes are needed for the i.v. administered drug to penetrate the brain and reach its receptors, and because the i.v. injection is given over a 5 minute period or longer, the brain and plasma are viewed as a single compartment. The effect occurs almost instantly, and declines with the decline in plasma concentrations. Thus, perhaps surprisingly, for thiopental, a highly lipid-soluble drug showing multi-compartment pharmacokinetics, and a high apparent volume of distribution, the receptors are considered to be in the central compartment, and equations closely relating the effect to the concentration in plasma are in order. It is of interest to compare this, conceptually, with expectations for the chemically-related drug, phenobarbital, which is highly polar, which appears to show one-compartment pharmacokinetics from its plasma decay curve, and which has an apparent volume of distribution that is not much different from total body water. However, it crosses the blood–brain barrier slowly, in relatively small quantities, in accord with the existence of a small and specific peripheral compartment that is in fact its site of action. In this case, the effect on the brain develops slowly after single doses, including those given by i.v. injection. In modelling terms an alternative model is needed for this drug, with an 'effect' compartment (the brain) that involves amounts of drug that are not large enough to purturb the plasma concentrations (see later sections).

14.2.6   Receptors in the peripheral compartment models

14.2.6.1   LSD

Wagner and colleagues (Wagner, 1968; Wagner *et al.*, 1968) showed that, after i.v. injections of lysergic acid diethylamide (LSD), the concentrations of LSD showed the typical bi-exponential decay associated with distribution through a two-compartment system. During the initial α-phase of decay, during which the drug was transferring from the plasma and other parts of the central compartment to the peripheral compartment, the effect, assessed using an arithmetical performance test, was rising. At the same time, the calculated concentrations in the peripheral compartment were rising. The apparent volume of the central compartment was $0.163 \, \mathrm{L \, kg^{-1}}$ (approximating to extracellular fluid) while the apparent volume of the peripheral compartment was $0.155 \, \mathrm{L \, kg^{-1}}$ – this compartment presumably included the intracellular fluid of the brain. There was a linear relationship between effect and the concentration in the peripheral compartment, with a baseline performance score. The data are shown in Figure 14.5. The model of Equation 14.2 would be appropriate in this case, with a clockwise hysteresis curve with no saturation.

14.2.6.2   Digoxin

There have been several studies of the multi-compartment nature of the distribution of digoxin after i.v. doses, and the delay in onset of useful effect after such doses is well-known to clinicians. Historically, there was a practice of digitalization – administration of the maximum possible tolerated dose followed by reduction of the dose to a more acceptable level, and there has been a need to devise a more logical approach

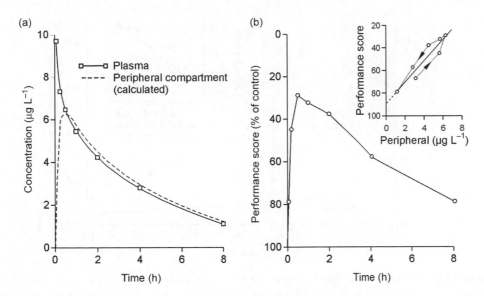

Figure 14.5 (a) Mean plasma concentrations of LSD in five individuals following i.v. bolus doses of $2\,\mu g\,kg^{-1}$, together with calculated concentrations in the peripheral compartment. (b) Performance score based on arithmetic tests in the same experiment. The insert in (b) shows the performance score plotted against concentration in the peripheral compartment, indicating a straight line relationship with evidence of modest hysteresis. (Redrawn after Wagner *et al.*, 1968.)

to determination of the optimum dose. PK/PD has helped to devise an answer. Typically such studies involve the inotropic effect being measured as systolic time interval.

Reuning *et al.* (1973) used a two-compartment model to describe the kinetics of digoxin, showing that the inotropic effect related more closely to the concentrations in the peripheral compartment than to those in plasma. In later studies, other investigators demonstrated the use of three-compartment models showing that there was a slightly better correlation between effect and a third, 'deeper' (slowly distributing) compartment.

Because the receptors for digoxin are in the heart, this provides an example of slow penetration to receptors in the most highly perfused tissue in the body. The sarcoplasmic reticulum of the cardiac cell is presumed to be the site of the receptors. The equilibrium takes as much as 5 hours to be achieved, which is many multiples of the half-life of the α-phase. At equilibrium, the majority of the body content of the drug is in the peripheral compartment. This was checked with studies of renal elimination to ensure that the drug thought to be in the body had not been excreted. It may be that this presaged the concept of an 'effect' compartment, but it should be noted that the proposal for digoxin kinetics arose from the analysis of concentrations in plasma (see later).

A graph of $\Delta(Q - S_2)$ against fraction in the tissue compartment was linear, and the intercept was not significantly different from zero (Figure 14.6). Thus an appropriate model to relate drug in the peripheral compartment with effect is Equation 14.1. The data are insufficient to determine whether or not there was a hysteresis effect in this study.

Reuning *et al.*, (1973) discussed the post-equilibrium phase, noting that up to 13 hours could be needed to reach this point. After this point, changes in effect parallel plasma concentrations. This work had practical consequences:

- The recognition that *V* is lower in renal failure patients and that they need lower doses.
- The use of a lower loading dose in such patients, instead of digitalization.
- The taking of clinical monitoring samples 12 hours after the dose.

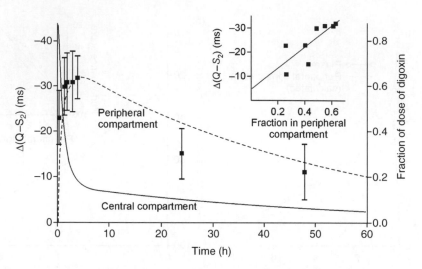

Figure 14.6   Relationship of the change in the $Q - S_2$ interval, $\Delta(Q - S_2)$ to computer-simulated curves for fraction of digoxin dose in the central and peripheral compartments as a function of time after a single intravenous dose of digoxin. Data are mean $\pm$ SEM from six normal subjects. Inset: plot of intensity of pharmacological effect as function of fraction of drug in peripheral compartment.

More recently, interest in the pharmacokinetics and pharmacodynamics has turned to the possibility of there being a specific 'effect' compartment, in addition to the distribution compartments.

14.2.7   Effect compartment models

These models are sometimes called 'link models' or 'effect-distribution models'. They facilitate estimation. of the *in vivo* pharmacodynamic effect from non-steady-state effect (E) versus time and concentration (C) versus time data, within which potential exists for observed E and C to display temporal displacement with respect to each other (Segre, 1968; Wagner, 1968; Dahlstrom *et al.,* 1978; Sheiner *et al.,* 1979). The rate of change of amount of drug (A_e) in a hypothetical effect compartment can be expressed as:

$$\frac{\mathrm{d}A_e}{\mathrm{d}t} = k_{1e} A_1 - k_{e0} A_e \tag{14.7}$$

where A_1 is the amount of drug in the central compartment of a pharmacokinetic model, linked to the effect compartment, with first-order rate constants k_{1e} and k_{e0} (Figure 14.7). Note the use of the same convention of labelling the parameters, e represents the effect compartment.

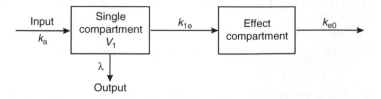

Figure 14.7   The one compartment distribution model of earlier chapters shown with the addition of an effect compartment.

The corresponding expression for the amount of drug in the effect compartment, for a one-compartment model with bolus input of dose (D) is:

$$A_e = \frac{k_{1e} D}{k_{e0} - \lambda} [\exp(-\lambda t) - \exp(-k_{e0} t)] \tag{14.8}$$

where λ is the elimination rate constant. The concentration of drug in the effect compartment, C_e, is obtained by dividing A_e by the effect compartment volume V_e:

$$C_e = \frac{k_{1e} D}{V_e (k_{e0} - \lambda)} [\exp(-\lambda t) - \exp(-k_{e0} t)] \tag{14.9}$$

At equilibrium, the rates of drug transfer between the compartments are equal:

$$k_{1e} A_1 = k_{e0} A_e \tag{14.10}$$

$$k_{1e} V_1 C_1 = k_{e0} V_e C_e \tag{14.11}$$

If the partition coefficient, K_p, equals C_e / C_1 at equilibrium (steady-state), then rearranging Equation 14.11 gives:

$$V_e = \frac{k_{1e} V_1}{K_p k_{e0}} \tag{14.12}$$

Substituting Equation 14.12 for V_e in Equation 14.9 yields:

$$C_e = \frac{k_{e0} D K_p}{V_1 (k_{e0} - \lambda)} [\exp(-\lambda t) - \exp(-k_{e0} t)] \tag{14.13}$$

At equilibrium, C will be equal to C_e / K_p, so dividing both sides by K_p and substituting C:

$$C_1 = \frac{k_{e0} D}{V_1 (k_{e0} - \lambda)} [\exp(-\lambda t) - \exp(-k_{e0} t)] \tag{14.14}$$

This is how the link-model relates the kinetics in plasma to the kinetics of drug in the effect compartment. When used together with the E_{max} model for estimation of the maximal drug-induced effect, the concentration at half-maximal effect (apparent EC_{50}) and the rate constant of the disappearance of the effect (k_{e0}):

$$E = E_0 + \frac{E_{max} C_e}{EC_{50} + C_e} \tag{14.15}$$

assuming the sigmoidicity factor (γ) to be equal to unity. Computer fitting of the equations to the effect data allows estimation of the rate constant for the disappearance of the effect, k_{e0}, EC_{50} and E_{max}.

At steady state, C_e is directly proportional to the plasma concentration (C), as $C_e = K_p C$. Consequently, the potency (EC_{50}) obtained by regressing the last two equations represents the steady-state plasma concentration producing 50% of E_{max}.

Note that the effect rate constant (k_{1e}) may be viewed as a first-order distribution rate constant. It assesses the rate of delivery of the drug to a specific tissue, determined by, for example, tissue perfusion rate, apparent volume of the tissue and eventual diffusion into the tissue. It should be emphasized that the link model applies to situations where a small but pharmacologically significant subset of the dosed molecules penetrates a separate compartment with its own input-output characteristics, distinct from anything observable in the plasma concentration decay curve or available for calculation as in the case with the pharmacokinetic distribution models (contrast LSD, and digoxin as an example of a compartment distribution model application). Literally hundreds, perhaps thousands, of cases, especially in regard to drugs acting within the cardiovascular system, such as depicted in Figure 14.8 have been described.

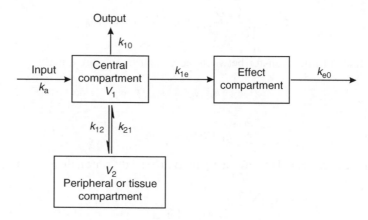

Figure 14.8   Representation of a link model applied to a drug that confers the characteristics of two-compartment model.

A tabulation of some of the times to equilibrium reported appears as Table 14.2. This model is now established as indicating a fundamental mechanism of pharmacology.

Table 14.2   Equilibration half-lives determined using the effect compartment model

Drug	Equilibration half-time (minutes)	Pharmacological effect
Disopyramide	2	QT prolongation
Verapamil	2	PR prolongation
N-Acetylprocainamide	6.4	QT prolongation
Quinidine	8	QT prolongation
Digoxin	214	LVET shortening
Ergotamine	594	Vasoconstriction

QT and PR are electrocardiographic intervals; LVET is left ventricular ejection time.

It is of interest that digoxin, a drug that has been extensively studied in this context, has blurred the distinction between the distribution compartment models and the effect compartment models. A similar comment could be made in relation to phenobarbital. The effect compartment model has been very successful in exploring a large number of pharmacological observations, and it has achieved a position of near universality in its position as describing a fundamental mechanism of pharmacology, finding application with single and multicompartment drugs with long equilibration half-times, to drugs with such short equilibration half-times that they are effectively zero.

Lorazepam has provided us with a remarkable demonstration of the power of PK/PD modelling with an effect compartment. Gupta *et al.* (1990) used a battery of psychomotor and cognitive tests in a group of subjects given oral lorazepam and compared the data with plasma concentrations of the drug. The plasma concentrations were best described by a two-compartment model with first-order absorption. There was an anticlockwise hysteresis loop for each effect. Fitting the time-course of the concentrations in an integrated PK/PD model required an effect compartment with a finite equilibrium rate constant between it and the plasma compartment. The magnitude of the temporal lag was quantified by the half-life of equilibration, and the CNS effect was characterized by a mean estimate of maximum predicted effect. Sigmoid concentration–effect curves were generated relating the concentration in the effect compartment and the measured pharmacological effects (Figure 14.9).

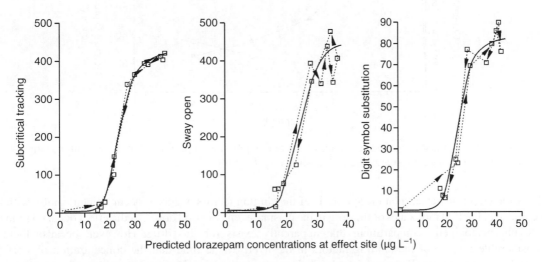

Figure 14.9   Relationship between predicted effect site concentration (C_e) of lorazepam ($\square$) and three psychomotor effects. Solid-lines fitted to the PK/PD model as a function of C_e; arrows show hysteresis. (Redrawn from Gupta *et al.*,1990.)

14.2.8   Warfarin type models

Effect compartment and link models are limited by their applicability to situations in which the equilibrium between plasma and response is due to drug distributional phenomena. In reality, there is often a delay between occurrence of maximum drug concentrations at the site of action and maximum intensity of effect, caused by slow development of the effect rather than slow distribution to the site of action. In this situation, indirect or 'physiological substance' models are more appropriate (Dayneka *et al.*, 1993; Levy, 1994; Sharma and Jusko, 1997). Warfarin is a good example. This drug inhibits the prothrombin complex activity (PCA), by its action on vitamin-K dependent activation of clotting Factors II, VII, IX and X, formed in the liver. The relationship between concentrations in plasma and effect is illustrated in Figure 14.10, which shows data for a single patient from among a group of five patients given oral doses of racemic warfarin (1.5 mg kg^{-1}). The peak plasma warfarin concentration was approximately 6 hours after the dose whereas the percent decrease in PCA, was maximal at approximately the 2-day point. Note that PCA, (or prothrombin time), is measured *in vitro*, as the time for blood to clot under controlled conditions, and normal values are in the region of 120 s. The therapeutic objective is commonly to increase this by approximately 70%, to ~200 s. This does not reflect return to normality, but the achievement of a modified physiological state in which the risk of blood clotting *in vivo* is reduced. It therefore increases the risk of bleeding, and management of warfarin dosing involves careful manipulation and control of these two risks.

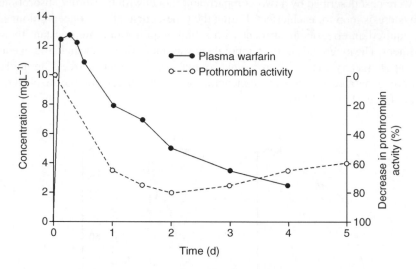

Figure 14.10   Plasma warfarin concentrations and depression of prothrombin activity after a single dose in one subject. (Redrawn after Nagashima *et al.*, 1969.)

Thus the effect of warfarin on the synthesis of the clotting factors begins to occur coincidentally with the appearance of the first molecules of the drug in the plasma, even after oral doses, as the effect occurs within the single pharmacokinetic compartment that apparently exists for this drug. However, because there is a considerable reserve of the clotting factors in the liver, by the time that the concentration of warfarin has reached its maximum (Figure 14.10), the decline of the reserve of clotting factors has only occurred to an extent of approximately 10%, as shown by the 10% increase in prothrombin time at this point, even though the drug concentration has been well above that needed to cause an effect for a considerable time. For the next 36–48 hours the synthesis remains blocked, and the clotting factor concentrations decline, causing further increases in the prothrombin time, after which the warfarin concentration declines further. The drug concentrations are now below the effective concentration, and the clotting factor concentrations begin to recover. This has major clinical implications, as anticoagulant activity is required as quickly as possible during a medical emergency, and so intravenous heparin is given as the first intervention, followed by oral warfarin.

In an investigation analogous to that of that of Figure 14.10, but using the more active (*S*)-enantiomer at a somewhat lower i.v. bolus dose (~10 mg), the concentrations of (*S*)-warfarin were described by the following exponential expression:

$$C = 1.05 \exp(-0.0288t) \tag{14.16}$$

where C was measured in mg L^{-1} and t was measured in hours. Thus the (*S*)-warfarin half-life in this case was approximately 24 h. The effect was again maximal at 36–48 h, with a 60–70% reduction in PCA. Recovery was somewhat faster than in the example in Figure 14.10 – it was 80% complete after about 5 days. Note that the bioavailability of racemic warfarin is over 90%, and that the oral and i.v. doses result in the same time course of effect, in spite of very different time courses of drug concentrations in plasma. The apparent volume of distribution of racemic warfarin is in the region of 0.14 L kg^{-1} and the half-life varies over a range from 20 to 50 hours, with the half-life of (*S*)-warfarin and (*R*)-warfarin being, on average, approximately 32 and 43 h, respectively. Accordingly, the clearance of the (*R*)-isomer is less than that of the (*S*)-isomer. The (*S*)-isomer is approximately three to five times more potent than the (*R*)-isomer.

Two further observations aid the understanding of the relationship between the pharmacokinetic and pharmacodynamic properties of warfarin. The dose in Figure 14.10 was a clotting factor synthesis blocking dose. In this circumstance, the changes in PCA can be used to evaluate the kinetics of the decline in concentrations of the clotting factors. It has been shown that this decline occurs monoexponentially, with a half-life, coincidentally with the warfarin half-life sometimes measured, of approximately 22 h, reaching a maximum of approximately 80% block at 48 h – no further increases in inhibition are seen, either with higher doses or at later time points [Figure 14.11(a)]. This figure is effectively a logarithmic transformation of the clotting data of Figure 14.10 over the first 36 h of the investigation.

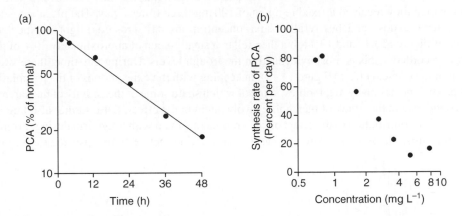

Figure 14.11 (a) Prothrombin complex activity declining exponentially over the first 48 hours after a synthesis blocking dose of warfarin sodium orally in a human subject. (b) Relationship between synthesis rate of the prothrombin complex activity (% d^{-1}) and plasma concentration of warfarin sodium. (Redrawn from Levy *et al.*, 1970.)

The second observation concerns the rate of synthesis of the clotting factors. A log(plasma concentration)–effect relationship with an IC_{50} of approximately 1.75 mg L^{-1} has been demonstrated for the racemate [Figure 14.11(b)]., where the effect is 'synthesis rate' of PCA (% day^{-1}). Data of this type are obtainable by dosing warfarin to the pharmacokinetic steady-state using different dose levels and measuring PCA once the steady-state is reached. The IC_{50} for (*S*)-warfarin is approximately 0.35 mg L^{-1}.

The PCA at steady state (PCA$_{ss}$) has been shown to be obtainable from:

$$PCA_{ss} = \frac{K_{in}}{k_d} \tag{14.17}$$

where K_{in} is the zero-order synthesis rate of the clotting factors, and k_d is the first-order rate constant of the decay ($\sim$0.03 h^{-1}) in Figure 14.11(a).

Equations have been derived to describe the turnover of the concentrations of the clotting factor(s) (d*P*/d*t*) (Nagashima *et al.*, 1969; Pitsiu *et al.*, 1993). Such equations incorporate k_d, and are complicated by the fact that PCA is increased when the synthesis rate of the clotting factors is reduced by the drug. The turnover rate of the clotting factors, assessed by means of PCA measurements, was 0.3 s h^{-1}. Warfarin, either as the racemate or the (*S*)-isomer reaches pharmacokinetic steady-state concentrations of approximately two to three times those achieved after single doses, within 2–3 days of initiation of treatment. This arises from the standard pharmacokinetic relationships between single and multiple doses discussed in Chapter 4. Thus, a reasonable target for steady-state plasma concentrations of racemic warfarin will be approximately

0.5–1 mg L^{-1}, achieving a 20–50% block of clotting factor synthesis, with the full effect evident within 2–4 days of initiation of therapy. This is achieved with long-term dosing with a wide range of oral doses in the range 1–10 mg day^{-1}. Monitoring of this scenario is usually achieved by means of PCA measurements, rather than drug concentration measurements.

14.2.9 Other examples

Two further examples of concentration–effect relationships provide insight into the complexity of quantitative pharmacology *in vivo*.

In the case of chlorpromazine, a group of newly diagnosed patients was observed over a period of forty days, during which they received a fixed regimen of 100 mg three times a day. The plasma concentrations rose and fell after each dose, and the overall plasma concentrations at first rose, as might be expected from the basic models of Figures 13.10 and 13.11, but then fell to a steady-state of approximately 30% of the highest concentration recorded – this is most evident from the trough levels. During this time, there was a steady development of the clinical effect (Figure 14.12), in keeping with the expectations of the psychiatrists at the time. Chlorpromazine has many metabolites, some of which share some of the activity of the parent drug, but there was no evidence in this study of metabolite involvement in the effect. Later studies also failed to reveal a robust metabolite contribution to the effect of chlorpromazine even with long-term dosing. From about the eighth day onwards, there was a strong *negative* correlation between the psychiatric score and the

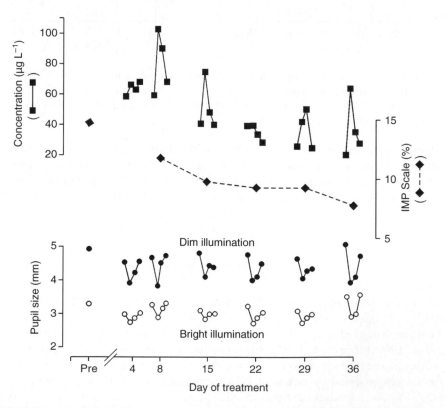

Figure 14.12 Mean plasma chlorpromazine concentrations, change in Inpatient Multidimensional Psychiatric Scale (% of maximum possible) and pupil size in 10 newly diagnosed patients receiving 100 mg every 8 hours. (Adapted from Sakalis *et al.*, 1972.)

concentrations in plasma. Pupil size is presented as an example of an autonomic nervous system effect which did show changes roughly in parallel with the drug concentrations.

Thus, Figure 14.12 illustrates several points already discussed, including enzyme induction and the difficulty of correlating effects in psychiatric patients to plasma concentrations, in part because of delayed effects and large contributions from placebo effects. A strong correlation between concentration and effect should not be expected for drugs designed to induce changes in cognition over relatively long periods of time.

The second example considers the treatment of bacterial infections with what Brodie described as 'a rational regimen of dosage', and to which Kruger-Thiemer applied a 'pharmacokinetic theory of dosage regimen' for sulfonamides. These concepts involve a loading dose followed by maintenance doses, designed to induce and maintain bacteriostatic concentrations for as long as is needed. This regimen yields a concentration curve at the site of action of the type shown in Figure 14.13.

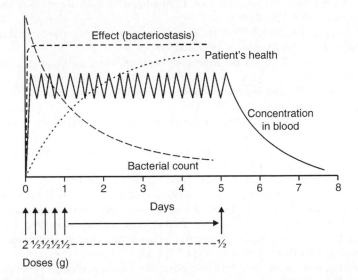

Figure 14.13 Time course of bacteriostatic chemotherapy. The figure shows drug concentrations in plasma induced by a loading dose, followed my maintenance doses for 5 days. The effect, bacteriostasis, the bacterial count, and the patient's feeling of well-being all show different time courses in this model.

While bacterial growth may be instantly inhibited, the bacterial count comes down more slowly, and the patient's feeling of improved health develops over a period of days. This figure is drawn as if for a sulfonamide with a loading dose of 2 g., and maintenance doses of 0.5 g. A basic feature of this diagram is that, with the dosing interval equal to the half-life of the drug, the concentration will fluctuate by 50% of the maximum. This type of regimen is very much used with drugs of all kinds. For example, treatment of joint pain with naproxen often involves a loading dose of two tablets followed by maintenance doses for several days of one tablet. Also, the monoclonal antibody, trastuzumab (Herceptin), is commonly used intravenously with a loading dose, followed by 3-weekly maintenance doses, for 6 months in order to suppress and keep suppressed any cancer cells that may have metastasized.

References and further reading

Belz GG, Kirch W, Kleinbloesem CH. Angiotensin-converting enzyme inhibitors. Relationship between pharmacodynamics and pharmacokinetics. *Clin Pharmacokinet* 1988; 15: 295–318.

Cohen BM, Tsuneizumi T, Baldessarini RJ, Campbell A, Babb SM. Differences between antipsychotic drugs in persistence of brain levels and behavioral effects. *Psychopharmacology (Berl)* 1992; 108: 338–44.

Colburn WA. Simultaneous pharmacokinetic and pharmacodynamic modeling. *J Pharmacokinet Biopharm* 1981; 9: 367–88.

Conolly ME, Davies DS, Dollery CT, Morgan CD, Paterson JW, Sandler M. Metabolism of isoprenaline in dog and man. *Br J Pharmacol* 1972; 46: 458–72.

Curry SH. *Drug Disposition and Pharmacokinetics*. 3rd edn. Oxford: Blackwell Scientific, 1980.

Dahlstrom BE, Paalzow LK, Segre G, Agren AJ. Relation between morphine pharmacokinetics and analgesia. *J Pharmacokinet Biopharm* 1978; 6: 41–53.

Dayneka NL, Garg V, Jusko WJ. Comparison of four basic models of indirect pharmacodynamic responses. *J Pharmacokinet Biopharm* 1993; 21: 457–78.

Derendorf H, Hochhaus G. (Eds) *Pharmacokinetic/Pharmacodynamic Correlation*. Boca Raton: CRC Press, 1995.

Donnelly R, Elliott HL, Meredith PA, Kelman AW, Reid JL. Nifedipine: individual responses and concentration-effect relationships. *Hypertension* 1988; 12: 443–9.

Donnelly R, Elliott HL, Meredith PA, Reid JL. Concentration–effect relationships and individual responses to doxazosin in essential hypertension. *Br J Clin Pharmacol* 1989; 28: 517–26.

Donnelly R, Meredith PA, Elliott HL, Reid JL. Kinetic–dynamic relations and individual responses to enalapril. *Hypertension* 1990; 15: 301–9.

Donnelly R, Elliott HL, Meredith PA. Antihypertensive drugs: individualized analysis and clinical relevance of kinetic-dynamic relationships. *Pharmacol Ther* 1992; 53: 67–79.

Elliott HL, Donnelly R, Meredith PA, Reid JL. Predictability of antihypertensive responsiveness and alpha-adrenoceptor antagonism during prazosin treatment. *Clin Pharmacol Ther* 1989; 46: 576–83.

Forester W, Lewis RP, Weissler AM, Wilke TA. The onset and magnitude of the contractile response to commonly used digitalis glycosides in normal subjects. *Circulation* 1974; 49: 517–21.

Gibaldi M, Perrier D. *Pharmacokinetics*, 2nd edn. New York: Marcel Dekker, 1982.

Girard P, Boissel JP. Clockwise hysteresis or proteresis. *J Pharmacokinet Biopharm* 1989; 17: 401–2.

Gupta SK, Ellinwood EH, Nikaido AM, Heatherly DG. Simultaneous modeling of the pharmacokinetic and pharmacodynamic properties of benzodiazepines. I: Lorazepam. *J Pharmacokinet Biopharm* 1990; 18: 89–102.

Holford NH, Sheiner LB. Understanding the dose–effect relationship: clinical application of pharmacokinetic-pharmacodynamic models. *Clin Pharmacokinet* 1981; 6: 429–53.

Jones BM, Vega A. Cognitive performance measured on the ascending and descending limb of the blood alcohol curve. *Psychopharmacologia* 1972; 23: 99–114.

Jusko WJ. Pharmacodynamics of chemotherapeutic effects: dose-time-response relationships for phase-nonspecific agents. *J Pharm Sci* 1971; 60: 892–5.

La Londe RL. Pharmacokinetic-Pharmacodynamic Relationships of Cardiovascular Drugs. In: Derendorf H and Hochhaus G, editors. *Handbook of Pharmacokinetic/Pharmacodynamic Correlation*. Boca Raton: CRC Press, 1995: 197–226.

Levy G. Relationship between elimination rate of drugs and rate of decline of their pharmacologic effects. *J Pharm Sci* 1964; 53: 342–3.

Levy G. Mechanism-based pharmacodynamic modeling. *Clin Pharmacol Ther* 1994; 56: 356–8.

Levy G, O'Reilly RA, Aggeler PM, Keech GM. Parmacokinetic analysis of the effect of barbiturate on the anticoagulant action of warfarin in man. *Clin Pharmacol Ther* 1970; 11: 372–7.

Macphee GJ, Howie CA, Meredith PA, Elliott HL. The effects of age on the pharmacokinetics, antihypertensive efficacy and general tolerability of dilevalol. *Br J Clin Pharmacol* 1991; 32: 591–7.

Meredith PA, Kelman AW, Elliott HL, Reid JL. Pharmacokinetic and pharmacodynamic modelling of trimazosin and its major metabolite. *J Pharmacokinet Biopharm* 1983; 11: 323–35.

Nagashima R, O'Reilly RA, Levy G. Kinetics of pharmacologic effects in man: the anticoagulant action of warfarin. *Clin Pharmacol Ther* 1969; 10: 22–35.

Olsen CK, Brennum LT, Kreilgaard M. Using pharmacokinetic-pharmacodynamic modelling as a tool for prediction of therapeutic effective plasma levels of antipsychotics. *Eur J Pharmacol* 2008; 584: 318–27.

Porchet HC, Benowitz NL, Sheiner LB. Pharmacodynamic model of tolerance: application to nicotine. *J Pharmacol Exp Ther* 1988; 244: 231–6.

Porchet HC, Benowitz NL, Sheiner LB, Copeland JR. Apparent tolerance to the acute effect of nicotine results in part from distribution kinetics. *J Clin Invest* 1987; 80: 1466–71.

Pitsiu M, Parker EM, Aarons L, Rowland M. Population pharmacokinetics and pharmacodynamics of warfarin in healthy young adults. *Eur J Pharm Sci* 1993; 1: 151–7.

Remington G, Shammi CM, Sethna R, Lawrence R. Antipsychotic dosing patterns for schizophrenia in three treatment settings. *Psychiatr Serv* 2001; 52: 96–8.

Reuning RH, Sams RA, Notari RE. Role of pharmacokinetics in drug dosage adjustment. I. Pharmacologic effect kinetics and apparent volume of distribution of digoxin. *J Clin Pharmacol New Drugs* 1973; 13: 127–41.

Robertson SA, Lascelles BD, Taylor PM, Sear JW. PK-PD modeling of buprenorphine in cats: intravenous and oral transmucosal administration. *J Vet Pharmacol Ther* 2005; 28: 453–60.

Ross EM. Pharmacodynamics: mechanisms of drug action and the relationship between drug concentration and effect. In: *Goodman and Gilman's Pharmacological Basis of Therapeutics*, 10th edn. New York: Pergamon, 1996.

Rubin PC, Butters L, Kelman AW, Fitzsimons C, Reid JL. Labetalol disposition and concentration-effect relationships during pregnancy. *Br J Clin Pharmacol* 1983; 15: 465–70.

Sakalis G, Curry SH, Mould GP, Lader MH. Physiologic and clinical effects of chlorpromazine and their relationship to plasma level. *Clin Pharmacol Ther* 1972; 13: 931–46.

Segre G. Kinetics of interaction between drugs and biological systems. *Farmaco Sci* 1968; 23: 907–18.

Sharma A, Jusko WJ. Characterization of four basic models of indirect pharmacodynamic responses. *J Pharmacokinet Biopharm* 1996; 24: 611–35.

Sheiner LB, Stanski DR, Vozeh S, Miller RD, Ham J. Simultaneous modeling of pharmacokinetics and pharmacodynamics: application to d-tubocurarine. *Clin Pharmacol Ther* 1979; 25: 358–71.

Smolen VF. Quantitative determination of drug bioavailability and biokinetic behavior from pharmacological data for ophthalmic and oral administrations of a mydriatic drug. *J Pharm Sci* 1971; 60: 354–65.

Smolen VF. Theoretical and computational basis for drug bioavailability determinations using pharmacological data. I. General considerations and procedures. *J Pharmacokinet Biopharm* 1976; 4: 337–53.

Tfelt-Hansen P, Paalzow L. Intramuscular ergotamine: plasma levels and dynamic activity. *Clin Pharmacol Ther* 1985; 37: 29–35.

Van Putten T, Marder SR, Wirshing WC, Aravagiri M, Chabert N. Neuroleptic plasma levels. *Schizophr Bull* 1991; 17: 197–216.

Vincent J, Elliott HL, Meredith PA, Reid JL. Doxazosin, an alpha 1-adrenoceptor antagonist: pharmacokinetics and concentration-effect relationships in man. *Br J Clin Pharmacol* 1983; 15: 719–25.

Vincent J, Liminana R, Meredith PA, Reid JL. The pharmacokinetics, antihistamine and concentration-effect relationship of ebastine in healthy subjects. *Br. J. Clin Pharmacol.* 1988; 39: 497–501.

Wagner JG. Kinetics of pharmacologic response. I. Proposed relationships between response and drug concentration in the intact animal and man. *J Theor Biol* 1968; 20: 173–201.

Wagner JG, Aghajanian GK, Bing OH. Correlation of performance test scores with 'tissue concentration' of lysergic acid diethylamide in human subjects. *Clin Pharmacol Ther* 1968; 9: 635–8.

Woo S, Jusko WJ. Interspecies comparisons of pharmacokinetics and pharmacodynamics of recombinant human erythropoietin. *Drug Metab Dispos* 2007; 35: 1672–8.

Wu B, Joshi A, Ren S, Ng C. The application of mechanism-based PK/PD modeling in pharmacodynamic-based dose selection of muM17, a surrogate monoclonal antibody for efalizumab. *J Pharm Sci* 2006; 95: 1258–68.

15

Extrapolation from Animals to Human Beings and Translational Science

15.1 Introduction

Historically, pharmacologists have conducted extensive laboratory studies of drugs *in vitro* reproducing physiological and neurochemical functions, and modifying them with drugs, in isolated animal tissues in organ baths. The qualitative and quantitative phenomena studied in this way were also studied in whole animals. Analogies would be sought in human therapeutics. New drugs were especially studied in mice, rats, and dogs, and a judgement would be made that the risk of initial exposure of humans, volunteers or patients, was warranted, with no pharmacokinetic data.

In the field of drug disposition and pharmacokinetics, a vast amount of valuable information has been discovered using isolated enzymes and functioning cells, liver and other tissue homogenates, perfused organs, and laboratory animals *in vivo*. However, it is rare that our need is to understand the disposition of a drug in the rat or other laboratory animals *per se*, and comparative pharmacokinetics is especially important in the discovery of new therapeutic agents. The transition from animals to human beings remains a critical step, and in its modern form it is facilitated by pharmacokinetics.

When a new chemical entity (NCE) intended for investigation in man as a potential therapeutic agent leaves the laboratory for the clinic, there will be a body of knowledge about its disposition and fate that is considerable in size but limited in scope. Types of information available will commonly include:

- Physical and chemical properties (solubility, pk_a, stability, etc.), and preliminary information on absorption potential, from molecular modelling or simple experiments
- Metabolic reactions in several species including, in part, human, *in vitro*, from enzyme, hepatocyte, and isolated organ (animals only) studies
- Protein binding in several species *in vitro* including human
- Renal clearance from *in vivo* studies in animals in which blood and urine samples were collected in parallel
- Microsomal intrinsic clearance in animals and humans, sometimes called 'metabolic stability'
- *In vivo* single-dose animal pharmacokinetics, sometimes including comparison of oral and i.v. doses, and including calculation of whole body half-life and clearance
- Limited data on pharmacokinetics during long-term dosing from toxicokinetic studies lasting up to 3 months
- Observations on pharmacological effects.

The challenge is to choose doses and blood sampling sequences for the initial human investigations, such that the safety of the exposed subjects is protected, and the maximum amount of data available is obtained from a

Drug Disposition and Pharmacokinetics, By Stephen H. Curry and Robin Whelpton
© 2011 John Wiley & Sons, Ltd

limited number of human exposures. While all of the data that exists is taken into account at this point in the process, the major emphasis is placed on the *in vivo* pharmacokinetics, and an approach called allometric scaling is a valuable tool in this context.

15.2   Allometric scaling

Allometry is the study of the relationships between size and shape. Allometric scaling in biology assumes that because most mammals have similar shapes and utilize the same physiological processes there must be some definable relationship between these processes in different mammalian species. For example, on average, a mammal's heart beats approximately four times for every breath taken no matter how large the animal. Note it is the *ratio* that is constant and this is so because both the breathing rate, *BR*, and heart rate *HR*, can be scaled to body weight, *W*:

$$BR = 0.169W^{-0.28} \qquad (15.1)$$

$$HR = 0.0428W^{-0.28} \qquad (15.2)$$

using the same power function, or exponent of -0.28. The same exponent has been used to relate life expectancy to body weight suggesting that all mammals will have the same total number of heart beats during a lifetime, although this must be a gross oversimplification. So, generally speaking, smaller animals live their lives at a faster pace. Mice, rats, dogs, monkeys and humans all absorb nutrients and drugs from their stomachs and intestines, they all deliver drug and nutrient molecules to the liver, brain, lungs and other organs, dissolved or solubilized in blood, they all metabolize chemicals in their livers, and they all excrete drugs and their metabolites in bile, urine and other fluids. However, drugs are eliminated much more rapidly in mice (which can rapidly metabolize weight-adjusted amounts of barbiturates that would kill a human being) and by different routes in rabbits and dogs (some rabbits can consume belladonna – deadly nightshade – with impunity without the cardiovascular effects that can kill humans).

Allometric scaling attempts to relate differences in pharmacokinetic parameters, such as systemic clearance (*CL*), volume of distribution (*V*) and elimination half-life ($t_{1/2}$) to differences in body weight between species. Equations 15.1 and 15.2 can be written in a general form:

$$Y = aW^b \qquad (15.3)$$

where *Y* is the physiological or pharmacological function, *W* is body weight, *b* is the 'allometric exponent' and *a* is the 'allometric coefficient'. Relatively little use is made of *a*. Logarithmic transformation of Equation 15.3 gives:

$$\log Y = \log a + b \log W \qquad (15.4)$$

showing that a graph of log *Y* versus log *W* is a straight line with a slope *b* and intercept log *a*. In many cases the slope is positive between 0 and unity, but it can be 0 and it can be negative. Thus:

• A slope of zero indicates that the parameter is the same across the species (e.g. number of eyes).
• A slope of unity indicates that the parameter increases in direct proportion to body weight.
• A positive slope between 0 and unity indicates that the parameter increases with body weight but not in direct proportion to body weight.
• A negative slope indicates that the parameter decreases with increases in body weight.

Body surface area has been related to body weight by the formula:

$$\text{surface area} = 0.1 \, W^{0.67} \tag{15.5}$$

and log–log transformations give the expected straight line (Figure 15.1). Drugs, particularly cytotoxic agents used in cancer treatment, are often dosed according to surface area, or in paediatric medicine. If Equation 15.5 is used then one is actually scaling to body weight. More complex formulae for dosage adjustment in human beings usually include height as well as weight.

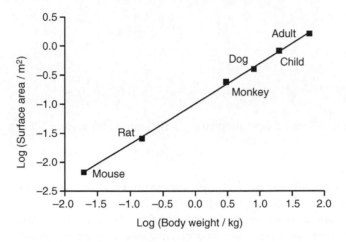

Figure 15.1 Allometric relationship between body surface area and species body weight. (Redrawn from Curry *et al.*, 2007.)

Table 15.1 shows a number of allometric exponents for some physiological and pharmacokinetic phenomena. Drug half-life, systemic clearance of drugs *in vivo*, renal clearance of drugs, and apparent volume of distribution of drugs all increase with body weight, but not proportionally. The rate constant of the terminal phase of elimination, being inversely related to half-life, decreases with body weight, and has a negative exponent.

Table 15.1 Some examples of allometric relationships

Parameter	Power
Heart rate	−0.28
Rate constant of elimination (λ_z)	−0.24
Terminal half-life	0.24
Systematic clearance *in vivo*	0.70
Renal clearance (of drugs)	0.78
Apparent volume of distribution	0.92

The exponent of the apparent volume of distribution is often close to 1, showing as might be expected, that this parameter correlates directly with body weight. Exponents for clearance are often approximately 0.75 whilst those for elimination half-life are about −0.25. This arises because:

$$t_{1/2} = 0.693 \frac{V}{CL} \tag{4.7}$$

For volume of distribution, Equation 15.3 becomes:

$$V = aW^1 \tag{15.6}$$

and for clearance it is:

$$CL = a'W^{0.75} \tag{15.7}$$

so substitution into Equation 15.7 gives:

$$t_{1/2} = 0.693 \frac{aW^1}{a'W^{0.75}} \tag{15.8}$$

Simplifying:

$$t_{1/2} = 0.693 \frac{aW^{(1-0.75)}}{a'} = 0.693 \frac{aW^{0.25}}{a'} \tag{15.9}$$

Combining, a, a' and 0.693 into a new allometric coefficient, A, Equation 15.9 can be written:

$$t_{1/2} = AW^{0.25} \tag{15.10}$$

which explains why the sum of the exponents for clearance and half-life should equal 1, provided of course that the allomeric exponent for volume of distribution is 1.

Clearly, a research team interested in predicting the half-life of a new drug in humans can obtain data in a variety of species, determine the allometric exponent, and calculate the expected half-life in humans. This will permit the adequate design of a Phase I experiment from the point of view of blood sampling times and overall duration of the experiment. It does not determine the safe dose; the initial dosing level is chosen after multiple-dose safety studies in two mammalian species.

Figure 15.2 shows a set of data of this kind for bosentan, an endothelin receptor antagonist for pulmonary hypertension. It is surprisingly difficult to find data of this kind in the public domain, although it exists

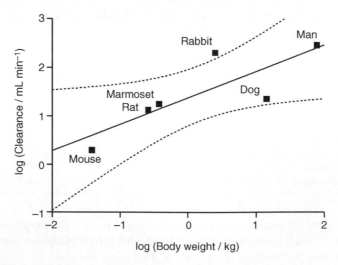

Figure 15.2   Allometric plot showing the relationship between systemic clearance and body weight in six species. The slope of line is the allometric exponent = 0.54. The broken lines represent the 95% confidence limits.

in plentiful supply in the files of research-based industry. In this case, data were collected in five species of laboratory animals. From these data, it would have been predicted that the total body clearance in humans would be in the region of 3 mL min^{-1}. The figure shows that *CL* recorded in humans was somewhat lower, at approximately 2.2 mL min^{-1}, but this would be considered a good result. Note that log–log plotting can give a false sense of validity to the prediction. The 95% confidence intervals of Figure 15.2 show that at high and low body weights the 95% intervals encompass more than 2 log units, that is more than two orders of magnitude in the parameter, *CL*. Regression analysis of the logarithmic values is akin to weighting the original data $1/x^2$ (Section 5.2.2.1).

15.2.1   Refinements to allometric scaling

15.2.1.1   Effect of neoteny

It has been proposed that through some evolutionary adaptation, human beings have an advantage over other mammals, resulting in increased longevity (modern medicine being excluded presumably). A larger brain and reduced metabolic rate (i.e. reduced mixed-function oxidase concentrations), have been suggested as possible reasons. Brain weight (*B*) has been used to compute a maximum life span potential (*MLP*):

$$MLP = a(W^b)(B^c) \qquad (15.11)$$

where $a = 185.4$, $b = -0.225$ and $c = 0.636$. Using *MLP* rather than chronological time improves the prediction of phenytoin clearance in man (Figure 15.3). A modification of Equation 15.11 in which *MLP* is replaced by *CL* has been used with drugs with low hepatic clearances, the assumption being that inclusion of a factor for brain weight is also a correction for low mixed-function oxygenase activity.

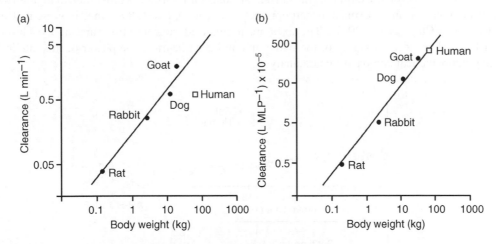

Figure 15.3   Comparison of allometric scaling of phenytoin clearance (a) using chronological time (b) using maximum life-span potential (MLP). See text for details. (Redrawn from the data of Campbell and Ings, 1988.)

15.2.1.2   Pharmacokinetic time

Just as it is possible to use maximum life span potential as a unit of time to calculate pharmacokinetic parameters, such as clearance (Figure 15.3) there is no reason why something analogous to *MLP* cannot be used as a scale for time. Thus, when comparing pharmacokinetic results from different species both dose and time would be scaled as appropriate. In their studies with methotrexate, Dedrick *et al.* (1970), converted chronological time into 'pharmacokinetic time' by dividing time by $W^{0.25}$. This has become known as the *Elementary Dedrick* plot where:

$$x\text{-}axis\ units = \frac{time}{W^{(1-b)}} \tag{15.12}$$

and the concentration is normalized for dose and body weight:

$$y\text{-}axis\ units = \frac{concentration}{dose/W} \tag{15.13}$$

In Equation 15.12, b is the allometric exponent for clearance whilst the exponent for volume of distribution is assumed to be 1 (see the derivation of Equation 15.10). No such assumption is made in the *Complex Dedrick* plot, where the time axis units are:

$$x\text{-}axis\ units = \frac{time}{W^{(c-b)}} \tag{15.14}$$

c being the exponent for volume of distribution. New terms, *kallynochrons* and *apolysichrons* have been introduced for the time units of elementary and complex plots, respectively. *Dienetichrons* include *MLP* so that the time axis units are:

$$x\text{-}axis\ units = \frac{time}{W^{(c-b)}} \frac{1}{MLP} \tag{15.15}$$

A complex Dedrick plot for an example of rat and human concentration data is illustrated in Figure 15.4. This graph reflects data for an experimental neuroprotective agent (AR-R 15896) that is presented in more detail elsewhere (Curry *et al.*, 2007). The axes are normalized concentration, and apolysichrons from Equation 15.13. The close proximity of the two graphing lines illustrates the power of this intellectually-stimulating mathematical approach to data analysis.

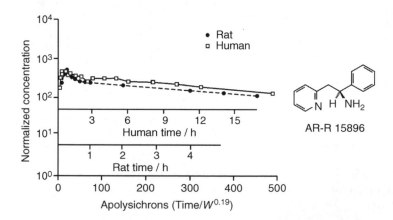

Figure 15.4   Complex Dedrick plot of rat and human data. (Redrawn from Curry *et al.*, 2007.)

However, simple allometric scaling, using Equation 15.3, also proved to be useful in this case in predicting human pharmacokinetic parameters from rat data (Table 15.2). Values of *b* were assigned for clearance (0.75), half-life (0.25) and volume of distribution (0.938). The intercept functions a, determined for each parameter from rat data, were then used to compute the pharmacokinetic parameters for a 70 kg man. Three methods were used to estimate the elimination half-life:

$$t_{\frac{1}{2}\text{ human}} = \frac{0.693 \, V}{CL} \tag{15.16}$$

$$t_{\frac{1}{2}\text{ human}} = t_{\frac{1}{2}\text{ rat}} \left(\frac{W_{\text{human}}}{W_{\text{rat}}}\right)^{y-x} \tag{15.17}$$

where y is the exponent for volume of distribution (0.938) and x is the exponent for clearance (0.75). Thus, $y - x = 0.188$.

$$t_{\frac{1}{2}\text{ human}} = \log a + \log W_{\text{human}} \tag{15.18}$$

Table 15.2 Predicted and actual pharmacokinetic parameters (from Curry *et al*, 2007)

Pharmacokinetic parameter	Predicted	Actual
Clearance ($\text{L h}^{-1}\text{kg}^{-1}$)	0.138	0.128
Half-life[a] (h)	14.5	13.6
V_1 (L kg^{-1})	1.01	1.02
V_{ss} (L kg^{-1})	2.14	2.11

[a] Average of three values calculated using Equations 15.16, 15.17 and 15.18.

15.2.2 Practical aspects of allometry

In general, the following holds true:

- Physical phenomena (e.g. tissue distribution measured by apparent volume of distribution) are relatively similar across species (exponent ~ 0.95).
- Processes dominated by physiological phenomena lead some drug related allometric relationships to reflect physiological allometric relationships (e.g. for drug metabolism when controlled by blood flow).
- Half-life and elimination rate constant tend to reflect a complex of factors, both physical and biological, including heart rate and tissue distribution, but also intrinsic metabolic clearance.
- Whole body clearance reflects a hybrid of factors in its allometric scaling, both physical and biological.
- Because *in vitro* clearance represents only a proportion of *in vivo* clearance, it will be expected to have allometric exponents that reflect isolated processes; allometric scaling is less commonly applied to *in vitro* data.

The scientific community is often asked about monkeys in this context. Intuitively, close analogies with humans might be expected. This is a vast topic, but while many scientists can cite individual examples of excellent predictions, as a general rule, data from monkeys are no better and no worse than data from mice, rats and dogs in the process of making predictions for humans. These data obviously can and do make a valuable contribution to the process of allometry, but they do no more than that. For example, some years ago, Chiou and Buehler (2002) showed that for a large number of compounds, by plotting log–log correlation diagrams of human half-life against half-life in monkeys, rats and dogs, success rates for human predictions were approximately the same for monkeys (87%) and rats (83%), and lower (72%) for dogs. One of the reasons why monkeys fail to give a more robust prediction for humans is that, like humans, and for that matter most dogs, they are heterogeneous – they are not as pure bred as are laboratory rats.

Thus, as a general observation, when following the traditional Phase I concept of choosing a single compound on the basis of preclinical data and preparing it for initial human testing, the investigator will, in all probability, want to predict the pharmacokinetic properties of the compound in human beings from animal data. This can be done quite well from allometric scaling if you first determine CL_{int}, CL, $t_{1/2}$, V, etc. in four to five other species. There are also some convenient rules of thumb based on experience:

- Human *in vivo* pharmacokinetic predictions can be obtained using the simple allometric scaling relationships of
  - CL (human) $= 40\,CL$ (rat) (L h^{-1}),
  - V_{ss} (human) $= 200\,V_{ss}$ (rat) (L),
  - $t_{1/2}$ (human) $= 4\,t_{1/2}$ (rat).
  (Note that units for CL and V_{ss} are not adjusted on a per kg basis in this rule.)
- Whole body clearance from plasma level decay (CL) is more predictable for compounds eliminated by renal (or biliary) excretion than by metabolism, because renal elimination has a larger physical (diffusion) component in its mechanisms.
- Drugs that are largely excreted unchanged will tend to have half-life values in humans three to four times those in rats; drugs that are largely dependent on metabolism for their elimination will have half-life values in humans that are 8–12 times those in rats.

15.3 Dose-ranging versus microdosing studies

The reality of choosing the dose for an initial human exposure is, however, a little different, because, as stated earlier, it is based on safety studies, not on pharmacokinetics. The first human exposure is likely to be made with a dose 1/50 or 1/100 of the lowest dose showing any effect, desired or undesired, in the laboratory animals, on a mg kg^{-1} basis. No effect is anticipated or sought in the human subjects.

Commonly, this dose is below the limit of analytical detection in plasma, preventing any pharmacokinetic studies with these initial doses. As the investigation proceeds, the dose is increased, slowly, step by step until detection in plasma becomes possible, and until pharmacological effects of significance, if any, occur. This is a slow, expensive process, with more exposures than is desirable in an expeditious program. Consequently, this leads us into consideration of the modern idea and practice of human microdosing.

Human microdosing is driven by developments in bioanalysis. Thanks to the use of such techniques as positron emission tomography (PET) and accelerator mass spectrometry (AMS) it is increasingly possible to administer minute doses of new drugs and to study their pharmacokinetic properties. These doses are sub-threshold in terms of interest by the medicine control agencies in them as potentially toxic. FDA and other such agencies permit the use of accelerated protocols for human microdosing, with what is called an Exploratory IND, after single doses followed by recovery studies in animals, rather than after one month of continuous exposure followed by necropsy.

This has some useful side effects. It reduces the use of animals, facilitates earlier human exposure, involves less CMC (chemistry and manufacturing controls) requirements, and makes it possible to conduct Phase I trials with less drug material. It becomes possible to compare multiple compounds in Phase I, and to choose the best compound for clinical trials from among several candidates on the basis of data in the ultimate test subject, the human. No longer are investigators optimizing new drugs for rats *per se*. Another objective of human microdosing (which is sometimes called Phase 0), can be to establish that the half-life of the drug is apparently the same over a broad range of doses, providing an early screening mechanism that can reduce the risk of a drug with the non-linear pharmacokinetic characteristics of phenytoin, reaching clinical trials, although this does not guarantee non-linearity at the larger doses likely to be needed in the clinic. Another valuable use of AMS technology is the measurement of concentrations of ^{14}C-labelled drug administered in very small doses intravenously, concurrently with conventional oral doses. Non-labelled drug is assayed by such techniques as high pressure liquid chromatography (HPLC) permitting calculation of absolute bioavailability using simultaneous i.v. and oral doses. However, PET and AMS, while reducing the ancillary costs in a variety of different ways, lead to individual experiments being very expensive.

15.4   Statistical approaches

Another developing idea is the use of statistical approaches in the dose and plasma concentration escalation process during Phase I. Whitehead and his colleagues (2007) have demonstrated how Bayesian decision-making procedures can be applied in choosing the dose to be used in the second and subsequent dosage rounds of the study. Bivariate observations are made of undesirable events and signs of therapeutic benefit (if any). Logistic regression models are used to study the two responses, which are assumed to take a binary form. Accurate definition of the therapeutic window is sought, and with each stage, optimal treatment of the subsequent cohort of subjects is achieved. Thus, using the first set of data for calibration, the probability of a particular response occurring at a particular dose, or plasma drug concentration, is calculated, and that dose, rather than some arbitrarily chosen dose in a predetermined ladder of small-step minor escalations is chosen. At its best, this process could apparently lead to one inactive dose, two individual doses focused separately on desired and undesired effects, and one dose focused on both types of effect – a total of four doses in all – being used to characterize the drug, thus achieving an accurate estimate of the therapeutic window at minimal cost. This has so far only been applied to the search for new anti-cancer drugs. Phase I studies in this field must incorporate a hope of therapeutic benefit and a tolerance of adverse effects not experienced in most other fields of medicine.

15.5   Translational science

'Translational science' is new terminology relevant to the transition from animals to man. It seeks to define a scientific track, not unlike that existing today, but making extensive use of biomarkers, and an administrative track of providing government funding for early clinical development of academic discoveries in academic centres. Biomarkers come in various shapes and sizes. They include imaging (e.g. PET), and assays of biochemical and physiological function including disease, with heavy emphasis on DNA-based gene sequence tests and proteomics. A reasonable objective would be discovery of *in vitro* biomarkers that predict the pharmacokinetic properties of drugs *in vivo* in humans with the target disease. Then, perhaps, it would be possible to define the properties of the drug that we want in human beings, and back-calculate through reverse allometric scaling (prediction of the desired properties in rats from the specification for man and reverse scale-up (from *in vivo* back to *in vitro*), permitting the choice of the right drug from candidate compounds, in the preclinical stage of the process, for the target humans from *in vitro* data. As yet, we are not there.

References and further reading

Boxenbaum H, Ronfeld R. Interspecies pharmacokinetic scaling and the Dedrick plots. *Am J Physiol* 1983; 245: R768–75.

Caldwell GW, Masucci JA, Yan Z, Hageman W. Allometric scaling of pharmacokinetic parameters in drug discovery: can human CL, Vss and t1/2 be predicted from *in-vivo* rat data? *Eur J Drug Metab Pharmacokinet* 2004; 29: 133–43.

Campbell DB, Ings RM. New approaches to the use of pharmacokinetics in toxicology and drug development. *Hum Toxicol* 1988; 7: 469–79.

Chiou WL, Buehler PW. Comparison of oral absorption and bioavailability of drugs between monkey and human. *Pharm Res* 2002; 19: 868–74.

Curry SH, McCarthy D, DeCory HH, Marler M, Gabrielsson J. Phase I: The first opportunity for extrapolation from animal to human exposure. In: Edwards LD, Fletcher AJ, Fox AW, Stonier PD, editors. *Principles and Practice of Pharmaceutical Medicine*, 2nd edn. Chichester: John Wiley & Sons Ltd, 2007.

Curry SH. Translational science: past, present, and future. *Biotechniques* 2008; 44: ii–viii

Dedrick RL, Bischoff KB, Zaharko DS, Interspecies correlation of plasma concentration history of methotrexate. *Cancer Chemother Rep Part I* 1970; 54: 95–101.

Ings RM. Interspecies scaling and comparisons in drug development and toxicokinetics. *Xenobiotica* 1990; 20: 1201–31.

Lappin G, Stevens L. Biomedical accelerator mass spectrometry: recent applications in metabolism and pharmacokinetics. *Expert Opin Drug Metab Toxicol* 2008; 4: 1021–33.

Mahmood I, Yuan R. A comparative study of allometric scaling with plasma concentrations predicted by species-invariant time methods. *Biopharm Drug Dispos* 1999; 20: 137–44.

Tang H, Mayersohn M. A novel model for prediction of human drug clearance by allometric scaling. *Drug Metab Dispos* 2005; 33: 1297–303.

Whitehead J, Zhou Y, Hampson L, Ledent E, Pereira A, A Bayesian approach for dose-escalation in a Phase I clinical trial incorporating pharmacodynamic endpoints. *J Biopharm Stat* 2007; 17: 1117–29.

16

Peptides and Other Biological Molecules

16.1 Introduction

This chapter is about a collection of important compounds that are characterized as being basically 'biological' in origin. However, no single descriptive term applies to all of them. While many are new, representing the impact of biotechnology on modern medicine, not all of them are new – insulin has been established for several generations as a biological product used in therapeutics. In contrast, monoclonal antibody drugs are very new. Neither are all the drugs in this chapter macromolecules – they include small peptide hormones used as drugs as well as large proteins. They are not all peptides or proteins, as they include a polysaccharide – heparin. They do not include all of the drugs containing peptide bonds, as the 'pril' antihypertensive drugs (such as lisinopril) contain peptide bonds. And they do not include all of the endogenous compounds used as therapeutic agents – there will be no further references to the ions, amino acids, sugars, or vitamins that are both essential nutrients and therapeutic agents, or to non-peptide/polysaccharide small molecule hormones. Thus the most useful classification is as 'compounds of biological origin, either fully or in part,' which serves little purpose! Table 16.1 lists the types of molecule in this category, with some examples and some of their properties. These compounds as a group show all of the properties to be found in 'small molecule' drugs, and more. Some examples of special interest, exemplifying recurring themes, are the subject of Section 16.6.

16.2 Chemical principles

The compounds in this category range in relative molecular mass from approximately 200 (simple dipeptides) up to 30,000 (some of the proteins). They are mostly polar, but some non-polar compounds that can be administered orally are included. Larger molecules behave differently from the relatively low molecular mass (<500), lipophilic, weak electrolyte drugs that have been produced in large numbers in medicinal chemistry laboratories over the last 50–60 years. Peptide and protein drugs contain amide links, and although many of these drugs may contain amine and carboxylic acid groups, as do the smaller molecules, they do not, in the main, fit Lipinski's 'Rules of Five', which for a molecule to be suitable as a drug state that it should have:

- A molecular weight less than $500 \, \text{g mol}^{-1}$.
- No more than five hydrogen-bond donors (sum of –OHs and –NHs),
- Log $P \leq 5$,
- No more than 10 hydrogen-bond acceptors (sum of Ns and Os).

Drug Disposition and Pharmacokinetics, By Stephen H. Curry and Robin Whelpton
© 2011 John Wiley & Sons, Ltd

Table 16.1 Examples of peptides, proteins and polysaccharides used as drugs

Group	Example	Route	Uses	Comments
Hormones	Vasopressin and desmopressin	i.n.	Control of urine in diabetes insipidus	Nonapeptides given by any parenteral route; first intranasal drugs
	Insulin	s.c.	Diabetes mellitus	Original model for radioimmunoassay; polypeptide hormone of 51 amino acids
	Oxytocin	i.v.	Induction of labour	Nonapeptide, given by any parenteral route
	Human growth hormone	i.v.	Promoting growth in children	Somatrophin – single polypeptide chain of 191 amino acids
Enzymes	Digestive enzymes	Oral	Promoting digestion of food	Long-established home medicine agent
	Pancreatic enzymes	Oral	Promoting digestion of food	Enteric coated granules – used in cystic fibrosis
	Thrombolytic agents (e.g. streptokinase)	i.v.	Dissolving blood clots	Degrades fibrin and breaks up thrombus
	Asparaginase	i.v.	Leukaemia	Can be PEGylated
Synthetic peptides	Capaxone	s.c	Suppression of autoimmune encephalomyelitis, MS	Mixture of synthetic tetrapeptides
Botanical peptides	Ciclosporin	Oral	Preventing transplant rejection	Natural fungal cyclic peptide – non-polar and P-450 substrate
Protein bound drugs	Paclitaxel/protein complex	i.v./i.m.	Cancer treatment	Paclitaxel is highly insoluble, and must be solubilized to facilitate injection
Polysaccharides	Heparin	i.v.	Anticoagulation	Complex drug used in management of risk of blood clots in several medical conditions

Proteins (natural or modified)	Interferon	i.v.	MS and other conditions	Non-glycosylated proteins with several applications; some products PEGylated.
	Erythropoietin	i.v.	Anaemia of renal failure	Product of recombinant technology – standardized in international units
	Interleukin	i.v.	Prevention of abortion and/or miscarriage	$Rh_o(D)$ immune globulin
Monoclonal antibodies	Trastuzumab	i.v.	Breast cancer (HER2 type)	Model for principles of monoclonal antibody drug production
	Rituximab	i.v.	Non-Hodgkin's lymphoma	Targets phosphoprotein CD 52 on T- and B-lymphocytes
	Adalimumab	s.c	Crohn's disease; and rheumatoid arthritis	IgG_1 monoclonal antibody
	Infliximab	i.v.	Crohn's disease	Inhibits TNF-α
	Basiliximab	i.v.	Acute rejection of kidney transplants	Inhibits IL-2 on activated T-cells

i.m., intramuscular; i.n., intranasal; i.v., intravenous; s.c., subcutaneous; MS, multiple sclerosis; IL-2, interleukin-2.

The above are the specifications for successful oral absorption, and it is recognized that they are irrelevant to drugs that are substrates for biological transporters and/or are polar biological molecules.

Although many of these molecules are biochemically important endogenous compounds, they may be administered in pharmacological doses for therapeutic purposes. Some of them are 'humanized' proteins (Section 1.1.1), that is they are manufactured to resemble the endogenous human protein as closely as possible. Historically, animal proteins of the same type, but with small, relatively unimportant, differences in amino acid sequence, have been used as therapeutic agents; an obvious example being the use of porcine rather than human insulin for the control of diabetes mellitus. More recently, the biotechnology industry has been able to manufacture 'recombinant' versions of these proteins, with the human arrangement of the amino acids, after isolation of the genes responsible for their synthesis. These proteins for human use are manufactured outside the human body, often in suspensions of bacteria such as *Escherichia coli* (Box 16.1).

Box 16.1   Stages in the production of recombinant DNA and its use to manufacture a human polypeptide or protein (e.g. human insulin)

- Identify the human genes required for synthesis of the polypeptide/protein *in vivo*.
- 'Cut' DNA with endonucleases to liberate the genes required.
- Use 'vectors' (e.g. viruses) to transfer these desired DNA fragments into host cells (e.g. bacteria) to be 'sliced' on to plasmid DNA – plasmids are small circular pieces of DNA lying outside the main bacterial chromosome – they themselves are opened up using endonucleases, then re-closed to include the gene that was transferred – other activity of the plasmid is lost.
- This creates a 'hybrid plasmid'.
- Mix the hybrid plasmid with the bacterial host cells to form 'transformed cells'.
- Separate transformed cells from others and grow in cultures.
- Create a fermentation chamber containing cultured cells and amino acids as a 'factory' for production of the polypeptide/protein.
- Separate and purify the human polypeptide/protein.
- Formulate pharmaceutically for human use (e.g. 'Humulin,' replacing animal sourced insulin that was not identical with human insulin).

These drugs include the monoclonal antibody products of the modern biotechnology industry that are revolutionizing medicine in both scientific and financial ways. The stages in a typical production of a monoclonal antibody are listed in Box 16.2.

Box 16.2   Stages in the production of a monoclonal antibody for use against human disease

- Identify an antigenic component of a disease-inducing protein (e.g. the HER2 extracellular domain of one of the breast cancer-inducing proteins).
- Separate the antigen, and inject into mice – separate spleen cells that have developed antibodies.
- Fuse the spleen cells containing the antibodies with myeloma cells, creating an immortal hybridoma; these cells now synthesize purines by means of the 'salvage' pathway.
- Grow clones of these cells in microtitre wells, manufacturing greater quantities of the cells that developed the original antigen.
- Scale up using the peritoneal cavities of mice, or a culture medium (preferred) in a fermentation chamber.
- Separate and purify the antibodies.
- Formulate pharmaceutically for human use (e.g. trastuzumab for the treatment of HER2 breast cancer).

Thus, many, but not all, of the chemical principles documented in earlier chapters do not apply with these compounds. Some of these compounds are sometimes described as 'biological response modifiers'.

16.2.1 PEGylation

This word is used to describe the process by which a polyethylene glycol (PEG) polymer chain is added to another molecule, commonly a therapeutic protein. It is also applicable to peptides, and of course, therefore, to antibody drugs. The PEG is bonded covalently to a region of the molecule not involved in the pharmacological action, and although it increases the molecular weight, this is not significant with such large molecules. This has several beneficial results, including masking of the protein from the host's immune system, increasing size in solution, which can reduce renal clearance, and adding water solubility to hydrophobic drugs and proteins. It can change the conformation, and electrostatic binding of the protein.

Thus apart from improving drug solubility, PEGylation can reduce dose frequency, extend circulating life in the body, and increase drug stability by enhanced protection from proteolytic degradation. This has found greatest application with interferon drugs, L-asparaginase, and recombinant methionyl human granulocyte colony-stimulating factor. It has also been used in formation of PEGylated liposomes containing doxorubicin.

16.3 Assay methods

While it is possible to achieve chromatographic separations of peptides and other biological molecules, these drugs often lack sufficient aromatic structures to facilitate their ultraviolet or fluorescence detection (although some successes have been achieved with electrochemical detection). Consequently, these compounds often need derivatizing before they can be quantified by ultraviolet or fluorescence detection. As this introduces a step into the assay that is difficult to control, less conventional methods of analysis may be needed. Methods commonly employed with these compounds are:

- Chromatographic separation with derivatization to confer absorbance or fluorescence properties; a variant on this can be radioactive derivatization, quantitation being dependent on assessment of the amount of radioactivity incorporated.
- Radioimmunoassay (RIA) using the drug as the antigen to which antibodies are developed. Competition between the drug and radioactively-labelled drug is used for quantification. The very first published application of RIA was for the polypeptide drug, insulin.
- Enzyme-linked immunosorbent assay (ELISA), an antibody binding approach of broad applicability to proteins in particular, and can be calibrated in mass per volume or 'units' per volume terms if the molecular mass of the analyte is not clearly known.
- Biological assays measure effects to indicate the drug concentration present by comparison with appropriate calibration experiments. For example, heparin can be assayed on the basis of the amount of hexadimethrine needed to neutralize heparin activity using thrombin-induced coagulation as an endpoint.
- Imaging assays can be used for monoclonal antibodies. An example is trastuzumab which has been studied by adding 'a new trifunctional chelating agent containing a biotin residue and a radiometal chelation moiety,' known as [111]In-1033-BR96 to the molecule and studying the distribution of the radioactivity. Much as PEGylation does not change the pharmacological properties of a protein, combination with this reagent, even though it creates a different molecule, with different physical and chemical properties, is considered not to change the disposition and fate of the drug.

- Another imaging method, also used with trastuzumab, has involved simultaneous use of ^{90}Y-trastuzumab (a β^--emitter, for tissue uptake studies in dissected animals) and ^{86}Y-trastuzumab (a positron emitter) for non-invasive live imaging work using PET scanning.

16.4   Pharmacokinetic processes

16.4.1   Administration and dosage

In some cases, compounds in this category are not only standardized, but also dosed in terms of internationally accepted units, based on biological potency, rather than on the basis of mass. This is because different batches of the drug may vary in chemical constitution, while retaining the required potency. However, standardization and measurement of dose by mass is sought whenever possible.

Compounds in this category are mostly administered by injection because of their lack of potential for absorption from the gastrointestinal tract. This results from instability in the acid of the stomach, or in the presence of peptidases in the intestines and/or the liver during first passage through that organ, or from their relative polarity, molecular size and charges resulting in too little transcellular diffusion. Whether i.v., i.m. or s.c. routes of administration are used will depend on how fast the pharmacological effect is wanted, and also on convenience (insulin is given i.m. or s.c because that provides the most suitable speed of onset of effect, and because the patient can conveniently self-administer it). There may also be a component of procedural need. Intravenous infusions of trastuzumab are supervised by clinic staff, in case of dose-related emergencies such as an allergic response.

In contrast to the examples in the previous paragraph, digestive enzymes given orally for local effects in the gastrointestinal tract survive exposure to acid and to peptidases, as does botulinum toxin and, presumably, polio vaccine. Despite being a peptide, ciclosporin is lipophilic and very poorly soluble in water. It is resistant to acid and peptidases, and is given orally. Peptides in general, and vasopressin in particular, can be given by the intranasal route for systemic action. Insulin has been studied for decades, and chemically modified, in the hope of finding an effective preparation that can be given i.n., by inhalation, by means of suppositories, through the skin, or orally, with little, if any, impact on its use by patients.

The lymphatic system has a significant role in the absorption of proteins. This system is a network of vessels draining from the many tissues of the body into the left and right subclavian veins. Although it is a one-way system, together with the cardiovascular system, it serves to deliver nutrients and oxygen to tissues, and to remove metabolic waste products from them. Simply put, the lymphatic system returns fluid (lymph) and its contents to the circulatory system from intercellular tissue spaces. Because drainage from the intestinal tissue is *towards* the venous system, materials from the intestine become dissolved in lymph and absorbed into the cardiovascular system.

The lymphatic system by-passes the portal system of the liver, and it is important to the absorption of dietary lipids. It also undoubtedly plays a part in the absorption of all orally-administered drugs, although with so much potential for absorption into the cardiovascular system via the mesenteric blood capillaries and the hepatic portal system, this contribution is, in many cases, quite small. However, and in particular, part of the lymphatic system is made up of thin-walled lymphatic capillaries within the tissues of the gastrointestinal tract. These vessels are more permeable than are blood capillaries. Also, the lymphatic vessels are well adapted for the movement of large molecules, particularly proteins, and this includes any peptide and protein drugs that survive the acidity of the stomach and the proteases of the intestine. For example, polypeptides of size greater than $20,000 \, \text{g} \, \text{mol}^{-1}$ are able to reach the blood via the lymphatic system, even though they are unable to traverse blood capillary membranes. Nevertheless, movement through lymphatics is relatively slow, and absorption in this way is rarely sought in the search for absorption of new drugs of any type.

The lymphatic system also plays a significant part in the absorption of subcutaneously-administered drugs of all kinds.

16.4.2   *Bioequivalence*

Concepts of bioequivalence have to be applied carefully with these drugs. In some cases, chemical properties and methods of analysis allow standards similar to those applied to small molecules to be applied. However, even different batches within the processes of a single manufacturer can vary with protein drugs, let alone between manufacturers, and regulatory authorities permit a rational protocol of the best available practices in manufacturing, chemical analysis, and bioassay, to be used in ensuring batch-to-batch and manufacturer-to-manufacturer consistency.

16.4.3   *Distribution*

Because most compounds in this category are polar, their apparent volumes of distribution tend to be low; often similar to the volume of plasma, or at the highest, a little over $1 \, \text{L} \, \text{kg}^{-1}$, indicating occurrence of some specific tissue uptake. Some typical apparent volume of distribution values are: $0.058 \, \text{L} \, \text{kg}^{-1}$ (heparin); $0.08 \, \text{L} \, \text{kg}^{-1}$ (streptokinase); $0.055 \, \text{L} \, \text{kg}^{-1}$ (L-asparaginase); and $1.3 \, \text{L} \, \text{kg}^{-1}$ (ciclosporin).

Distribution can be very specific. After intranasal administration, cholecystokinin is found only at sites in the stomach associated with the neuronal pathways to the brain that are connected with feelings of satiety. Trastuzumab shows specific localization in tumour tissue, because it forms a complex with the extracellular domain of HER-2.

16.4.4   *Metabolism*

Peptides are mostly metabolized by peptidases in the liver and kidney. Half-life values vary from a few minutes to as much as 28 days, depending on the ease with which the peptide bonds interact with the peptidase active sites. Just exactly why particular peptides and proteins vary so much in their sensitivity to proteases has been the subject of a considerable amount of research. Proteases are broadly classified into endopeptidases, which cleave the internal peptide bonds in substrates, and exopeptidases, which cleave the terminal peptide bonds. Exopeptidases can be further subdivided into aminopeptidases and carboxypeptidases. There is a further subclassification based on models applicable to describing the physical nature of the interactions between the substrate and the active site of the enzyme. Proteases can also be classified as aspartic proteases, cysteine proteases, metalloproteases, serine proteases, and threonine proteases, depending on the nature of the active site. Whether or not a particular substrate interacts with a particular active site will depend on the degree to which the peptide bond(s) in the substrate lack(s) hindrance to a close fit, and whether or not the substrate and enzyme are to be found in close proximity in the same body fluids. Thus, small, straight chain peptides have little chance of survival in the presence of intestinal proteases, while the peptide bonds of ciclosporin are so hindered that this drug barely behaves like a peptide, and is primarily metabolized by cytochrome P-450 enzymes. The therapeutic proteins that are given by intravenous injection do not come into contact with intestinal proteases. Whenever possible, protein drugs will be designed to resist attack by proteases.

16.4.5   *Excretion*

The standard concepts apply (Section 3.3). In particular, insulin ($M_r \sim 6{,}000$) is filtered at the glomerulus and is then metabolized to its amino acids by enzymes in the brush border of the renal tubular lumen, and/or within luminal cells.

Superoxide dismutase shows saturation of renal excretion as a function of plasma concentration, a phenomenon that is unusual amongst drugs, but familiar to investigators of transport maxima in the kidney with endogenous compounds, as occurs with glucose reabsorption.

16.5   Plasma kinetics and pharmacodynamics

A variety of phenomena familiar to pharmacokineticists is seen with compounds in this category. Thus:

- Many quite conventional studies are conducted: measurement of half-life values, protein binding, growth and decay of plasma concentration curves, renal excretion of metabolites, etc.
- Heparin shows dose-dependent kinetics; this causes difficulty in controlling dosing.
- Trastuzumab also shows dose-dependent kinetics. This is so prominent that it has been turned to advantage in devising optimum dosage regimens for this drug.
- Several of the molecules in this class show bi- or tri-exponential plasma level decay after i.v. bolus or infusion doses. For example, trastuzumab data were fitted to a 'two-compartment linear model,' even though its apparent volume of distribution was just 2.95 L (~plasma volume).
- A close correlation between kinetics and effect does not always occur. For example, erythropoietin has a half-life after i.v. injection of 10 hours, but the response can be delayed for up to 2–6 weeks. Desmopressin shows a biphasic plasma concentration decay after i.v. dosing, with the half-life values of 6.5–9 minutes and 30–117 minutes, but its antidiuretic effect lasts for 6–20 hours. Insulin has a half-life of 5–6 minutes, and its time-course of effect is controlled by its pharmaceutical formulation. Somatotrophin (HGH) has a half-life of 20–30 minutes, although its therapeutic value is measured over a period of months or years.
- It would appear that PK/PD models with effect compartments, or invoking principles similar to those applicable with warfarin are most appropriate with drugs in this class.

16.6   Examples of particular interest

16.6.1   *Cholecystokinins*

Cholecystokinins are a family of hormones produced by post-translational metabolism of preprocholecystokinin. Individual products are identified by the number of amino acids: for example CCK58, CCK33. CCK8 (Figure 16.1). CCK8 is not a drug *per se*, but it is of importance as a probe for the study of satiety mechanisms. After intravenous or intranasal pharmacological doses it is to be found almost exclusively in the pyloric region of the stomach wall, where there are stretch receptors which send signals to the brain via the

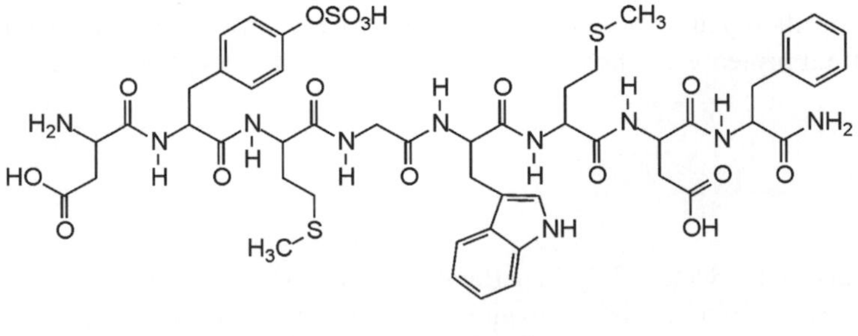

Figure 16.1   Formula of CCK-8.

vagus nerve to indicate that the stomach is full and that appetite should be shut down. This has been explored in human pharmacological investigations by Greenough *et al.* (1998) and others. Cholecystokinin has a half-life measured in minutes when incubated with kidney peptidase preparations. However, it is sufficiently stable *in vivo* for it to be possible to show a clear reduction in meal size in human volunteers given intranasal doses, thus shedding light on satiety mechanisms. There have been attempts to synthesize cholecystokinin analogues that are more stable, and have the same properties, in the search for anti-obesity drugs.

16.6.2   Ciclosporin

Ciclosporin (cyclosporine) is a cyclic polypeptide (Figure 16.2) of fungal origin, which is used as an immunosuppressant in transplant medicine. It is lipophilic and hydrophobic. It can be administered i.v., dissolved in a modified castor oil/ethanol mixture, or orally in a soft capsule when it has systemic availability of 20–50%. Soft capsules containing a micro-emulsion formulation have 10% greater availability. It has an apparent volume of distribution of $1.3 \, L \, kg^{-1}$, and approximately 30 known metabolites. It is a cytochrome P-450 substrate, and so it is prone to the drug interactions common with small molecule drugs. Its half-life is 5–6 hours. It is mostly excreted as metabolites in bile and hence faeces. Its therapeutic use requires close monitoring of concentrations in plasma.

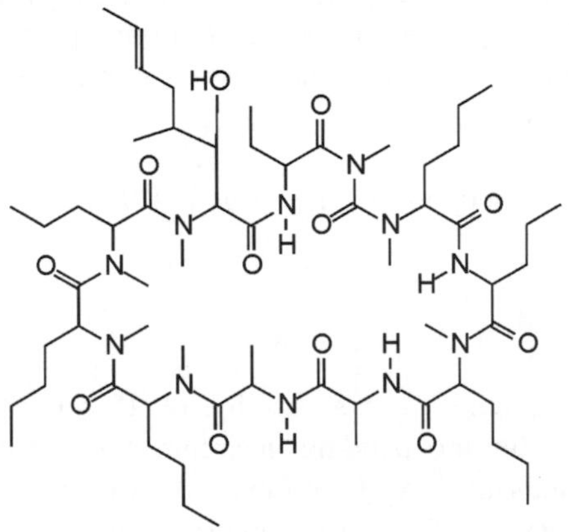

Figure 16.2   Structure of ciclosporin.

16.6.3   Heparin

Heparin (Figure 16.3) is a naturally-occurring negatively charged glycosoaminoglycan, and thus is a polymer of alternating D-glucuronic acid and *N*-acetyl-D-glucosamine residues. It is produced in the body in mast cells where it is stored in granules along with histamine. It is used as an anticoagulant, and it is obtained for

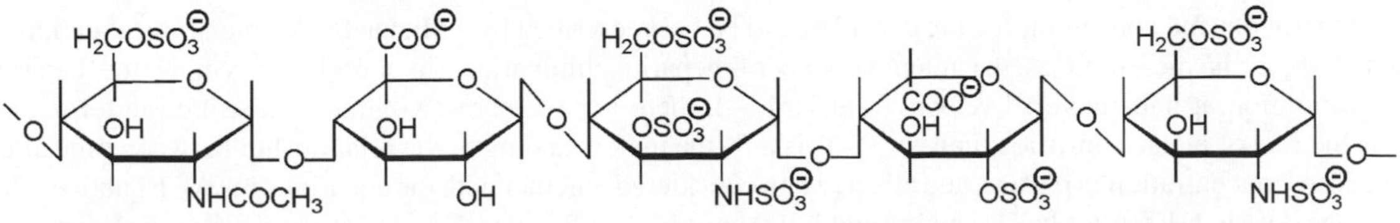

Figure 16.3   Representation of heparin. Heparin consists of repeating disaccharide units which may be sulfated to varying degrees.

medical purposes from biological sources as a by-product of the meat industry. Different batches are not consistent with each other. Its molecular weight varies from 4,000 to 30,000 g mol^{-1} depending on the degree of breakdown of its mucopolysaccharide side chains. The low molecular weight fractions and high molecular weight fractions have effects at different points in the blood clotting cascade.

Heparin is standardized and dosed in internationally recognized potency units, not in mass units, and even plasma concentration measurements use these units. It is given by i.v. injection for rapid effect, or by subcutaneous injection, when the onset time is 1–2 hours. Even with i.v. injection there is a short delay in onset time, and the effect lasts longer than would be expected from the plasma concentration–decay kinetics. This has led to suggestions that 'activation' is needed. The apparent volume of distribution is 0.058 L kg^{-1}. There is a short phase of fast decline of plasma concentrations after a bolus injection, with a half-life of approximately 5 minutes, followed by a slower phase with a half life in the range 1–5 hours. There has been extensive discussion about whether there is or is not dose-dependent decline of plasma concentrations, but the evidence favours dose dependence. Half-lives of the terminal phase of concentration decline in plasma are generally accepted to be 1, 2.5 and 5 h after doses of 100, 400 and 800 units kg^{-1} respectively, showing marked non-linearity in the elimination kinetics of the drug.

Part of the confusion over the kinetics of heparin has undoubtedly been caused by the methods of analysis. While direct chemical assay is desirable, several 'pharmacokinetic' studies have really been studies of the time course of effect, using methods described as the Lee–White clotting time, the activated clotting time, the thrombin clotting time, the anti Xa assay, and the activated thromboplastin time (APTT) assay. Also, there has been inconsistency concerning application of these assays to plasma and/or blood. The APTT assay is probably the most accepted. In this assay:

- Oxalated plasma + a partial thromboplastin time reagent + a surface activator are incubated for 3–5 minutes.
- Calcium chloride is added.
- The coagulation time is measured.

Heparin increases the coagulation time, from a baseline of 25–50 s. A plot of APTT versus heparin concentration may be linear, or log–linear depending on circumstances that are not always fully understood. The range of the assay is 0.1–0.8 units mL^{-1}, with a normal value of around 0.4, and a reasonable expectation of therapy resulting in a 50–75% increase. Thus the apparent non-linearity in kinetics may have been because of a lack of appreciation of the non-linearity of the assays used. However, there is broad acceptance now that a log–linear model is applicable to the effect of heparin, thus:

$$APTT = APTT_0 \exp(mC) \tag{16.1}$$

and as taking logarithms gives:

$$\ln(APTT) = mC + \ln(APTT_0) \tag{16.2}$$

m is the slope of the graph of $\ln(APTT)$ versus concentration, C.

Heparin plasma concentrations are considered to be best described by a Michaelis–Menten type equation in which V_{max} is measured as the maximal rate of heparin elimination (in units h^{-1}), Km is the heparin concentration at half-maximal velocity (units mL^{-1}), there is an apparent volume of distribution term, and R is the rate of heparin infusion (units h^{-1}). This aids the understanding of the relationship between time after the dose, concentration in plasma, and effect, when considered together with the equations for APTT above. The average dose is 1,400 units h^{-1} for males, and 1,100 units h^{-1} for females. V_{max} tends to be 40% greater in males than in females (3,555 ± 2,139 units h^{-1} in males). Km is 0.35 units mL^{-1} – in the lower part of the normal range for useful therapeutic effect.

A scatter diagram of *APTT* versus plasma concentration of heparin in a population of patients is truly a scatter diagram, showing no meaningful between-patient correlation. Heparin is a complex drug to understand, and a difficult drug to manage in the clinic.

16.6.4 Trastuzumab

This compound is a monoclonal antibody devised by the biotechnology industry for the specific treatment of HER-2-positive breast cancer. It shows marked non-linearity in its elimination kinetics, and as the aim of therapy is to maintain trough concentrations in plasma of $20\,\mu g\,L^{-1}$ over a 25–30 week period, it has been possible to replace the originally proposed dosing regimen of a $4\,mg\,kg^{-1}$ loading dose followed by $2\,mg\,kg^{-1}$ weekly, for about 6 months, with an $8\,mg\,kg^{-1}$ loading dose followed by $6\,mg\,kg^{-1}$ every three weeks for the same time. Because the drug must be given i.v. in an outpatient clinic setting, this has resulted in a much more convenient regimen, with the number of clinic visits reduced by two-thirds. The half-life of trastuzumab is similar to that of the endogenous IgGI immunoglobulin that constitutes the backbone of the drug. Of particular interest is the fact that it forms a complex with the extracellular domain of HER-2 and the clearance of the complex is greater than that of the drug – demonstrating a rational approach to removing an unwanted residue from the body.

16.6.5 Erythropoietin

This compound is typical of the protein drugs formed using recombinant DNA techniques to imitate the normal body constituent. It consists of 193 amino acid residues and has a molecular weight of approximately $30,000\,g\,mol^{-1}$. It is heavily glycosylated, and is used to treat the anaemia of chronic renal disease. There is always a baseline erythropoietin concentration in plasma, so therapy adds 'drug' to the endogenous amount. The recombinant product is epoetin alfa, and it can be given i.v. or s.c. Its half-life is 10 hours, although its t_{max} is often said to be between 5–24 hours; the response can be delayed for 2–6 weeks in some patients – success in treatment is measured over long periods of time.

16.6.6 Vasopressin and desmopressin

Vasopressin, antiduretic hormone (ADH), is a natural nonapeptide (Figure 16.4) that acts in the body to control urine production in the process of homeostasis. Desmopressin (Figure 16.5) is a synthetic peptide

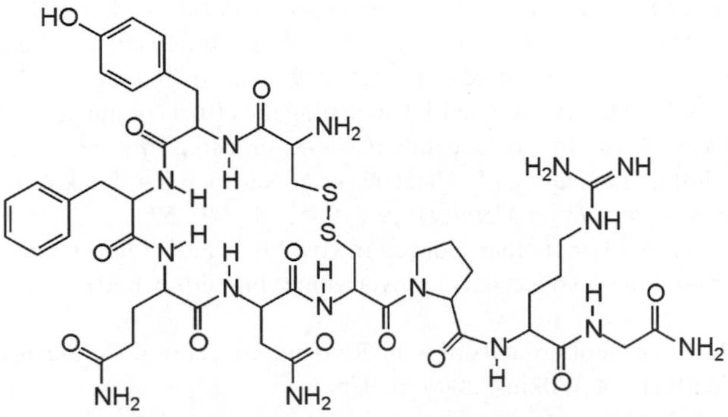

Figure 16.4 Structure of vasopressin.

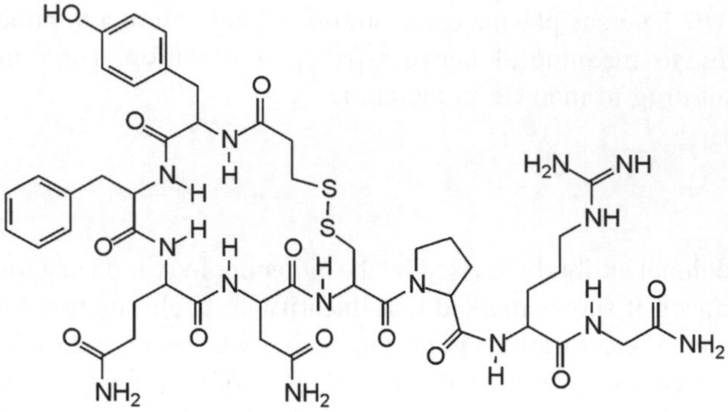

Figure 16.5   Structure of desmopressin.

analogue. Both compounds are used in pharmacological doses to control the excessive water loss of diabetes insipidus. Vasopressin has been used in this way for many years, and it was the first intranasally (i.n.) administered peptide, having been formulated as a powder which was known in the past as 'pitressin snuff'. This route of administration, while being the only feasible method for this drug, does provide a measure of control of dosing for what is a very difficult drug to manage.

16.7   Conclusion

The last 30 years have probably seen the birth of a brave new world of scientifically-conceived highly-specific 'biological' drugs. The examples quoted here are probably just the leading edge of what is to come, in the search for specificity in drug delivery and effect, in freedom from unwanted pharmacological effects, and in what is coming to be called 'personalized medicine.'

References and further reading

Chiang J, Gloff CA, Yoshizawa CN, Williams GJ. Pharmacokinetics of recombinant human interferon-beta ser in healthy volunteers and its effect on serum neopterin. *Pharm Res* 1993; 10: 567–72.

Curry SH, Schlosser MJ, Rawleigh S, Webborn P, Willson VJC, Wilkinson D, Logan CJ. Concentrations of the appetite-suppressing cholecystokinin analog FPL 15849KF in dogs after IV and intranasal doses. *Int J Obesity* 1995; 19 (Suppl. 2): 414.

Estes JW. Clinical pharmacokinetics of heparin. *Clin Pharmacokinet* 1980; 5: 204–20.

Fahr A. Cyclosporin clinical pharmacokinetics. *Clin Pharmacokinet* 1993; 24: 472–95.

Gemmill JD, Hogg KJ, Burns JM, Rae AP, Dunn FG, Fears R, *et al*. A comparison of the pharmacokinetic properties of streptokinase and anistreplase in acute myocardial infarction. *Br J Clin Pharmacol* 1991; 31: 143–7.

Greenough A, Cole G, Lewis J, Lockton A, Blundell J. Untangling the effects of hunger, anxiety, and nausea on energy intake during intravenous cholecystokinin octapeptide (CCK-8) infusion. *Physiol Behav* 1998; 65: 303–10.

Ho DH, Brown NS, Yen A, Holmes R, Keating M, Abuchowski A, Newman RA, Krakoff IH. Clinical pharmacology of polyethylene glycol-L-asparaginase. *Drug Metab Dispos* 1986; 14: 349–52.

Kidd JG. Regression of transplanted lymphomas induced in vivo by means of normal guinea pig serum. I. Course of transplanted cancers of various kinds in mice and rats given guinea pig serum, horse serum, or rabbit serum. *J Exp Med* 1953; 98: 565–82.

Krall D, DerMarderosion AH. Biotechnology and drugs. In: Remington G, editor. *The Science and Practice of Pharmacy*. Philadelphia, Lippincott Williams & Wilkins, 2006: p. 976.

McKeage K, Perry CM. Trastuzumab: a review of its use in the treatment of metastatic breast cancer overexpressing HER2. *Drugs* 2002; 62: 209–43.

Neuhaus O, Kieseier BC, Hartung HP. Pharmacokinetics and pharmacodynamics of the interferon-betas, glatiramer acetate, and mitoxantrone in multiple sclerosis. *J Neurol Sci* 2007; 259: 27–37.

Palm S, Enmon RM Jr., Matei C, Kolbert KS, Xu S, Zanzonico PB, Finn RL, Koutcher JA, Larson SM, Sgouros G. Pharmacokinetics and Biodistribution of (86)Y-Trastuzumab for (90)Y dosimetry in an ovarian carcinoma model: correlative MicroPET and MRI. *J Nucl Med* 2003; 44: 1148–55.

Perona JJ, Craik CS. Structural basis of substrate specificity in the serine proteases. *Protein Sci* 1995; 4: 337–60.

Plosker GL, Lyseng-Williamson KA. Adalimumab: in Crohn's disease. *BioDrugs* 2007; 21: 125–32; discussion 133-4.

Tokuda Y, Watanabe T, Omuro Y, Ando M, Katsumata N, Okumura A, *et al*. Dose escalation and pharmacokinetic study of a humanized anti-HER2 monoclonal antibody in patients with HER2/neu-overexpressing metastatic breast cancer. *Br J Cancer* 1999; 81: 1419–25.

Wang Z, Martensson L, Nilsson R, Bendahl PO, Lindgren L, Ohlsson T, *et al*. Blood pharmacokinetics of various monoclonal antibodies labeled with a new trifunctional chelating reagent for simultaneous conjugation with 1,4,7,10-tetraazacyclododecane-*N*,*N'*,*N''*,N--tetraacetic acid and biotin before radiolabeling. *Clin Cancer Res* 2005; 11: 7171s–7177s.

Wills RJ. Clinical pharmacokinetics of interferons. *Clin Pharmacokinet* 1990; 19: 390–9.

17

Drug Interactions

17.1 Introduction

When two or more drugs are used together, the pharmacological result is not necessarily the sum of the effects obtainable from the drugs used individually in the same doses. This is because one drug may affect the action of another. This is termed 'drug interaction'.

With thousands of drugs available, the statistical probability of interactions when drugs are used in combination is high. There are many mechanisms of drug interaction. Some are chemical, some are biochemical, and some are physiological/pharmacological. Many of the most easily understood drug interactions are mediated by drug-induced changes in drug absorption, distribution, metabolism and excretion. Certain drug interactions are of great clinical importance, usually involving a drug with a narrow therapeutic window, while others are no more than interesting pharmacological curiosities.

Sometimes a distinction is made between drug–drug interactions, drug–food (e.g. grapefruit juice) interactions, and drug–chemical (e.g. alcohol) interactions. A widespread example of a drug–chemical interaction is illustrated by the fact that pharmaceutical oral syrup preparations of most amine drugs produce an ugly precipitate with the tannin content of the beverage when, with good intent, these syrups are diluted with tea (Curry *et al.*, 1991). This chapter considers pharmacological, pharmacokinetic and pharmaceutical interactions, including examples of drug–drug, drug–food, and drug–chemical interactions. However, the influence on food *per se* on drug absorption in particular was considered in Chapter 9.

17.2 Terminology

When two (or more) interacting drugs have the same effect, for example central nervous system (CNS) depressant drugs and ethanol, the interaction is described as *homergic*. More often, only one of the drugs produces the effect being studied, although each drug in the combination may have an effect that is modified. This can be described as a *heterergic* interaction. Much discussion occurs in regard to whether drug interactions should be described under the headings of *antagonism*, *addition* (*or summation*), *potentiation* or *synergism*. Additionally, the use of the terms *supra-additive* and *infra-additive* has been proposed.

Antagonism is when one drug reduces the effect of another. This is clear for a heterergic inhibition interaction when the resultant effect is less than that recorded before the interaction. However, in a homergic interaction, the two doses together will generally produce an effect greater than that produced from either of the individual doses on its own. But if the new effect is less than that predicted from sum of the two individual contributions, then this is classified as antagonism (Section 17.6.4).

According to its Greek origins, the word '*synergism*' merely means 'working together', so it could be used to describe any interaction, but it has evolved in meaning to be synonymous with potentiation. Summation

Drug Disposition and Pharmacokinetics,   By Stephen H. Curry and Robin Whelpton
© 2011 John Wiley & Sons, Ltd

and addition also appear to be synonymous. If the effects of two drugs in a homergic interaction add together and lead to the effect expected by simple arithmetical addition, then summation and addition are appropriate terms. However, some writers differentiate between these two terms, on the basis of summation being used exclusively for a combined effect at the centre of the sigmoid log dose–response curve. Summation sometimes becomes a special form of addition in this argument, and other additive effects can be described as supra-additive if they are below the inflection of the curve, and infra-additive if they are above. Some authors stress the *mutual* nature of drug interactions, in that an effect of one drug on another is often reciprocated by an effect of the other on the one.

Life is obviously simpler with a heterergic interaction involving enhanced effect. If the effect of the one drug causing the effect is increased, then potentiation is the appropriate description. Fortunately, most interactions are heterergic. The drug affected in an interaction is sometimes called the 'victim' drug.

17.3   Time action considerations

It is easy to consider an interaction in relatively simple terms when thinking of it as occurring *in vitro* in a pharmacologist's organ bath, with an effect occurring immediately the drug is added to the bath and continuing until the bath is washed out. However, *in vivo* effects have onset, peak intensity and duration characteristics (Chapters 4, 13 and 14). These are functions of the drug concentrations in blood and tissues, and these concentrations are in turn controlled by the balance of absorption, tissue distribution, metabolism and excretion. This can introduce added terminology issues. In the introduction to Chapter 4 it was noted that modifications of drug absorption, metabolism and excretion could all affect a drug response differently. For example, an increase in the rate of absorption with no change in the degree of absorption, will lead to an earlier onset of effect and a greater and earlier maximum effect, but a shorter duration of effect. Slightly different, but completely analogous considerations are relevant to fluctuating concentrations during long-term treatment. Thus by reference to onset of effect and intensity of effect, an increased rate of absorption might be termed a potentiation, but the shorter duration of effect would be an antagonism. Given this set of circumstances, it is best to discuss the results of an interaction in terms of measured phenomena and pharmacological mechanisms, and be careful not to be too dogmatic in use of terminology. It is now recognized that every interaction of importance should be evaluated, where possible, in terms of the effect of the interacting drug on the absorption rate, bioavailability, time and height of maximum concentration, half-life, area under the curve, apparent volume of distribution, clearance, maximum and minimum concentrations in the fluctuating pattern of drug concentrations seen during long-term dosing, and in changes in the effect of the victim drug.

For the most part, the results of interaction studies tend to be expressed in terms of changes in the descriptive pharmacokinetic parameters listed in the previous paragraph, such as half-life and bioavailability. However, the more advanced clearance concepts discussed in Chapter 7 make it possible to perform calculations that provide greater insight into the underlying mechanism than is possible with descriptive pharmacokinetics alone. For example, propranolol has been shown to lower the hepatic clearance of lidocaine. This occurs because lidocaine has a high extraction ratio, and its hepatic clearance is very much affected by liver blood flow. Propranolol reduces cardiac output, which in turn reduces hepatic blood flow, reducing the clearance of lidocaine. This translates into a longer half-life of the lidocaine. Similarly, enzyme induction interactions are most important with low extraction ratio drugs, and protein binding interactions commonly involve drugs that demonstrate the saturability of protein binding within the clinical range of concentrations.

Similar approaches to the above permitted a perceptive study of the interaction of theophylline and enoxacin. Enoxacin inhibits theophylline metabolism in the liver, leading to nausea as an adverse effect of the increased theophylline concentrations. In the absence of enoxacin, the half-life of theophylline is approximately 9 hours, so steady-state conditions should be established in approximately 2 days in a long-term dosing regimen (Section 4.2.6). However, under the influence of enoxacin, 4 days were required in order to achieve the new steady-state, consistent with the half-life having doubled. The experimental design permitted calculation of the new half-life as 22 hours, consistent with a reduction in unbound clearance by 56%. Although the raised theophylline concentrations were shown in the experimental protocol to take place within the clinically accepted range, it was clear that the same interaction occurring in patients would cause the plasma concentrations of theophylline to rise above the upper limit of this range and hence result in nausea.

17.4 Interactions involving drug distribution and metabolism

As already stated, a great many interactions can be explained in terms of effects on drug absorption, plasma protein binding, metabolism and excretion. Others can be explained in terms of classical theories of drug–receptor interaction (Chapter 13). Yet others are at present unexplained. The three most important mechanisms involving changes in drug disposition and fate are associated with drug metabolism, both induction and inhibition, and with binding of drugs to plasma protein and sequestration in tissues.

17.4.1 Enzyme induction

Induction of drug metabolizing enzymes is probably the longest established of the three major pharmacokinetic mechanisms of drug interaction. This phenomenon was discovered when the influence of phenobarbital on the effects and plasma concentrations of anticoagulants, in particular, was first observed. Enhanced enzyme activity can be detected in animals after just two or three doses of phenobarbital. Affected drugs typically show shortened half-life values, reduced areas under the curve, lower steady-state concentrations on multiple dosing, increased metabolite formation, and reduced effects. In addition to phenobarbital and virtually all of the other barbiturates, enzyme inducers include 3-methylcholanthrene (a laboratory model), rifampicin (rifampin), carbamazepine, valproic acid and phenytoin, among many other compounds, and most of them induce microsomal P-450 activity (Table 3.1). With the passage of time the major clinical focus has changed from phenobarbital to rifampicin in this context.

17.4.2 Enzyme inhibition

Although mutual inhibition of metabolism by various substrates for the P-450 system had been known for many years previously, it was the observation in the early 1980s that cimetidine caused major changes in the responses to a wide variety of drugs that focussed attention on this mechanism of drug interactions. Other H_2-blocking drugs were found to have lesser abilities to inhibit the P-450 system, and this distinction was soon exploited in the marketing of ranitidine, in particular, as a relatively non-interacting drug. The list of drugs that inhibit CYP3A4 and/or CYP2D6 in particular, now includes ketoconazole, fluoxetine, fluvoxamine, quinidine, theophylline and erythromycin (Table 3.1), with several examples causing interactions that affect prescribing practices.

17.4.3 Plasma protein binding and tissue uptake

Competition between different compounds for the same binding sites on albumin and other proteins can lead to the incumbent drugs being displaced. In a balanced therapeutic situation, this in turn can lead to an increased amount of displaced drug being available in plasma water and in other body fluids, for pharmacological action. However, tissue uptake is also affected by concentrations in plasma water, so the overall effect may be a transient increased effect, but this is likely to be rapidly compensated by increased localization in tissues, and, sometimes, by enhanced elimination.

17.4.4   Mechanisms of enzyme induction

The mechanism of enzyme induction has received considerable attention. The mechanism of microsomal oxidation of drugs was considered in Chapter 3. When enzyme induction occurs, the liver increases in weight and in protein content, the smooth membranes of the endoplasmic reticulum (microsomes) increase in quantity and in content of protein, RNA and phospholipid (especially when phenobarbital is the inducing agent) and the microsomal messenger RNA and the rate of microsomal incorporation of amino acids from transfer RNA is increased (especially when 3-methylcholanthrene is the inducing agent). This activation is reduced by ethionine, puromycin and actinomycin D, all of which inhibit protein synthesis. In laboratory animals, enzyme induction can be demonstrated as a decrease in hexobarbital sleeping time, and an increase in biphenyl 4-hydroxylation, in ethylmorphine *N*-desmethylation, and in *p*-nitrobenzoate reductase activity. Hexobarbital sleeping times have been widely used to assess undifferentiated P-450 activity. The liver concentrations of reduced nicotinamide adenine dinucleotide phosphate (NADPH)-cytochrome *c* reductase, cytochromes P-450, and cytochromes b_5 increase. However, there is no effect on glucose-6-phosphatase dehydrogenase, 6-phosphogluconate dehydrogenase or glucose-6 phosphatase. This was taken as evidence that the mechanism of enzyme induction involves changes in the enzymes specifically present as part of the drug metabolizing microsomal system, rather than changes in rate of NADPH generation. It appears that enzyme induction involves an increase in the rate of synthesis rather than a decrease in the rate of destruction of drug metabolizing enzymes. As the enzymes are relatively non-specific, induction by one substrate often leads to other substrates being more rapidly metabolized, thus providing a means of one drug affecting the response of another. Microsomal enzymes are constantly undergoing synthesis and degradation, with half-lives of turnover varying from hours to days, so induction results in increased amount of enzyme and, consequently, increased V_{max} values for the metabolic reactions. The full effect of the inducing agent may be quite rapid, or delayed, depending on the turnover rate of the enzyme isoform involved.

In the case of CYP3A4 induction, the process is believed to involve ligand binding to the nuclear pregnane X receptor (PXR). The activated PXR complex then forms a heterodimer with the retinoid X receptor (RXR), which binds to the XREM region of the CYP3A4 gene. XREM is a regulatory region of the CYP3A4 gene, and binding causes a co-operative interaction with proximal promoter regions of the gene, resulting in increased transcription and expression of CYP3A4 mRNA and protein. Induction of CYP2D6 involves the CAR/RXR nuclear receptor heterodimer. This process can be viewed as a defence mechanism of the body; a method of preserving homeostasis, that involves the body recognizing the ligand (in the current context, a drug) as an unwanted chemical that might be toxic, and automatically responding by increasing the amount of enzyme that can help rid the body of that drug.

17.5   Extent of drug interactions

The common feature of all of the compounds shown to be potent enzyme inducers is that they are highly lipophilic and are substrates for the enzymes concerned. It seems that important enzyme inducing agents must persist in the extracellular fluid in sufficient quantities for sufficient time, so they tend to be compounds with relatively long half-lives (generally resulting from relatively slow metabolism and limited renal excretion as unchanged drug). In addition, they must bind to P-450 as substrates without destroying the enzyme. This is illustrated well by the barbiturates. While barbital, phenobarbital, allobarbital, pentobarbital, quinalbarbital and thiopental are all inducing agents, phenobarbital is the most potent, as barbital is relatively polar and can be excreted most easily without prior metabolism, allobarbital binds to P-450 but destroys the enzyme, pentobarbital and quinalbarbital are intermediate in potency and thiopental is weak because of its high degree of tissue localization reducing the amount available in extracellular fluid at any time.

The list of compounds affected by enzyme inducing agents is long, potentially including any compound that is a substrate for the enzymes. However, a change in the rate of metabolism does not necessarily bring about a change in drug response and any list of compounds for which an important change in response is likely, consequent on enzyme induction, is relatively short. Historically, the focus was on anticoagulants and anticonvulsants, and also chlorpromazine, for which enzyme induction has been shown to change autonomic nervous system responses without changing antipsychotic effects. Antipyrine has been used extensively in demonstrating enzyme induction. The following appear to be the conditions for maximal influence of enzyme induction, all of which are satisfied by the drugs mentioned here:

- The affected compound is slowly metabolized, often with dose-dependent kinetics.
- Termination of the effect is dependent on metabolism by microsomal enzymes, rather than on redistribution, excretion, or physiological effects.
- The effect is closely related to the concentration of the drug in the body.
- The compound affected has a low therapeutic index.

17.6   Key examples

In this section, four examples illustrating the most important pharmacokinetic mechanisms of drug interaction are described. The results of research into all four examples have, to a smaller or larger extent, affected the practice of medicine.

17.6.1   Warfarin

This coumarin anticoagulant, which can be considered as the quintessential victim drug, is mentioned in almost every drug interaction discussion because:

- It is a weak acid which is bound to plasma albumin, and is displaced from binding sites by other weak acids such as phenylbutazone; this interacting drug, obsolete as a therapeutic agent, has been a key compound in aiding our understanding of drug interaction mechanisms.
- It has a low volume of distribution and capacity-limited hepatic clearance.
- It is a substrate for the non-specific liver microsomal drug oxidase system, so its metabolism is accelerated by enzyme inducing agents such as phenobarbital, and its metabolism may be inhibited by competing substrates such as ketoconazole and imipramine.
- It has a low therapeutic index so small influences caused by a wide variety of drugs can assume great importance.
- It exerts its effect on a delicately balanced physiological control mechanism, thus maximizing the opportunity for clinical demonstration of any interaction.

However, it should be appreciated that while the effect of warfarin is closely related to the concentration of warfarin in plasma, or even more closely related to its concentration in plasma water, there is a 2–3 day lag between a change in concentration and a change in effect, except in situations of acute overdosage (Chapter 14).

The classical warfarin interactions are with phenobarbital and with various anti-inflammatory drugs, such as phenylbutazone. Table 17.1 shows a summary of the results of these interactions. It is of interest because of the fact that both the protein binding and the enzyme induction interactions lead to a reduction in total warfarin in plasma, but one of these interactions leads to an increased response and the other leads to a decreased response. A barbiturate added to a prevailing warfarin regimen will lead to accelerated warfarin metabolism. The result will be lower concentrations of warfarin in plasma, both free in plasma water and bound to protein, but with virtually the same proportion bound. This will lead to a slow change in warfarin concentrations and so to a reduced overall response. In contrast, addition of a displacing agent such as salicylate or phenylbutazone to a warfarin regimen will cause an immediate reduction in the amount and proportion of warfarin bound and a corresponding increase in unbound warfarin, with the total at first unchanged. If dosing with the displacing drug is continued then the total concentration of warfarin in plasma decreases as increased amounts of unbound drug are taken up into tissues (including, possibly liver, where the site of action of warfarin lies), until tissue/plasma concentrations equilibrate. The new equilibrium will then be: (i) increased tissue warfarin, (ii) decreased total and bound warfarin in plasma, (iii) increased unbound warfarin in plasma water, and (iv) consequent enhanced effects.

Table 17.1   Drug interactions and warfarin

Measure	Type of interaction	
	Protein binding	Enzyme induction
Total warfarin	Reduced	Reduced
Bound warfarin	Reduced	Reduced
Unbound warfarin	Increased	Reduced
Effect (short-term)	None	None
Effect (long-term)	Haemorrhage	Clotting[a]

[a] If barbiturate added to warfarin regimen.

The increased unbound fraction, fu, may result in increased metabolism of a drug such as warfarin, however the interaction with phenylbutazone is more complex than a simple displacement interaction. Warfarin is marketed as the racemate and the S-isomer, which is some three to four times more potent than the R-isomer, is metabolized to 7-hydroxywarfarin whilst the R-isomer is metabolized to warfarin alcohol. Normally the S-isomer is more rapidly metabolized than the R-, but in the presence of phenylbutazone the production of 7-hydroxywarfarin is reduced. In fact, phenylbutazone inhibits the clearance of S-warfarin but increases the clearance of the R-isomer so that the 'total' remains more or less the same (Figure 17.1).

In reality, the greatest clinical danger with the enzyme induction interaction has been when a barbiturate was withdrawn from an established regimen of barbiturate plus warfarin, such as occurred when patients admitted to hospital for anticoagulation therapy, were also given sleeping tablets, and became stabilized on the combination. On returning home, the patients no longer required the hypnotic drugs and their enzymes reverted to lower activity, leading to the warfarin then being metabolized more slowly. The increased concentrations of warfarin could lead to haemorrhage. Today's hypnotics do not affect anticoagulation therapy in the same way, but this example provides a lesson of lasting significance in the history of, and potential for, drug interactions.

Non-barbiturate drugs which can affect anticoagulants by virtue of their enzyme inducing properties include rifampicin, haloperidol and griseofluvin, as well as several obsolete compounds mentioned in older literature. Non-anticoagulant drugs affected by barbiturates are many in number, e.g. phenytoin. Other drugs which affect anticoagulants by protein binding effects include a variety of older drugs, plus aspirin.

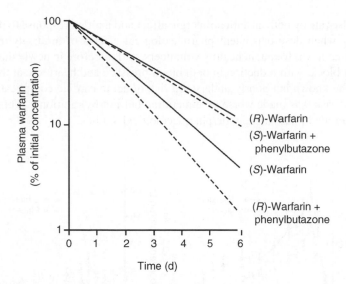

Figure 17.1   Schematic diagram of changes in plasma concentrations of warfarin enantiomers after single doses before and after phenylbutazone, 100 mg three times a day for 10 days. (Redrawn after Lewis *et al.*, 1974.)

17.6.2   Cimetidine and ketoconazole

These drugs are ideal examples of enzyme inhibitors. Although it has long been known that there is considerable potential for drug interactions caused by inhibition of isoforms of the P-450 system, and, now, grapefruit juice is known to have similar properties, it was only in the late 1970s and early 1980s, when cimetidine was introduced as the first of a series of highly innovative drugs that specifically inhibits gastrointestinal acid secretion (the H_2-blocking drugs), that clinically-important interactions related to this mechanism started to emerge. Cimetidine was found to inhibit the metabolism of a wide range of P-450 substrates, including theophylline, diazepam, tricyclic antidepressants, and metoprolol. Other inhibitors have since been found. These have included the unrelated antifungal drug, ketoconazole. One consequence of the observations with cimetidine was a vigorous search among its analogues and successors, especially ranitidine, for similar properties. These, basically, were not found, leading to vigorous competitive marketing, on the basis of these differences, among the makers of H_2-blocking drugs. As knowledge of the isoforms of P-450 emerged, it was seen that cimetidine, and later ketoconazole, particularly inhibited CYP3A4, the isoform especially responsible for first-pass metabolism, so that parent drugs were found to have enhanced bioavailability under the influence of cimetidine and ketoconazole, while the metabolic products of first-pass metabolism had reduced plasma concentrations.

   One of the cimetidine interaction studies of particular interest involved amitriptyline, the tricyclic anti-depressant at one time commonly co-prescribed with cimetidine in the hope of effectively treating both ends of the 'gut–brain' axis responsible, in part, for gastrointestinal ulcers. The first-pass desmethylation of amitriptyline to nortriptyline was inhibited in the interaction. Plasma amitriptyline concentrations were higher, and plasma nortriptyline concentrations were lower, as Figure 17.2 shows. Because this work was conducted with single doses of amitriptyline in healthy volunteers, it was possible to measure changes in blood pressure and heart rate, as well as performance in digit symbol substitution tests of brain function, in relation to plasma drug concentrations. The effects were greater when the amitriptyline concentrations were higher. However, this work also revealed that higher plasma amitriptyline concentrations resulted in more intense reductions in both heart rate and blood pressure, which was not, at first, explicable in terms of the known central and autonomic nervous system effects of the drug, which were expected to reduce blood pressure, but raise pulse rate by both an anticholinergic affect and a reflex response to the reduction in blood pressure. It was only when dose-dependent pharmacological effects of amitriptyline were studied in a separate experiment that it was found, in healthy volunteers, and therefore in newly diagnosed patients, that the drug caused heart block, with reduction in both blood pressure and heart rate at the same time, at least partially explaining the known but poorly understood risk of acute cardiac complications of amitriptyline in naïve subjects. This risk was made worse by cimetidine but not by ranitidine. This study illustrates the extreme need for integrated pharmacokinetic/pharmacological studies.

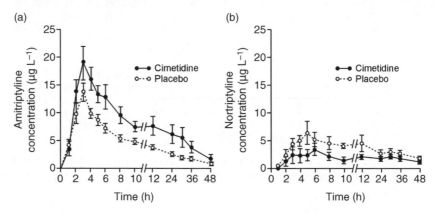

Figure 17.2 Interaction of cimetidine and amitryptiline. (a) Co-administration of cimetidine increased amitriptyline concentrations compared with placebo, whereas (b) the corresponding concentrations of the metabolite, nortriptyline were reduced. Note that the area under the curve for amitriptyline was increased, and that for nortriptyline was reduced, consistent with inhibition of CYP3A4 during presystemic metabolism of the oral doses. (Redrawn from Curry *et al.*, 1985.)

Ketoconazole, like the effect of the cimetidine on amitriptyline, inhibits the first-pass conversion of terfenadine to its metabolite fexofenadine by CYP3A4 (Figure 17.3). The raised terfenadine concentrations in plasma are sufficient to cause a severe toxic effect of terfenadine-induced prolongation of the QT interval

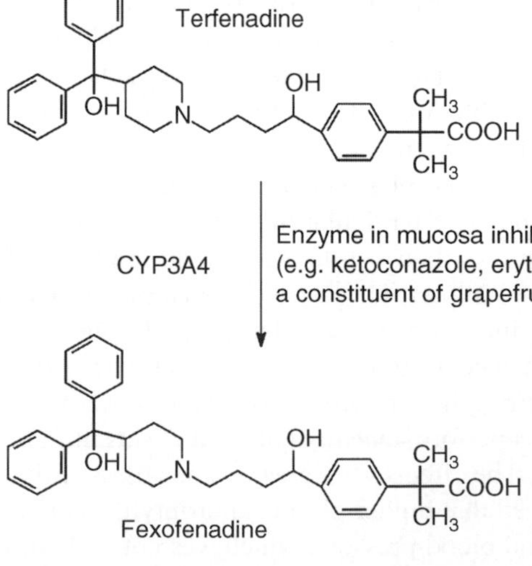

Terfenadine

CYP3A4

Enzyme in mucosa inhibited by drugs (e.g. ketoconazole, erythromycin) and a constituent of grapefruit juice

Fexofenadine

Figure 17.3 Inhibition of terfenadine prevents conversion to the antihistamine fexofenadine.

in the electrocardiogram. Both the C_{max} and the half-life of terfenadine are increased, greatly increasing the patient exposure to the drug (Table 17.2).

Table 17.2 Effect of ketoconazole on terfenadine and fexofenadine kinetics

Treatment	Pharmacokinetic parameters			
Terfenadine alone	$C_{max} < 10\,\text{ng mL}^{-1}$	Metabolite half-life	13.2 h
		Metabolite C_{max}	471 ng mL^{-1}
		Metabolite t_{max}	2.8 h
Terfenadine + ketoconazole	$C_{max}\ 27\,\text{ng mL}^{-1}$	Metabolite half-life	35.7 h
		Metabolite C_{max}	134 ng mL^{-1}
		Metabolite t_{max}	7.0 h

The pharmacokinetics of the metabolite, fexofenadine, are also affected, partly because of its slower formation from terfenadine, but also because fexofenadine itself is a substrate for the affected P-450 isoforms. Fexofenadine formed from terfenadine experienced a delay in t_{max}, an increase in C_{max}, and a longer $t_{1/2}$. The interaction with terfenadine was deemed to be so significant that it had two dramatic consequences: (i) terfenadine was withdrawn from the market, and replaced with fexofenadine, which, although sensitive to the effect of ketoconazole, has much lesser potential, if any, to cause prolongation of the QT interval; and (ii) testing of the QT interval has become routine in the safety testing of new drugs, both in preclinical studies and in the Phase I and Phase II human studies, and in Phase III pivotal clinical trials, creating a whole new testing culture and industry at the same time.

Ketoconazole also inhibits first-pass metabolism of ciclosporin (cyclosporine), enhancing its effect.

17.6.3 Digoxin and quinidine

The interaction of quinidine with digoxin illustrates two important issues: (i) inhibition of tissue uptake, and (ii) that drugs that are routinely co-prescribed may indeed interact. The clinical problem was that quinidine elevates plasma concentrations of digoxin two- to threefold, inducing arrhythmias as an adverse effect of the digoxin, in patients being treated for arrhythmias by the two drugs in combination. Quinidine is a potent inhibitor of P-glycoprotein (P-gp; Section 2.3.1.2) and in cultured cell lines containing P-pg was shown to inhibit transport of digoxin by 57%. Mice in which the gene expressing P-gp was disrupted ('knock-out', KO) were compared with wild-type mice. The quinidine doses were reduced in the study in KO mice to allow for impaired quinidine transport, so that exposure in both groups was the same. Quinidine increased digoxin plasma concentrations by 73% in wild-type mice, but only 19.5% in KO mice. Quinidine also increased digoxin concentrations in the brains of wild-type mice, by 73.2%, but decreased brain concentrations in KO mice by 30.7%. It was concluded that quinidine inhibits the transport of digoxin, both *in vitro* and *in vivo*, including at the blood–brain barrier, decreasing the apparent volume of distribution of digoxin, and raising concentrations in plasma in the wild-type mice. The effect on brain concentrations in the KO mice was not fully explained.

Rifampicin also has effects on digoxin, not through enzyme induction, but by a mechanism involving induction of P-gp. Plasma concentrations of digoxin are reduced, assessed by *AUC* measurements, with oral doses being affected more than i.v. doses. This has been investigated using i.v. and oral doses, with and without rifampicin, in human volunteers who underwent duodenal biopsies for P-pg assessment. It was found that rifampicin increased intestinal P-gp threefold, with no effects on renal clearance or the half-life of digoxin. Thus the interaction is induction of digoxin efflux in intestinal tissue, reducing the bioavailability. A similar mechanism may apply to St John's wort, which reduced digoxin concentrations by approximately 20% over 16 days when 900 mg day^{-1} of hypericum (St John's wort) was co-administered. Hypericum is also a CYP3A4 inducer, with potentially catastrophic effects on ciclosporin in particular, because of its popularity as an over-the-counter treatment for the depression which is common in post-transplant patients.

17.6.4   Alcohol and other depressants (notably barbiturates)

This is the homergic interaction par excellence with almost infinite possibilities for discussion of addition, potentiation, etc. (see earlier). A number of facts are pertinent:

- Ethanol can accelerate or retard the absorption of other drugs.
- Other drugs can accelerate or retard the absorption of alcohol.
- Alcohol can accelerate or retard the metabolism of other drugs.
- Other drugs can accelerate or retard the metabolism of alcohol.

Interactions with alcohol often lead to concentration changes in plasma and appear to occur to some extent with alcohol and barbiturates, other depressant drugs, such as phenothiazines and antihistamines, phenytoin, and some antidepressants. There is a significant interaction with benzodiazepine drugs, but any changes resulting from mutual influences on the plasma concentrations of benzodiazepines and ethanol appear to be minimal, although the pharmacological consequences of this interaction are considerable. In regard to changes in plasma concentrations the arguments of earlier sections must be borne in mind, in that different conclusions may be drawn depending on whether the rate of rise or fall of the concentration is measured or whether the area under the curve is measured.

In addition to the above:

- Certain doses of drugs are sub-threshold (Chapter 13) and a dose may be sub-threshold for sedation but above threshold for, say, anticonvulsant therapy.
- Sub-threshold doses of two depressant drugs can combine to form an above threshold response – which will be an effect of unexpected magnitude.

The above considerations combined lead to the conclusion that an all-embracing description of the combination of alcohol and other depressant drugs is inappropriate, and precise terms such as addition, potentiation or antagonism should only be applied to precisely defined circumstances. Some other statements of fact reinforce this opinion:

- As already stated, diazepam and alcohol exert relatively small mutual effects on their various plasma concentrations (very little or no pharmacokinetic interaction).
- Sub-threshold doses of diazepam and alcohol cause a recordable effect when combined (potentiation or synergy, or addition within the limits of a common dose–response relation).
- Chlorpromazine increased arterial alcohol concentrations (potentiation of absorption).
- Chlorpromazine enhanced the effects of alcohol (potentiation, or addition within the limits of a common dose–response relationship).
- Phenobarbital and alcohol given in combination at 50% of doses which individually had shown responses gave a much enhanced response (potentiation).

- An analysis of deaths involving alcohol and barbiturates showed that death from barbiturates alone occurred at a mean concentration of $3.67\,\mu g\,mL^{-1}$, death from alcohol alone occurred at $65\,g\,L^{-1}$, and death from the combination occurred at 70% of the blood barbiturate concentration $(2.55\,\mu g\,mL^{-1})$ plus 27% of the ethanol concentration $(17.5\,g\,L^{-1})$ which is very close to an additive response.

Thus, it seems that the interaction is basically additive, but this will show as enhancement of effect or addition of effect depending on the interplay of dose–response and time–action factors.

One particular technique especially applicable to homergic interactions is isobolography. An isobol is a line linking equipotent doses of two drugs, in a graph in which the two axes are doses of the two drugs involved in the interaction. Thus, it is necessary to define an 'end-point' or 'criterion-effect' such as ED_{50}, or minimum inhibitory concentration, in the case of antimicrobial agents. Figure 17.4(a) is an isobologram for the interaction of diazepam and alcohol, studied by means of a digit symbol substitution test. Based on preliminary information, the protocol included high and low doses of the two drugs individually, a combination dose that gave an effect between those of these high and low doses, and placebo. Two-point dose–response graphs were thus established for the two drugs alone. The combination dose established the criterion-effect for the isobologram. The doses of the individual drugs that were expected to induce the criterion effect were then calculated, and the three-point isobologram was constructed using these calculated doses and the combination dose. This line was straight, indicating addition of the contributions from the two drugs. A concave line would have indicated potentiation, while a convex line would have indicated antagonism [Figure 17.4(b).] More complex isobolograms have been constructed in situations where the pharmacology was more complicated, and more data were available.

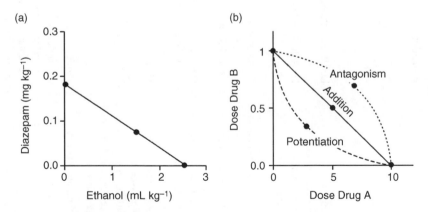

Figure 17.4   (a) Isobologram depicting effects of diazepam, ethanol and a combination of the two on digit substitution test (from Curry *et al.*, 1985). (b) Isobologram showing differentiation of potentiation, addition and antagonism for homergic drug interactions.

17.7   Further examples and mechanisms of a wide range of drug interactions

These are listed in detail in Table 17.3. The classification is by site of interaction and each interaction is considered in regard to whether it modifies the onset, intensity and/or duration of the effect being recorded. In some cases, no entry is made under these headings because of a lack of relevant information and the impossibility of making a useful prediction. The reader is recommended to refer to the time–action principles

Table 17.3   A table of drug interactions

Group	Site of interaction	Drug affecting the status quo (A)	Drug affected (B)	Overall effect of A on B	Modification of effect of B		
					Onset	Intensity	Duration
I	In pharmaceutical preparation of intravenous infusion fluids[a]	Tetracycline drugs	Penicillin drugs	Inactivated			
II	At the site of administration (a) Parenteral[b]	Adrenaline/other vasoconstrictors	Local anaesthetics	Potentiation			Prolonged
	(b) In the gastrointestinal tract[b]	Liquid paraffin (mineral oil)	Fat soluble vitamins	Reduced absorption	Delayed	Decreased	Shortened
		Liquid paraffin (mineral oil)	Fat soluble drugs	Reduced absorption	Delayed	Decreased	Shortened
		Arachis oil and dietary lipid	Griseofulvin	Accelerated absorption	Earlier	Increased	Shortened(?)
		Monoamine oxidase inhibitors (e.g. tranylcypromine)	Dietary amines (e.g. tyramine in cheese)	Increased absorption	Earlier	Increased	Prolonged
		Smooth muscle depressants (e.g. barbiturates, atropine-like drugs and opium alkaloids)	Many drugs	Decreased absorption	Delayed	Decreased	Shortened
		Calcium and/or iron	Tetracycline	Reduced absorption		Reduced	Shortened
III	Distribution/protein binding[c]	Salicylates	(a) Anticoagulants (warfarin and other coumarins)	Increase in clotting time	Earlier	Increased	Prolonged
			(b) Tolbutamide	Hypoglycaemia	Earlier	Increased	Prolonged

	Phenylbutazone	Warfarin	Increase in clotting time	Earlier	Increased	Prolonged
	Sulfonamides	(a) Tolbutamide	Hypoglycaemia			
		(b) Bilirubin in neonates	Kernicterus			
	Quinidine	Digoxin	Arrhythmias	Earlier	Increased	Prolonged
IV Drug metabolism[d]						
(a) Induction	Phenobarbital/rifampin	(a) Phenytoin	Reduced effect	Delayed	Decreased	Shortened
		(b) Warfarin	Reduced effect	Delayed	Decreased	Shortened
(b) Inhibition	Phenylbutazone	Tolbutamide	Hypoglycaemia	Earlier	Increased	Prolonged
	Anticholinesterases	Suxamethonium	Paralysis			
	Tranylcypromine	Pethidine and morphine	Convulsions	Earlier	Increased	Prolonged
	Cimetidine and ketoconazole	Many drugs (e.g. amitriptyline and terfenadine)	Enhanced effects	Earlier	Increased	Prolonged
V Drug excretion[e]	Probenecid	Penicillin	Prolonged effect			Prolonged
	Ammonium chloride	(a) Amphetamine	Reduced effect		Reduced	Shortened
		(b) Phenobarbital and salicylates	Increased effect		Increased	Prolonged
	Sodium bicarbonate	(a) Amphetamine	Increased effect		Increased	Prolonged
		(b) Phenobarbital and salicylates	Reduced effect		Reduced	Shortened
	Verapamil	Digoxin	Inhibits biliary excretion		Increased	
VI Pharmacological; classical drug receptor studies[f]	(a) Atropine	Cholinergic drugs	Inhibited		Abolished	
	(b) Anticholinesterases	Muscle relaxants (curare-type)	Reversal of effect		Abolished	
	(c) Amphetamine	Guanethidine	Reversal of hypotensive effect			
	(d) Tranylcypromine	Imipramine	Hypertension			
	(e) Desipramine	Guanethidine	Reversal of hypotensive effect			

(continued)

Table 17.3 (*Continued*)

Group	Site of interaction	Drug affecting the status quo (A)	Drug affected (B)	Overall effect of A on B	Modification of effect of B		
					Onset	Intensity	Duration
VII	Pharmacological: unclassified[g]	(a) Salicylate	Insulin	Hypoglycaemia		Increased	Prolonged
		(b) Thiazide diuretics	Digitalis glycosides	Cardiac effects increased		Increased	
		(c) Tranylcypromine	Amphetamine	Hypertension	Earlier	Increased	
		(d) Corticosteroids	Anaesthetics	Hypotension			
VIII	Pharmacological: CNS[h]	(a) Amphetamine	Barbiturates	Euphoria			
		(b) Ethanol	Many drugs with central depressant actions				

[a] Chemical mechanism of pharmaceutical incompatibility; interactions of this type can also occur in body fluids through inadvertent co-prescribing.

[b] (a) Adrenaline causes vasoconstriction, so the rate of removal from the site of administration (which is also the site of action) is reduced. (b) Vitamins do not have acute effects; prolonged interaction leads to vitamin deficiency. Griseofulvin is unusual in that it is absorbed intimately mixed with dietary lipid. Normally, tyramine is extensively metabolized by monoamine oxidase (MAO) in the intestinal lumen and mucosa and during first passage through the liver. Tranylcypromine inhibits MAO and, hence first-pass metabolism of tyramine which enters the general circulation and displaces noradrenaline from stores in adrenergic neurons. Divalent metal ions (Ca^{2+}, Mg^{2+}, Fe^{2+}) react with tetracyclines to form less easily absorbed complexes.

[c] Drugs A displace drugs B from binding sites on plasma protein. This increases the amount of unbound B available for action at receptors. This effect may be short-lived, as the increased amount of unbound B can diffuse to tissue binding sites. Thiazides increase protein binding of pempidine but the significance of this is uncertain.

[d] Barbiturates (phenobarbital is the one most studied) and many other drugs stimulate the liver microsomal enzymes to greater activity in metabolizing warfarin, phenytoin and a wide variety of other drugs. Interestingly, phenobarbital and phenytoin are still used in combination. Problems with warfarin arise when patients are taken off barbiturates and their enzymes return to normal. For cimetidine and ketoconazole details see Sections 17.4.2 and 17.6.2.

[e] Probenecid blocks penicillin excretion by active transport into the proximal tubule – this has been used to prolong the action of penicillin. Ammonium chloride reduces reabsorption of amphetamine and other amines in the renal tubule by reducing pH; it increases reabsorption of phenobarbital, salicylic acid, and other weak acids. Sodium bicarbonate causes effects opposite to those of ammonium chloride, by increasing urine pH.

[f] (a) Inhibition of acetylcholine at cholinergic receptors; (b) Accumulation of acetylcholine displaces competitive neuromuscular blocking drugs; (c) Guanethidine is displaced from tissue binding sites; (d) Inhibition of metabolism of noradrenaline by tranylcypromine plus inhibition of reuptake of noradrenaline into adrenergic nerve endings by imipramine leads to excessive effects of endogenous noradrenaline.

[g] Salicylates exert a hypoglycaemic effect by enhancing beta-cell sensitivity to glucose and potentiating insulin secretion.

[h] (a) Appears as a qualitatively different CNS effect, but it is probably the result of partial pharmacological inhibition of the effects of each drug. (b) Special problems arise when two drugs have similar actions; opinions differ concerning whether to describe this interaction as addition, potentiation, or synergy.

discussed at the beginning of Chapter 4 in his or her assessment of an interaction. Readers may also find it useful to refer to electronic databases of information on particular interactions.

17.8   When are drug interactions important?

There is an immense literature on drug interactions. Almost any combination of drugs can be employed to demonstrate an interaction provided enough ingenuity is used in designing the experiments concerned. Thus the literature is packed with reports that one drug affects the plasma levels of another in man, the metabolism of another in mouse liver homogenates, the concentrations of another in plasma, the action of another in isolated enzyme preparations, etc. Many of these interactions are benign. For example, it was shown many years ago that of a group of 237 warfarin-treated patients who received chloral hydrate only about 10% showed clear potentiation of the effects of warfarin. As already partly discussed in relation to enzyme induction in particular, in practice drug interactions seem to be important when:

* One or more of the drugs concerned has a low therapeutic index.
* The effects of two of the compounds involved are similar (in contrast with the situation of an interacting drug with its own action of a type totally different from that of the drug affected).
* The compounds are interacting to influence delicately balanced physiological control mechanisms, e.g. those concerned with the heart, maintenance of blood pressure, blood coagulation, etc.

Thus, the interactions most obviously affecting the prescription of drugs are those involving anticoagulants, cardioactive drugs, antihypertensive drugs, antidepressants, cold cures, local vasoconstrictors, older hypnotics, anticonvulsants and ethanol.

In addition, it should be realized that a number of clinical procedures apparently involve auto-adaptation to the presence of interacting drugs (e.g. the combined use of phenobarbital and phenytoin, and digoxin, diuretics and potassium supplements).

Borrowing terminology from classical pharmacology Rowland and Tozer (1995) have stressed the 'graded,' rather than 'quantal,' nature of enzyme induction, (and also the opposite, metabolic inhibition), changes in enzyme concentration that occur during induction, occur in proportion to the concentration of the inducing ligand, and over a period of time, thus rarely, if ever, creating an *acute* medical emergency. Rather, they cause a gradual change in drug response that can be recognized and controlled in the patient by appropriate dosage adjustment. However, it should be noted that there have been cases of acute haemorrhage after addition of phenylbutazone to a warfarin regimen.

17.9   Desirable drug–drug interactions

The term 'drug interaction' is often associated with alarm, as it engenders feelings related to potential lack of control over the effects of drugs in the body. There is some validity in this, as some drug interactions are, indeed, the cause of great difficulty in therapeutics, as we have observed in the previous sections. However, it should not be forgotten that some interactions can be turned to advantage, and far beyond the use of combination therapy. For example:

* The use of vasoconstrictors, both adrenaline and octapressin, in prolonging the effect of local anaesthetics thus allowing larger doses to be given safely.
* The use of carbidopa to enhance the effect of L-dopa, allowing lower doses and reduced unwanted effects
* The use of probenecid to prolong the effect of penicillin.
* The effect of cilastin to reduce the renal metabolism of imipenem, enhancing its duration of action and preventing formation of nephrotoxic metabolites.

17.10   Predicting the risk of future drug interactions with new chemical entities

With so many drugs in use, and with others in development, there is an obvious need for accurate and efficient evaluation of whether or not a new chemical entity (NCE) is likely to have significant effects on the actions of existing drugs when it is co-prescribed, and/or whether the NCE itself will cause unexpected effects in patients because of drug interactions. In recent years, several new drugs, as with terfenadine, have been withdrawn from the market because of drug interactions not anticipated before their introduction, discovered only through post-marketing surveillance, highlighting the risk of loss of the investment of time, money and hope represented by the NCE. It is not possible to conduct a comprehensive laboratory or clinical evaluation of this risk before approval for marketing, for reasons of both practicality and protection of the human subjects used in clinical pharmacological testing, yet the need is recognized by both research personnel and regulatory authorities. Three approaches to this problem exist: (i) *in vitro* experiments, (ii) use of drug mixtures and phenotyped subjects in Phase I studies conducted by clinical pharmacologists, and (iii) limited Phase III trials of relevant drug combinations.

17.10.1   *In vitro* prediction of drug interactions

It has been known since the 1980s that enzyme inhibitors, such as cimetidine, can be introduced into *in vitro* CYP 450 incubations, to determine K_i values, using conventional enzymology approaches. Also, using phenobarbital as an example, it is well-established that pretreatment of rats *in vivo* for as little as 2 days, with an enzyme-inducing agent, leads to an increase in microsomal intrinsic clearance of other compounds when measured *in vitro* using the livers from the pretreated animals. Some inducing agents take longer to exert an effect. Both inhibition and inducing techniques are widely used. The need is to utilize such information predicatively, as there is just as much risk of rejection of a useful NCE with a benign level of interaction risk, as there is of acceptance for further development of a NCE carrying with it serious future risk. Most of the current work in this regard is with the potential for enzyme inhibition, evaluated using K_i values.

Inhibition can be reversible or irreversible, competitive (competition for the active sites on the enzyme), non-competitive (such as induction of an allosteric change in the enzyme molecule reducing its activity), or 'un-competitive.' These three categories are differentiated by their different effects on the Km and V_{max} values, as determined using Lineweaver–Burk plots relating the rate of reaction (v) and substrate concentration (C). For competitive inhibition, the rate equation becomes:

$$v = \frac{V_{max}C}{Km(1 + I/K_i)} \tag{17.1}$$

in which I is the inhibitor concentration and K_i is the inhibition constant.

At one time the doctrine was simple. An inhibitor of microsomal P-450 was considered 'weak' if it showed K_i values against various substrates of more than $100 \, mol \, L^{-1}$, 'intermediate' if the K_i values were <100 but $>10 \, mol \, L^{-1}$, and 'potent' if the K_i values were $<10 \, mol \, L^{-1}$. It was considered in the 1990s that a K_i of $>10 \, mol \, L^{-1}$ was unlikely to lead to a clinically-important drug interaction. More recently, it has been recognized that several considerations other than *in vitro* potency must be taken into account, including:

- The concentration of inhibitors likely to occur *in vivo*.
- The concentration range of potential substrates (victim drugs) occurring *in vivo*.
- The therapeutic index of the victim drug.
- Whether the inhibitor and victim were likely to be co-prescribed.

As a result, *in vitro* experiments of this type have become more complex, with, for example:

- Recognition that inhibitor concentrations occurring at the active enzyme site should be incorporated into the calculations.
- Recognition that non-specific binding to the liver material can affect the active concentrations in the biophase.
- Particular focus on CYP2C9, CYP2D6 and CYP3A4 as these are especially important drug-metabolizing isoforms of P-450.
- Recognition that the fraction of the dose metabolized by the particular CYP isoform will affect the risk of the interaction studied being significant *in vivo*.
- Incorporation of the absorption rate constant of the inhibitor into risk assessment calculations because rate of absorption can affect whether or not the inhibitor and victim drugs are likely to interact with CYP3A4 at the same time.
- The average systemic plasma concentration of the inhibitor should be considered for interactions involving CYP2D6 in particular, incorporating evaluation of risk resulting from parallel drug elimination pathways for inhibitor and victim drug.

The result of this attention to detail in the prediction process has been a marked improvement in the *in vitro* predictability of CYP P-450 interactions in laboratory animals *in vivo* assessed using *AUC* values. In contrast, attempts to allow for potential protein binding and metabolism by intestinal mucosa in P-450 inhibition interactions have been less successful. Obviously, there can be no data on whether a NCE rejected in development would have caused interactions in patients, and there are no data at present on whether these more modern approaches are leading to drugs in clinical trials with lesser incidences of inhibition-based drug interactions. In reality, it is probably true to say that an inhibitor of CYP2D6 or CYP3A4 is unlikely to go forward into clinical trials, unless it is of such exceptional value that any risk of interactions would be seen as acceptable in relation to the potential clinical benefit.

An interesting case study in this context was presented by Chien and colleagues (2006) on the interaction between ketoconazole (the inhibitor) and midazolam (the victim drug), with focus on non-linear and non-steady state conditions such as first-pass effects and accumulation of inhibitor during multiple dosing regimens. This work used statistical probability, sensitivity analysis, and uncertainty concepts, and integrated kinetic models of the two drugs, and tested the *in vitro* prediction against five published *in vivo* studies. These investigators were able to:

- Reveal the optimal dose(s) and regimen for achieving inhibition (a regimen to avoid).
- Show that the most influential variable was the fractional clearance of midazolam by CYP3A4 – basically the proportion subjected to a first-pass effect.
- Demonstrate a saturable ketoconazole efflux process from the site of enzyme inhibition in the liver, as a key mechanistic component of the interaction.

17.10.2 In vivo *predictions in human subjects*

At one time, the half-life of antipyrine (phenazone) was commonly used to demonstrate enzyme induction in human subjects. Antipyrine is an obsolete analgesic drug that has rapid oral absorption, an oral systemic availability of close to unity, an apparent volume of distribution of total body water, and it is a substrate for human P-450. Thus changes in its half-life will be a result of changes in enzyme activity (clearance) rather than changes in volume of distribution. Test doses are safe, and it was popular to show that the antipyrine half-life became shorter in subjects pretreated with enzyme-inducing drugs such as phenobarbital.

There are numerous examples of drugs that have been tested with antipyrine, however dichloralphenazone, a drug product that is now virtually obsolete, is a curious, and probably unique, example to consider. This product combined the volatile compound chloral, which is still in use, with the (mostly) benign antipyrine, in a once popular sleeping tablet. Chloral is an enzyme inducer, and is metabolized to trichloroacetic acid, which displaces warfarin from its binding sites. Also, antipyrine, although pharmacologically of no significance in this product, induces P-450 activity after long-term treatment, shortening its own half-life. Thus dichloralphenazone, in one product, illustrated two major mechanisms of interaction, and provided a test method for demonstrating the consequences of its own properties!

Antipyrine was thus used as a non-specific test of P-450 activity. It has been superseded by compounds specific to particular isoforms. For example, a team of investigators at the Indiana University School of Medicine is credited with 'invention' of the 'Indiana Cocktail' comprising five test drug dosages that are notable test substrates for particular isozymes of cytochrome P-450 (Table 10.2). This has been used in the study of St. John's wort (Wang *et al.*, 2001) and clarithromycin (Bruce *et al.*, 2001) in particular. In the case of the clarithromycin study, the cocktail included tolbutamide, caffeine and dextromethorphan administered simultaneously. In the case of St. John's wort, it was shown that short-term administration had no effect on enzyme activities. Long-term administration caused a significant increase in the clearance of oral midazolam, and a lesser increase in the clearance of intravenous midazolam. The activities CYP1A2, CYP2C9 or CYP2D6 were unaffected. The clinical implications for management of patients using oral substrates of CYP3A4 are considerable.

Another clinical pharmacology approach to the study of specificity in drug interactions is the use of human subjects previously phenotyped for their levels of the various CYP isozymes. Thus, for an evaluation of a drug for its ability to inhibit CYP2D6 in particular is sought, then subjects with known CYP2D6 hepatic activity can be used, ensuring optimal experimental design. The human subjects with such a level of hepatic calibration would seem to possess a priceless commodity!

17.10.3 Innovative phase III clinical trial designs

Protocols that allow a drug to be tested as both inhibitor and inducer, and as both interacting drug and victim drug, in complementary investigations have been designed to produce comprehensive data for the purpose of risk assessment during clinical development of NCEs (Fox and van Troostenburg de Bruyn, 2007). One such design is shown in Figure 17.5. Human microdosing may also have a role in this work, eventually displacing

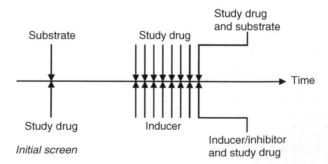

Figure 17.5 Two typical designs for a Phase I drug interaction trial. The horizontal line represents time, over several days. Above the line the study drug is being tested for any interaction, e.g. enzyme inhibition or induction, with a known isoenzyme substrate. Below the line the test drug is being examined for susceptibility to inhibition or induction by some other drug; note that in this protocol the roles of the drug inducing the effect, and the victim drug, are reversed. (Redrawn from Fox & van Troostenburg de Bruyn , 2007.)

the need for pharmaceutical innovators to conduct as many as 5–10 full-scale drug interaction clinical trials in the target population during Phase III development. These trials, historically, have focused on the drugs most likely to be co-prescribed in the patients with the indication to be treated with the NCE, and so have little value in predicting the risk of interaction with the thousands of other interacting drugs that might emerge after the drug has been approved for marketing.

References and further reading

Bruce MA, Hall SD, Haehner-Daniels BD, Gorski JC. In vivo effect of clarithromycin on multiple cytochrome P450s. *Drug Metab Dispos* 2001; 29: 1023–8.

Burk O, Koch I, Raucy J, Hustert E, Eichelbaum M, Brockmoller J, Zanger UM, Wojnowski L. The induction of cytochrome P450 3A5 (CYP3A5) in the human liver and intestine is mediated by the xenobiotic sensors pregnane X receptor (PXR) and constitutively activated receptor (CAR). *J Biol Chem* 2004; 279: 38379–85.

Chien JY, Lucksiri A, Ernest CS, 2nd, Gorski JC, Wrighton SA, Hall SD. Stochastic prediction of CYP3A-mediated inhibition of midazolam clearance by ketoconazole. *Drug Metab Dispos* 2006; 34: 1208–19.

Curry SH, DeVane CL, Wolfe MM. Cimetidine interaction with amitriptyline. *Eur J Clin Pharmacol* 1985; 29: 429–33.

Curry SH, Smith CM. Diazepam-ethanol interaction in humans: addition or potentiation? *Commun Psychopharmacol* 1979; 3: 101–13.

Curry ML, Curry SH, Marroum P. Interaction of phenothiazine and related drugs and caffeinated beverages. *DICP The Annals of Pharmacotherapy* 1991; 25: 437–8.

Fox AW, van Troostenburg de Bruyn A-R. Drug interactions. In: Edwards LD, Fletcher AJ, Fox AW, and Stonier PD, editors. *Principles and Practice of Pharmaceutical Medicine*, 2nd edn. Chichester: John Wiley & Sons, 2007.

Gibson GG, Plant NJ, Swales KE, Ayrton A, El-Sankary W. Receptor-dependent transcriptional activation of cytochrome P4503A genes: induction mechanisms, species differences and interindividual variation in man. *Xenobiotica* 2002; 32: 165–206.

Hussar DA. Drug interactions. In: Remington G, editor. *The Science and Practice of Pharmacy*. Philadelphia: Lippincott, Williams & Wilkins, 2006: 1889–1902.

Lewis RJ, Trager WF, Chan KK, Breckenridge A, Orme M, Roland M, Schary W, Warfarin. Stereochemical aspects of its metabolism and the interaction with phenylbutazone. *J Clin Invest* 1974; 53: 1607–17.

Rowland M, and Tozer TN. *Clinical Pharmacokinetics: Concepts and Applications*. 3rd. edn. (1995. Media, Pennsylvania, Lippincott, Williams & Wilkins.

Walsky RL, Boldt SE. In vitro cytochrome P450 inhibition and induction. *Curr Drug Metab* 2008; 9: 928–39.

Wang Z, Gorski JC, Hamman MA, Huang SM, Lesko LJ, Hall SD. The effects of St John's wort (*Hypericum perforatum*) on human cytochrome P450 activity. *Clin Pharmacol Ther* 2001; 70: 317–26.

18

Drug Metabolism and Pharmacokinetics in Toxicology

18.1 Introduction

The study of the drug metabolism and pharmacokinetic (DMPK) properties of xenobiotics provides critical support to the science of toxicology, from the point of view of (i) evaluation of exposure, (ii) determination of mechanisms, and (iii) guidance on treatment. At one time, in industrial drug discovery, DMPK studies were conducted at the same time as safety evaluation. Nowadays, they are a component of the new drug discovery process itself, conducted in parallel with preclinical chemical and biological studies. 'Toxicokinetics' is the study of pharmacokinetics in the context of exposure to doses associated with unwanted effects, in both animals and human beings. This includes the design, conduct and interpretation of safety evaluation protocols, and also validation of dosing in such studies. Toxicokinetics also plays a part in the process of extrapolation of data from animals to humans. 'Clinical toxicology' is the study of xenobiotic toxicity in human beings, especially patients, but also non-patient populations, exposed to therapeutic agents and environmental toxicants, at trace, therapeutic and excessive exposures. 'Analytical toxicology' focuses on methods of measurement of toxicants in biological material.

18.2 Terminology

An 'unwanted' drug effect is any effect occurring in the wrong circumstances. It can be an extension of the normal response, or something completely different. An unwanted effect can occur at the same or a higher dose than that inducing the wanted effect, or even, in special circumstances, at a lower dose. Almost any pharmacological mechanism can be involved in an unwanted effect, including that of the wanted effect. Thus, for example, unwanted effects can be ascribed to unusual drug metabolic routes and/or pharmacokinetics, unusual responses mediated through the nervous or endocrine systems, and many other events. However, there are some mechanisms that are exclusively associated with unwanted effects, and these are the subject of this chapter, illustrated by a number of notable examples. The objective of this chapter is to introduce the reader to some of the more important concepts of toxicology, as they relate to the DMPK properties of toxic molecules. It should be appreciated that all unwanted effects are related in some way to exposure, controlled by dosage and pharmacokinetic influences.

Various terms are used in relation to toxicology. 'Unwanted effect' is deliberately non-specific, but does allude to the fact that drugs have 'effects', which may be 'wanted' or 'unwanted'. Furthermore, an unwanted effect in one situation may be wanted in another. So, for example, constipation arising from the use of

Drug Disposition and Pharmacokinetics,   By Stephen H. Curry and Robin Whelpton
© 2011 John Wiley & Sons, Ltd

morphine postoperatively to treat pain is an unwanted effect, particularly after bowel surgery but the same effect is a wanted effect of morphine in antidiarrhoeal preparations. The term 'side effect' is similarly all inclusive, and frequently used, but it has the disadvantage that it suggests that effects and side effects are different and can be separated – this is often not the case. The term 'adverse reaction' has a connotation of greater severity than does unwanted effect, but that is not always intended when this term is used. 'Toxicity', similarly and often, but not always, implies a greater problem than does side effect or unwanted effect. An idiosyncratic unwanted effect is usually thought of as occurring in a definable subgroup of the population. The non-specific term 'secondary effect' is also sometimes found in scientific literature.

There have been many attempts to classify unwanted drug effects, based variously on mechanisms, clinical manifestations and other approaches. These have at times led to arguments about terminology and meanings of words that have not been helpful to scientific understanding. One relatively simple and useful classification is as follows:

1. *Pharmacokinetic unwanted effects.* Effects arising from the presence of excessive concentrations of molecules that exert only desirable effects in normal quantities. This group makes allowance for overdose, be it deliberate or accidental, acute or cumulative, absolute or relative, and resulting from anything to do with DMPK that can be described as idiosyncrasy or intolerance.
2. *Pharmacodynamic unwanted effects.* Effects occurring by pharmacodynamic mechanisms at normal clinical dosage, acutely or after prolonged exposure. This group allows for anything describable as a side effect, a secondary effect, or an idiosyncrasy or intolerance where physiological or biochemical sensitivity is a factor.
3. *Drug allergy.* Hypersensitivity with an immunological basis.
4. *Toxic drug reactions.* Effects caused basically by means of covalent chemical reaction between pharmacologically active foreign molecules and biologically important materials. This includes blood dyscrasias and tissue necrosis.
5. *Drug interactions.*

In this classification, groups 1 and 2 are considered to be reversible, in the sense that conventional dose–response relationships apply to both wanted and unwanted effects, so that reduction in concentrations consequent on removal of the drug and/or its metabolites from the body leads to the dissipation of the effects. Numbers 3 and 4 are thought of as irreversible, because, although they can clearly be reversed by a variety of medical interventions, and by the normal processes of body healing, covalent chemical reactions are key to their origins, and removal from the body of the offending molecules and the molecules with which they have reacted is presumed to be needed. Group 5 was the subject of Chapter 17. It should be appreciated that the clinical outcome with any particular drug can invoke a combination of these processes.

18.3 Dose–response and time–action with special reference to toxicology

Interpretation of toxicity studies is conducted, in part, in ways already discussed in regard to studies of the desired effects of drugs. Thus appropriate recognition is always given to dose–response relationships, and the growth and decay curves underlying the time–action relationships at the core of pharmacological science. So, the time course and dose relationships discussed in the opening paragraphs of Chapter 4, and in later chapters, are crucial. Similarly, pharmaceutical factors, special population factors, and disease factors potentially apply equally in toxicology as in pharmacology. There are, however, some variations on this theme in toxicology. For example, it is not uncommon for unwanted drug effects to arise only in subgroups of the treated population, and the characteristics of those subgroups become very important. Also, toxicity is often discovered only through epidemiological studies. For example, the most dramatic and arguably most tragic drug toxicology disaster involved exposure of foetuses to thalidomide given to their mothers. Its occurrence involved disposition mechanisms such as placental transfer hitherto ignored in new drug discovery. Its exemplification was a rare congenital malformation already known to the medical professions, then linked to thalidomide when its incidence increased dramatically in the new-born children of the treated mothers. Similarly, the cardiac toxicity of the COX-2 inhibitors (anti-inflammatory analgesics) only emerged after patients in vastly larger numbers than those involved in the clinical trials had been prescribed the drugs.

18.4 Safety studies in new drug discovery

Safety studies in laboratory animals are a critical component of new drug discovery and preparation of new chemical entities for initial human exposure. These studies use various designs, including single dose with recovery, and two-week, one-month, three-month and six-month exposure regimens followed by autopsy. They commonly utilize one rodent (e.g. rat) and one non-rodent (e.g. dog) species in each project. In order to meet the requirements of good laboratory practice (GLP), exposure must be confirmed using a validated method of analysis specific for the compound under investigation. Generally, samples for toxicokinetic analysis are taken from a subgroup of animals from within the treatment group, but a parallel group is permitted if it is considered that sampling from the test group would unduly influence the outcome of the safety evaluation. Commonly, plasma samples for toxicokinetics are taken at the beginning of the dosing time, such as after the first dose, and after the last dose. The regulatory objective is solely to prove that there was exposure to the drug, but these studies commonly provide the first opportunity to investigate doses other than the single doses typically used in pre-clinical studies, and useful multiple-dose pharmacokinetic data can be obtained. For example, toxicokinetics can detect whether the long-term concentrations predicted from the single doses are reached, and whether there are any differences in metabolic handling as the result of long term dosing. Toxicokinetics can also test whether long term effects are more closely related to C_{max} concentrations or to overall exposure, and whether there is evidence of tissue accumulation.

Safety pharmacology is often restricted to single doses, and involves the study of general organ systems, such as heart, kidney and lung function, regardless of whether the drug is designed to have an effect on one of these organs. Because safety pharmacology is generally conducted to GLP, toxicokinetic studies are required. Doses are increased beyond those required to demonstrate the desired effect, so toxicokinetics can provide the opportunity to check for deviations from linearity in the pharmacokinetic properties of the compound in question. However, LD_{50} (the dose that kills 50% of a population of laboratory animals) values are no longer required or scientifically justifiable – modern emphasis is on such parameters as the maximum repeatable dose (MRD). Determination of the MRD is an iterative process. Laboratory animals are first given increasing doses up to a point where unwanted effects are seen. Dosing then continues at a level just below this toxic dose, for a period of time sufficient to show that it can be repeated without induction of toxicity.

18.5 Examples

18.5.1 Isoniazid

Primary amines and hydrazines undergo condensation reactions with aldehydes to form imines (Schiff's bases):

$$R\text{-}CHO + R'NH_2 \rightarrow R\text{-}CH{=}N\text{-}R' + H_2O$$

An analogous reaction occurs between isoniazid and pyridoxal and/or pyridoxal phosphate, the active form of vitamin B6 (Figure 18.1). Pyridoxal phosphate is an important cofactor for the synthesis of monoamine neurotransmitters and a reduction in its availability can produces a peripheral neuropathy and central nervous

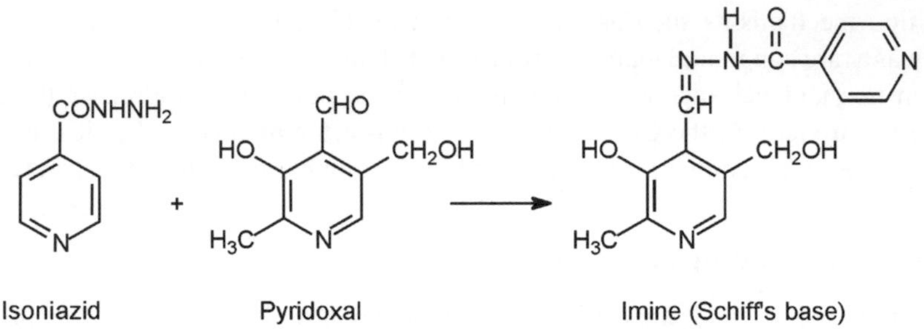

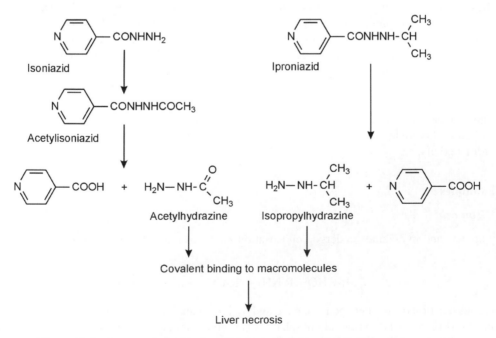

Figure 18.1   Condensation of pyridoxal with isoniazid.

system (CNS) effects. The neuropathy can be reversed by removal of the isoniazid, administration of pyridoxine, or both. It occurs most commonly in slow-acetylators (Section 10.3.1). These patients have unusually high plasma concentrations of isoniazid when treated with conventional doses. It has been suggested that the response of tuberculosis patients to isoniazid is associated more with the magnitude of peak concentrations in plasma than with overall exposure during daily dosing. The drug is activated by bacterial catalase which results in inhibition of the synthesis of mycolic acid required for mycobacterial cell walls.

Isoniazid is also hepatotoxic in certain individuals, shown by raised serum liver enzyme and bilirubin concentrations. In this case the toxicity is more prevalent in fast acetylators. The chemically-related monoamine oxidase inhibitor, iproniazid, causes similar toxicity, although this drug has a substituent on its terminal nitrogen atom, preventing metabolic acetylation (Figure 18.2). Phenobarbital pretreatment leads to increased incidence of this liver toxicity in rats treated with either isoniazid or iproniazid, and this toxicity can be prevented by treatment with inhibitors of drug metabolism. When [^{14}C-acetyl]-labelled acetylisoniazid was administered to rats, a large amount of covalently bound radioactivity was

Figure 18.2   Proposed mechanism of hepatotoxicity of isoniazid and iproniazid.

found in the liver. No covalent binding of radioactivity resulted when the acetylisoniazid was ring-labelled. The conclusion was drawn that hepatic necrosis results in this case from cleavage of the substituted compounds, acetylisoniazid and iproniazid, producing acetylhydrazine and isopropylhydrazine respectively. These two products are highly toxic. Metabolic activation, induced by phenobarbital, and leading to reactive species, is needed to facilitate their toxicity by means of covalent binding (Figure 18.2).

Interestingly, isoniazid demonstrates most of the general ideas involved in drug toxicity, in that there is a basis for the use of the various terms including unwanted effect and adverse reaction: the toxicity occurs in a subgroup of the population thereby qualifying to be described as idiosyncratic; there is a pharmacokinetic influence under genetic control; there are drug metabolism sequences involved in its toxicity; there is a plausible physiological/biochemical explanation for the toxicity providing a basis for methods of successful treatment of the toxicity; and covalent chemical reactions are involved. Isoniazid therefore invokes mechanisms 1 and 4 of the classification listed above (Section 18.2).

18.5.2 Paracetamol and phenacetin

The story of these two drugs and related toxic chemicals epitomizes just about the entire scope of drug toxicity studies. Phenacetin (acetophenetidin) had been in use for some 50 years before Brodie and Axelrod discovered a metabolite, paracetamol (acetaminophen), of even greater potential value as an analgaesic. Phenacetin was notorious for its propensity to cause renal damage, and paracetamol provided hope that there could be a related compound that did not cause this toxicity. Paracetamol has since become the most widely used pain killer worldwide, and also, coincidentally, the most important single cause of fulminant liver, but not renal, failure in Western countries. These two compounds, together with related molecules, provide a fascinating case study in mechanisms of toxicity, the role of metabolic and other pharmacokinetic processes, and the application of plasma concentration measurement as an aid to the diagnosis and treatment of drug toxicity.

Phenacetin is chemically related to acetanilide, which is metabolized to aniline and then to phenylhydroxylamine (Figure 18.3). *In vivo*, acetanilide causes haemolysis and methaemoglobinaemia. Phenylhydroxylamine, is highly toxic if administered directly.

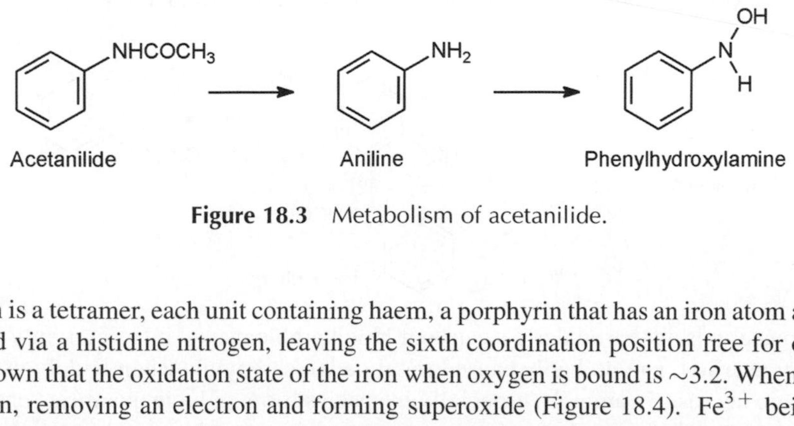

Figure 18.3 Metabolism of acetanilide.

Haemoglobin is a tetramer, each unit containing haem, a porphyrin that has an iron atom at its centre. The protein is bound via a histidine nitrogen, leaving the sixth coordination position free for oxygen binding. Studies have shown that the oxidation state of the iron when oxygen is bound is ~3.2. When oxygen binds it oxidizes the iron, removing an electron and forming superoxide (Figure 18.4). Fe^{3+} being smaller than Fe^{2+} moves closer to the centre of the porphyrin ring causing the allosteric change that has been observed. This results in co-operative binding of O_2 to the other three haem groups. Because molecular oxygen can only bind to haemoglobin, the Fe^{3+} in methaemoglobin must be reduced back to Fe^{2+}. Several enzymes are involved, the chief one being cytochrome b_5-reductase [reduced nicotinamide adenine dinucleotide

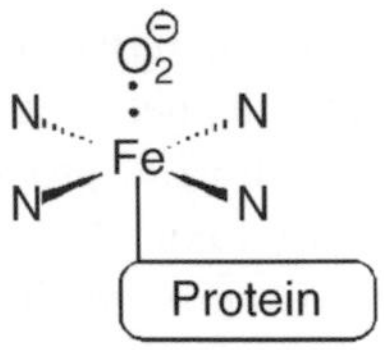

Figure 18.4   Simple representation of oxyhaemoglobin. The porphyrin nitrogens are depicted and the globular protein binds via a histidine nitrogen. Oxyhaemoglobin is diamagnetic, i.e. there are no unpaired electrons, so it is postulated that one electron from iron (giving Fe^{3+}) is paired with one from oxygen, giving superoxide.

(NADH), others being NADPH-methaemoglobin reductase and the involvement of ascorbate and reduced glutathione (GSH).

Thus, oxidizing agents including oxidized metabolites, for example phenylhydroxylamine which can be oxidized to a nitroso derivative and reduced back to aniline, can promote the formation of methaemoglobin from haemoglobin, either by direct oxidation or by depleting the cofactors required for the reduction pathways referred to above.

Haemolysis caused by aromatic amines, such as aniline, has been traced to the same chemistry, but with variations. The demonstration that haemolysis only occurs *in vivo* provides evidence that metabolic activation is needed. The *N*-acetylated precursors of amines (e.g. acetanilide) cause haemolysis to a lesser degree than do their respective amines, and this toxicity is prevented by inhibitors of deacetylation. These inhibitors have no effect on amine toxicity as such. Thus it is apparent that the anilides require metabolic transformation in two steps in order to cause haemolysis, the two steps being deacetylation followed by *N*-hydroxylation (a *C*-hydroxylation analogue is possible). In spite of this, the two toxicities, haemolysis and methaemoblobinaemia are differently affected by phenobarbital pretreatment, which increases methaemoglobinaemia but decreases haemolysis. Thus, although the same basic chemistry is involved in the two reactions, it seems likely that different active intermediates are involved. In fact, the existence of a reactive intermediate in the sense of a free radical alkylating agent is not needed to explain the methaemoblobinaemia, whereas such an intermediate is postulated for the haemolysis reaction.

The metabolic reactions of phenacetin can include hydroxylamine production.

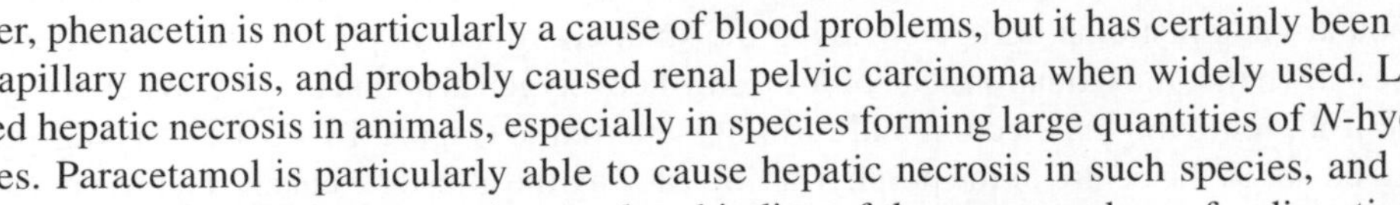

However, phenacetin is not particularly a cause of blood problems, but it has certainly been implicated in renal papillary necrosis, and probably caused renal pelvic carcinoma when widely used. Large doses also caused hepatic necrosis in animals, especially in species forming large quantities of *N*-hydroxylated metabolites. Paracetamol is particularly able to cause hepatic necrosis in such species, and in humans taking acute high doses. With phenacetin, covalent binding of the compound, or of radioactivity derived from it paralleled hepatic necrosis. This necrosis is associated with GSH depletion, and when caused by paracetamol in humans it can be successfully treated with glutathione precursors, or SH-donors, including *N*-acetylcysteine, L-cysteine, cysteamine, L-methionine, D-penicillamine, and dimercaprol. *N*-Acetylcysteine has emerged as the treatment of choice. Glutathione itself fails to penetrate hepatocytes. These compounds do not affect paracetamol metabolism as such. With phenacetin, the intact molecule is involved in the covalent binding, shown by the use of ring-labelled and acetyl-labelled derivatives which bind equally. However, elucidating a definitive mechanism is complicated by the fact these drugs may be metabolically deacetylated and reacetylated (Nicholls *et al.*, 2006).

In the search for the understanding of mechanisms, it was originally suggested that an additional hydroxylated metabolite of phenacetin (*N*-hydroxy-4-ethoxyacetanilide) might be involved, first losing its ethyl group non-enzymatically, forming a reactive imidoquinone [*N*-acetyl-4-benzoimidoquinone or *N*-acetyl-*p*-benzoquinone imine (NAPQI)]. This product is now known to be formed from paracetamol, without the requirement for the desethylation step that is needed with phenacetin. The imidoquinone can undergo nucleophilic addition at the 3-position of its benzene ring, reacting with functional –SH groups in cells. This explains the finding that the glutathione conjugate of phenacetin appears in the urine as a mercapturic acid conjugate of a 3-acetyl-cysteine-4-hydroxy derivative (see Figure 3.16). The reactive imidoquinone is capable of mediating both liver and kidney toxicity, but only if it is formed in quantities great enough to exceed the glutathione availability, as the toxic metabolite only reacts with key tissue constituents once the glutathione pool has been depleted by ∼30%. The formation of a mercapturic acid conjugate shows that at some stage the unstable arylating intermediate acts as an electrophilic reactant, having a fractional positive charge (indicated in Figure 18.5 as δ^+). This will react with nucleophilic cell macromolecules.

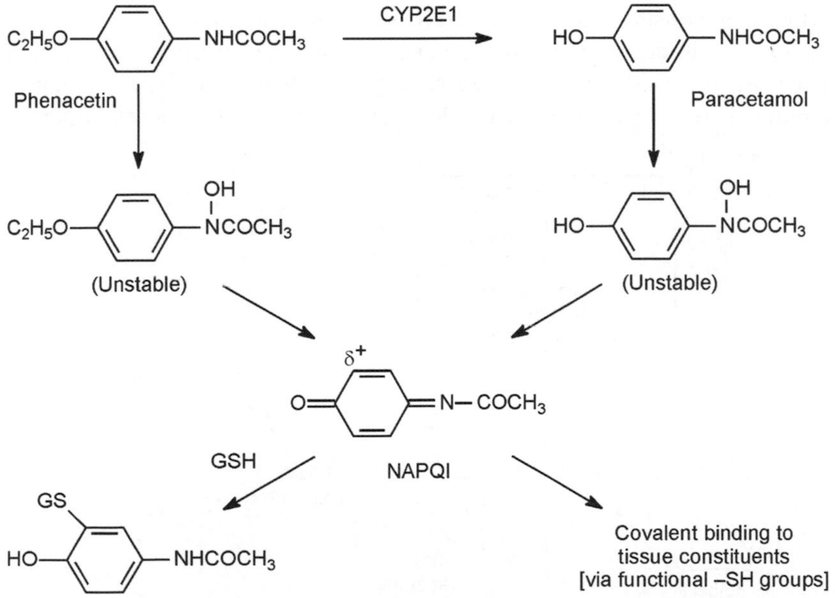

Figure 18.5 Proposed scheme for production of toxic metabolites of phenacetin and paracetamol.

The question arises as to why phenacetin and paracetamol, with such similar chemistry, show different patterns of toxicity. They differ in polarity of course, and therefore in affinity as substrates for P-450, which is responsible for *N*-hydroxylation and presumably for conversion of phenacetin to paracetamol. The *N*-oxidation can occur in both liver and kidney, and can be induced by phenobarbital pretreatment, and reduced by inhibitors of drug metabolism. It may occur to an increased proportional extent in one tissue at high doses, or after prolonged exposure, the former affecting paracetamol, the latter phenacetin. So the quantities and location of the active material can easily differ between the two compounds. It has been suggested that paracetamol fails to produce acute renal effects because it is not metabolized by kidney cells at a rate sufficient to exceed the glutathione capacity for conjugation. Because the formation of the reactive product from the *N*-oxides is promoted by acidic conditions, which occur more readily in the kidney, this could affect phenacetin more than paracetamol in some way. It has also been suggested that the *N*-hydroxylated phenacetin but not the *N*-hydroxylated paracetamol might be sufficiently stable after formation in the liver to reach the kidney in sufficient quantities to be toxic. Finally, phenacetin might affect the renal papilla and cause pelvic carcinoma, but not affect the proximal tubule, because of some drug concentration effect, or some difference in local concentrations of P-450. It is remarkable that two compounds could be so similar yet so different, and that one should have been rejected clinically and the other adopted with such enthusiasm as a substitute.

It is now established that the initial step in paracetamol toxicity is metabolism to NAPQI, which leads to depletion of GSH and covalent adduct formation of 'acetaminophen-cysteine adducts'. Immunochemical studies have shown that the cellular sites of covalent binding correlate with the toxicity. It has been shown that nitrated tyrosine is formed in hepatic centrilobular cells, and these adducts colocalize in cells containing the acetaminophen-cysteine adducts. Peroxynitrite, which is a highly reactive nitrating and oxidizing species formed by rapid reaction of nitric oxide and superoxide, produces nitrated tyrosine. Normally, GSH detoxifies peroxynitrite, but after GSH depletion by NAPQI, peroxynitrite nitrates protein tyrosine. It has also been hypothesized that lipid peroxidation may play a role. Thus nitration of tyrosine correlates with necrosis (Figure 18.6).

Paracetamol is rapidly absorbed from the small intestine, with its $C_{\max}$ occurring within 1–2 hours of tablet dosing. Normal $C_{\max}$ values are around $20\,\mathrm{mg\,L^{-1}}$. There is 20% presystemic metabolism by sulfation in the gut wall. Further metabolism occurs in the liver and the half-life in plasma is 1.5–3 hours. About 90% is metabolized to inactive sulfate and glucuronide conjugates that are excreted in urine, but the remainder is metabolized to the highly reactive intermediate NAPQI, which is immediately bound by intracellular glutathione and eliminated as mercapturic acid adducts in the urine. High doses of paracetamol produce

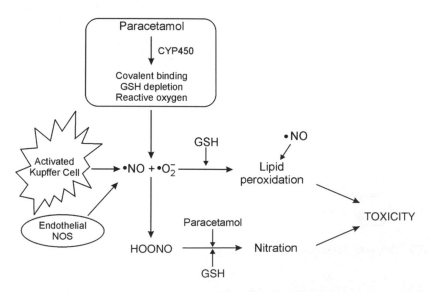

Figure 18.6 Postulated mechanisms of paracetamol-induced hepatotoxicity. Hepatocytes, Kupffer cells and endothelial cells participate in production of reactive nitrogen and oxygen species. The relative levels of nitric oxide ($\cdot$NO) and superoxide ($\cdot O^-_2$) determine whether mechanism of hepatic necrosis is dependent on protein nitrosylation or lipid peroxidation. (Redrawn from Jaeschke *et al.*, 2002.)

greater quantities of NAPQI and cause toxic depletion of glutathione stores, and also binds to other proteins causing damage to liver cells. The hepatotoxic dose is generally thought to be about 150 mg kg^{-1} but there is considerable variation between individuals.

The treatment of paracetamol overdose is guided either by dose consumed (when ingestion is staggered over a period of time) or by plasma concentrations of the drug (when the dose was in one single event at an identifiable time). The plasma concentration approach uses various nomograms linking points on a semilogarithmic graph of concentration between four hours post-overdose to 11–20 hours later. The early versions of these nomograms were devised by Prescott (see Prescott, 1996) and Rumack and Matthew (1975). Thus, if the plasma concentration is less than a point on a line between 200 mg L^{-1} at the 4 hour point and 30 mg L^{-1} at the 15 hour point, little more than supportive treatment is needed. This can include the use of ipecacuanha to induce vomiting, gastric washout (lavage) of unabsorbed drug, and activated charcoal to reduce any further absorption, but note that charcoal will inhibit the absorption of orally administered antidotes such as methionine. The nomogram should not be confused with the decline of the plasma concentration of the paracetamol – it reflects experience with various approaches to treatment and the probability of significant quantities of the toxic metabolite being present, and the fact that the need for the antidote relates to the dose consumed and the time since consumption. This nomogram line has been revised lower over the years, with one group of authors favouring use of a 24-hour data point, and others favouring a 25% reduction in the threshold concentrations to comply with US Food and Drug Administration guidelines designed to provide a better safety margin, albeit increasing the use of *N*-acetylcysteine which is not without complications. A further lowering is favoured by some physicians in vulnerable ('high-risk') populations such as those on enzyme inducing drugs, chronic alcohol abusers, fasting patients, and dehydrated patients, who, at least in some cases, are likely to produce greater proportional conversion to the toxic product.

Figure 18.7 illustrates the clinical approach to paracetamol poisoning. This figure relates plasma paracetamol concentration at diagnosis to time post-ingestion. It shows the three nomogram lines, normal, US and high-risk, and the time course of plasma paracetamol in three different reports. In all three cases, proactive treatment with *N*-acetylcysteine was warranted because the concentrations at one time or

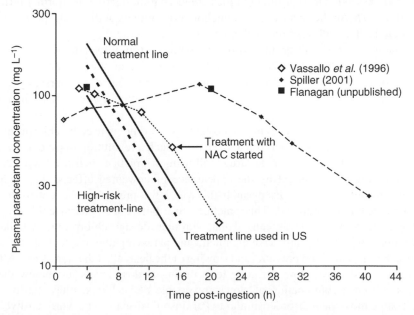

Figure 18.7 Paracetamol concentrations after paracetamol self-poisoning. (From Flanagan *et al.*, 2008.)

another were to the right of the nomograms. Note that in the 'Spiller' case, absorption of the paracetamol was delayed such that a single early plasma paracetamol concentration would have indicated no need for antidote dosing, and even in the other two cases the initial plasma concentration could have been misleading.

Finally, perhaps the single biggest risk factor is that there are patients treated with combination products of paracetamol plus narcotic analgaesics (such as 'Darvocet') who then add over the counter paracetamol to their dosing regimen without realizing that they are exposing themselves to potentially lethal doses of the paracetamol in the process.

18.5.3   *Toxicity associated with prolonged exposure to therapeutic doses*

Some of the most serious toxic reactions are associated with the normal pharmacological properties of drugs, experienced over a prolonged period of time, presumably at steady-state concentrations, and possibly involving long-term exposure to metabolites of the drugs concerned. No particular pharmacokinetic observations have been made in association with such problems, except perhaps, the general guideline that patients should receive the lowest possible dosage over the shortest possible time consistent with a useful therapeutic effect.

One such problem is 'serotonin syndrome'. This is associated with prolonged use of selective serotonin reuptake inhibitors (SSRIs) and, to lesser extent, with tricyclic antidepressants (TCAs) which increase the concentrations of serotonin in the brain. The clinical picture involves confusion, autonomic instability, and neuromuscular abnormalities. It is most often seen in patients taking two or more drugs at therapeutic doses that increase CNS serotoninergic activity by different mechanisms, such as monoamine oxidase inhibitors or pethidine, with SSRIs or TCAs. The combination of monoamine oxidase inhibitors and SSRIs or TCAs is discouraged but not unknown. The serotonin syndrome can be precipitated by single doses, but usually at single doses above the therapeutic range. Additionally, at least with amitriptyline, there can be a life-threatening combination of hypotension and heart block (produced by peripheral anti-adrenergic effects on the heart and blood vessels), parasympathetic block (rendering ineffective the normal reflex responses to reduction in heart rate, and/or the occurrence of anticholinergic effects of the amitryptiline on the heart) and CNS depression associated with respiratory depression.

A complex situation occurs with prolonged opioid exposure (opioid syndrome), in which the clinical manifestations are again bradycardia, CNS depression, hypotension and respiratory depression, plus, with some opioids, miosis and decreased gastrointestinal motility, complicated, of course, by addiction. Again, acute overdose can be life-threatening, more, in this case, from respiratory depression than from peripheral nervous system effects.

One of the oldest of these syndromes is the 'neuroleptic syndrome' caused by chlorpromazine and other phenothiazine antipsychotic drugs. Chlorpromazine caused such a revolution in psychiatry that prescribers became overenthusiastic in their use of the drug, following a practice then used with digoxin (Section 14.2.6.2), in first dosing to a limit set by the appearance of unwanted effects, then reducing the dose sufficiently for the unwanted effects to disappear. In this way, it was presumed, the probability of trouble-free therapy being achieved was optimized. This practice arose before any pharmacokinetic data had been obtained with these drugs. In time, it was realized that, with acute dosage, chlorpromazine, like amitriptyline, can cause nerve block, with CNS, cardiac, sympathetic and parasympathetic effects combining to produce life-threatening sedation, low blood pressure, and slowing of the heart all at the same time. After acute doses, recovery is complete and rapid. However, long term dosing at dose levels just below the threshold for unwanted effects caused a combination of slowly reversible nerve block, plus, 'tardive dyskinesia', a seemingly irreversible movement disorder similar to Parkinson's disease. The word 'tardive' indicates 'late in onset'. This syndrome was at first thought to be a characteristic of the schizophrenia that was being treated, and in a number of cases this led to an *increase* in dosing. When it became possible to measure chlorpromazine plasma concentrations it was found that reducing the dose and thus plasma concentrations, to the minimum consistent with useful therapeutic effects, the risk of developing of neuroleptic syndrome and tardive dyskinesia was greatly reduced. This work did not lead to monitoring of plasma concentrations in the way now common with drugs with narrow therapeutic indices (Chapter 19), but it did have a considerable impact on the treatment of schizophrenia, which evolved from being a long-term problem treated with high doses of drugs over periods of years in psychiatric hospitals, to an acute condition treated with smaller doses in the context of outpatient care.

As already stated, there is no unifying pharmacokinetic theory for these syndromes, characterized by the occurrence of a combination of normal and abnormal pharmacology with, mostly, long-term exposure to drugs and their metabolites. All can involve 'normal' doses, and/or both acute and chronic overdosing, especially, when long-term, possibly with the metabolites of the drugs involved. Treatment will almost always involve sampling for identification and quantitative assessment of the concentrations of the causative agents in the blood. Treatment of acute toxicity will involve reduction in exposure, by means of gastric lavage, dosing with activated charcoal to adsorb unabsorbed drug, and observation while the normal elimination processes take effect. Sometimes proactive means such as adjustment of urinary pH, or plasmapheresis, will be used to reduce body content. Thus, while drug concentrations will be especially useful in diagnosis of problems in asymptomatic patients at risk, it is rare for drug concentrations to dictate patient management decisions.

18.5.4 *Salicylate*

Poisoning by salicylate and other non-steroidal anti-inflammatory drugs is a common problem in emergency medicine. This syndrome is characterized by haematemesis, tachypnoea, hyperpnoea, dyspnoea, tinnitus, deafness, lethargy, seizures and confusion. Clinical toxicity is primarily a function of the degree and duration of acid–base disturbance, resulting from the depression of respiratory centre and consequent carbon dioxide exchange. Aspirin toxicity can result from both acute and chronic overexposure, and the pharmacology of the toxicity can be related to or unrelated to the mechanism of analgesia, anti-inflammation, and antipyresis. At therapeutic doses aspirin can cause gastric irritation, as well as nausea and vomiting. Occasionally, the effect of aspirin on blood clotting through its inhibition of platelet aggregation can cause bruising. Very rarely, there can be an allergic response exhibited as urticaria or anaphylactic shock.

Acute or chronic overexposure causes ringing in the ears and deafness as well as a broad variety of CNS effects ranging from headache to coma. Stimulation by salicylate of the respiratory centre in the medulla causes hyperventilation and hyperpyrexia, which in turn lead to sweating and dehydration. Metabolically there is an increase in oxygen consumption and carbon dioxide production, and thus metabolic acidosis plus respiratory alkalosis. The metabolic acidosis reduces the ionization so that more salicylate enters tissues including the brain, thereby enhancing its CNS toxicity. Also, the kidney excretes more bicarbonate (along with sodium and potassium). The bicarbonate in the urine causes the urine pH to rise, which in fact accelerates the excretion of the salicylate. Thus aspirin in overdose exhibits two of the important pH effects on distribution and elimination discussed in Chapter 2.

Treatment involves whole body cooling, plus cardiovascular and respiratory support, and correction of the acid-base abnormalities. Toxicokinetic intervention involves sampling to identify the toxic chemical(s) involved, administration of activated charcoal to adsorb any unabsorbed aspirin in the gastrointestinal tract, and administration of sodium bicarbonate to alkalinize the urine further in order to speed the elimination of the absorbed dose. The bicarbonate also assists in the reversal of the metabolic acidosis. Forced diuresis along with the bicarbonate is no longer recommended, as it has been found that the additional fluid load involved can complicate the stabilization of the electrolyte balance. Dialysis can be used as a treatment but it is not often needed.

Concentrations of salicylate in plasma are predictive of severity of intoxication and can define treatment. Thus concentrations above $4\,\mathrm{mmol\,L^{-1}}$ dictate that intensive care unit monitoring be employed, while concentrations above $4.45\,\mathrm{mmol\,L^{-1}}$ in chronic intoxication, and above $9.4\,\mathrm{mmol\,L^{-1}}$ in acute ingestion, can indicate a need for the use of haemodialysis.

18.5.5   Toxic chemicals

This group is made up of non-therapeutic toxic chemicals for which there may be or may not be an underlying pharmacokinetic theory in treatment. The toxic alcohols are a case in point.

18.5.5.1   Methanol and ethylene glycol

Methanol and ethylene glycol lead to accumulation of aldehyde and acid metabolites, and from promotion of lactic acid formation due to reduced NAD/NADH ratios. This is assessed as the anion gap:

$$\text{Anion gap} = ([\mathrm{Na^+}] + [\mathrm{K^+}]) - ([\mathrm{HCO_3^{2-}}] + [\mathrm{Cl^-}])$$

If the anion gap exceeds $14\,\mathrm{mmol\,L^{-1}}$, action is required. Ethylene glycol or methanol concentrations of $> 500\,\mathrm{mg\,L^{-1}}$ are a possible indication for pro-active treatment, with formepizole (4-methylpyrazole, 4MP), or intravenous ethanol, and/or haemodialysis. The use of ethanol as a treatment is based on the fact that ethylene glycol and methanol poisoning are mediated by metabolites. If the formation of these metabolites can be slowed down, then the alternative pathway of renal excretion will dominate the elimination. Ethanol does just that. In being a competing substrate for the enzyme alcohol dehydrogenase, ethanol slows down the metabolism of ethylene glycol and methanol, thus prolonging the treatment (which may seem counterintuitive) but reducing exposure to the toxic metabolites. Ethanol itself is metabolized to less toxic metabolites. The pharmacokinetic data relevant to this approach is summarized in Table 18.1. Formepizole is an inhibitor of alcohol dehydrogenase that achieves a result similar to that achieved with ethanol, without the ethanol effects. However, ethanol is inexpensive and usually readily available in hospitals.

Table 18.1   Mean elimination half-lives in hours of ethylene glycol and methanol (data from Chu *et al.*, 2002)

	Treatment			
	Drug alone	Ethanol	Formepizole	Ethanol + haemodialysis
Ethylene glycol	2.5–4.5[a]	17	19.7	2.6
Methanol	3.01	43	54	3.5

[a]Low level infusions in normal subjects, longer at toxic concentrations.

18.5.5.2   Halobenzenes

Bromobenzene and related compounds cause hepatic necrosis. Virtually all of the metabolism of bromobenzene is through the formation of 3,4-bromobenzene epoxide, which undergoes further metabolism by a variety of routes depending on the concentration available. At low doses, conjugation with glutathione and eventual excretion as a mercapturic acid conjugate is favoured. At higher doses, the concentrations of the epoxide rise, and formation of 4-bromophenol and arylation of microsomal protein occurs. Another possible route, employing epoxide hydrase, completes the picture without direct involvement in hepatotoxicity (Figure 18.8). Bromobenzene is a model for several widely used drugs, such as phenytoin (diphenylhydantoin) and

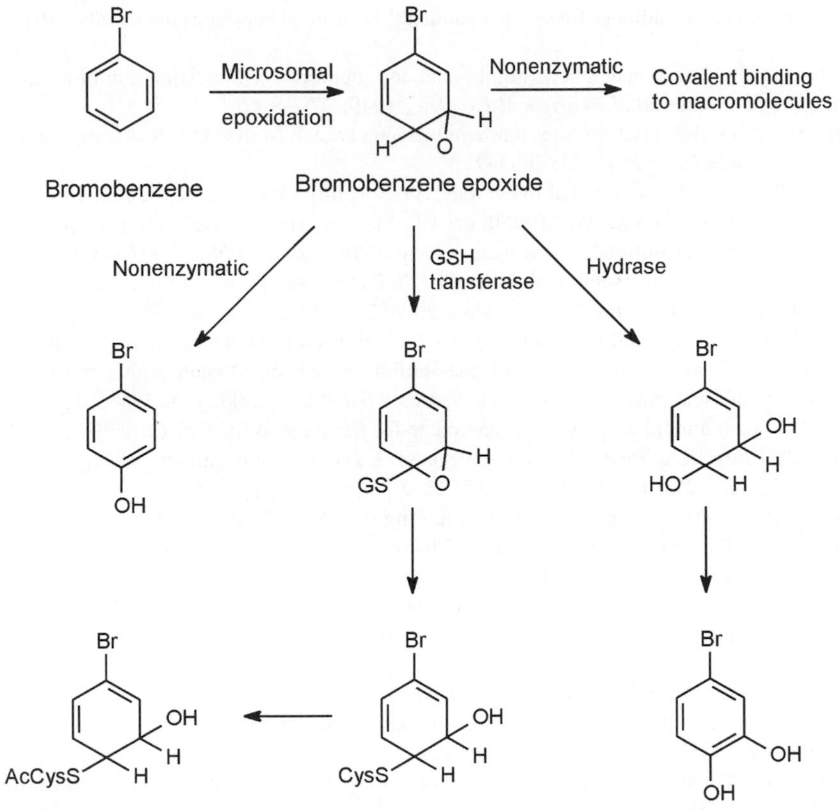

Figure 18.8 Enzymatic and non-enzymatic transformations of bromobenzene.

phenobarbital, which are metabolized to vicinal diol derivatives indicating the formation of epoxide intermediates. However, formation of such intermediates does not necessarily imply that liver toxicity will result.

18.5.5.3 Miscellaneous

Further examples of covalent chemistry involved in toxicology include carcinogenicity (such as with 2-naphthylamine which is an aromatic amine that is converted to a hydroxylamine), chemical damage, teratology (such as with thalidomide) and mutagenicity (such as with mustine hydrochloride, a cancer fighting drug), and with drug allergy (such as with penicillin). Heavy metals provide yet another area of toxicology in which principles of drug disposition and pharmacokinetics are relevant. Thus the world of toxicology provides a fascinating area of application of most, perhaps all, of the principles involved in this book, and, while treatment relies more on history, clinical assessment and interpretation of ancillary investigations, measurement and interpretation of drug concentrations remain critical in providing baseline data, and understanding of mechanisms.

References and further reading

Alapat PM, Zimmerman JL. Toxicology in the critical care unit. *Chest* 2008; 133: 1006–13.
Bertolini A, Ferrari A, Ottani A, Guerzoni S, Tacchi R, Leone S. Paracetamol: new vistas of an old drug. *CNS Drug Rev* 2006; 12: 250–75.

Boger RH. Renal impairment: a challenge for opioid treatment? The role of buprenorphine. *Palliat Med* 2006; 20 Suppl 1: s17–23.

Brodie BB, Axelrod J. The fate of acetophenetidin in man and methods for the estimation of acetophenetidin and its metabolites in biological material. *J Pharmacol Exp Ther* 1949; 97: 58–67.

Brune K. Persistence of NSAIDs at effect sites and rapid disappearance from side-effect compartments contributes to tolerability. *Curr Med Res Opin* 2007; 23: 2985–95.

Chu J, Wang RY, Hill NS. Update in clinical toxicology. *Am J Respir Crit Care Med* 2002; 166: 9–15.

Chyka PA, Erdman AR, Christianson G, Wax PM, Booze LL, Manoguerra AS, *et al*. Salicylate poisoning: an evidence-based consensus guideline for out-of-hospital management. *Clin Toxicol (Phila)* 2007; 45: 95–131.

Clark DW, Layton D, Shakir SA. Do some inhibitors of COX-2 increase the risk of thromboembolic events?: Linking pharmacology with pharmacoepidemiology. *Drug Saf* 2004; 27: 427–56.

Daly FF, Fountain JS, Murray L, Graudins A, Buckley NA. Guidelines for the management of paracetamol poisoning in Australia and New Zealand – explanation and elaboration. A consensus statement from clinical toxicologists consulting to the Australasian poisons information centres. *Med J Aust* 2008; 188: 296–301.

Dargan PI, Jones AL. Acetaminophen poisoning: an update for the intensivist. *Crit Care* 2002; 6: 108–10.

Davies NM, Good RL, Roupe KA, Yanez JA. Cyclooxygenase-3: axiom, dogma, anomaly, enigma or splice error? –Not as easy as 1, 2, 3. *J Pharm Pharm Sci* 2004; 7: 217–26.

Dawson AH, Whyte IM. Therapeutic drug monitoring in drug overdose. *Br J Clin Pharmacol* 1999; 48: 278–83.

Day RO, McLachlan AJ, Graham GG, Williams KM. Pharmacokinetics of nonsteroidal anti-inflammatory drugs in synovial fluid. *Clin Pharmacokinet* 1999; 36: 191–210.

Flanagan RJ. Fatal toxicity of drugs used in psychiatry. *Hum Psychopharmacol* 2008; 23Suppl 1: 43–51.

Flanagan RJ, Taylor A, Watson ID, Whelpton R. *Fundamentals of Analytical Toxicology.* Chichester: John Wiley & Sons, 2008.

Hinz B, Brune K. Cyclooxygenase-2–10 years later. *J Pharmacol Exp Ther* 2002; 300: 367–75.

Isbister GK, Bowe SJ, Dawson A, Whyte IM. Relative toxicity of selective serotonin reuptake inhibitors (SSRIs) in overdose. *J Toxicol Clin Toxicol* 2004; 42: 277–85.

Isbister GK, Buckley NA. The pathophysiology of serotonin toxicity in animals and humans: implications for diagnosis and treatment. *Clin Neuropharmacol* 2005; 28: 205–14.

Isbister GK, Dawson AH, Whyte IM. Comment: serotonin syndrome induced by fluvoxamine and mirtazapine. *Ann Pharmacother* 2001; 35: 1674–5.

Isbister GK, Prior FH, Foy A. Citalopram-induced bradycardia and presyncope. *Ann Pharmacother* 2001; 35: 1552–5.

Jaeschke H, Gores GJ, Cederbaum AI, Hinson JA, Pessayre D, Lemasters JJ. Mechanisms of hepatotoxicity. *Toxicol Sci* 2002; 65: 166–76.

Kaplowitz N. Idiosyncratic drug hepatotoxicity. *Nat Rev Drug Discov* 2005; 4: 489–99.

Knight TR, Ho YS, Farhood A, Jaeschke H. Peroxynitrite is a critical mediator of acetaminophen hepatotoxicity in murine livers: protection by glutathione. *J Pharmacol Exp Ther* 2002; 303: 468–75.

Maurer HH, Sauer C, Theobald DS. Toxicokinetics of drugs of abuse: current knowledge of the isoenzymes involved in the human metabolism of tetrahydrocannabinol, cocaine, heroin, morphine, and codeine. *Ther Drug Monit* 2006; 28: 447–53.

Mazer M, Perrone J. Acetaminophen-induced nephrotoxicity: pathophysiology, clinical manifestations, and management. *J Med Toxicol* 2008; 4: 2–6.

Nicholls AW, Wilson ID, Godejohann M, Nicholson JK, Shockcor JP. Identification of phenacetin metabolites in human urine after administration of phenacetin-C^2H_3: measurement of futile metabolic deacetylation via HPLC/MS-SPE-NMR and HPLC-ToF MS. *Xenobiotica* 2006; 36: 615–29.

Prescott LF. *Paracetamol (Acetaminophen) A critical bibliographic review.* London: Taylor & Francis, 1996.

Rischitelli DG, Karbowicz SH. Safety and efficacy of controlled-release oxycodone: a systematic literature review. *Pharmacotherapy* 2002; 22: 898–904.

Rollason V, Samer C, Piguet V, Dayer P, Desmeules J. Pharmacogenetics of analgesics: toward the individualization of prescription. *Pharmacogenomics* 2008; 9: 905–33.

Rumack BH, Matthew H. Acetaminophen poisoning and toxicity. *Pediatrics* 1975; 55: 871–6.

Spiller HA. Persistently elevated acetaminophen concentrations for two days after an initial four-hour non-toxic concentration. *Vet Hum Toxicol* 2001; 43: 218–9.

Vassallo S, Khan AN, Howland MA. Use of the Rumack-Matthew nomogram in cases of extended-release acetamino-phen toxicity. *Ann Intern Med* 1996; 125: 940.

Whyte IM, Dawson AH, Buckley NA. Relative toxicity of venlafaxine and selective serotonin reuptake inhibitors in overdose compared to tricyclic antidepressants. *QJM* 2003; 96: 369–74.

Woolf AD, Erdman AR, Nelson LS, Caravati EM, Cobaugh DJ, Booze LL, *et al*. Tricyclic antidepressant poisoning: an evidence-based consensus guideline for out-of-hospital management. *Clin Toxicol (Phila)* 2007; 45: 203–33.

19

Drug Monitoring in Therapeutics

19.1 Introduction

The basic tenet of drug monitoring is that drug therapy is more satisfactorily controlled if patients receive doses adjusted to give optimum plasma concentrations or optimum effects, whichever is the easier to measure. The aim is to provide objective assessments that are more likely to lead to successful therapy than clinical judgment alone. Having said that, many drugs can be used without a need for the kind of monitoring as described in this Chapter. It is those drugs for which the margin between adequate dosage and potentially toxic dosage is small, those that show latent toxicity, and those for which a direct pharmacological response cannot be measured, that should be monitored.

For some drugs it is relatively easy to measure the pharmacological (physiological or biochemical) effects. Examples include:

- Blood glucose for antidiabetic drugs such as insulin and oral hypoglycaemic drugs
- Blood lipids for 'statins' and other hypolipidaemic agents
- Blood pressure for antihypertensive drugs
- Electrocardiogram for antiarrhythmic drugs
- Prothrombin time for anticoagulants such as warfarin.

However, for those drugs for which it is difficult to predict effects from the size of the dose, then measurement of drug concentrations in plasma, or some other suitable fluid, may be appropriate. Thus, the combination of therapeutic drug monitoring (TDM) using chemical assays and the discipline of pharmacokinetics is useful for only a small number of compounds, and generally should not be contemplated unless there is:

- No direct method of measuring the pharmacological effect of the drug
- A poor or non-existent correlation between dose and effect
- A narrow therapeutic window
- Large inter- and intra-individual differences in drug disposition.

Furthermore, TDM requires:

- That there is a correlation between the measured concentration of drug and/or its metabolite(s) and the pharmacological effects of the drug.
- That reliable analytical methods for measuring the drug and/or its active metabolite concentrations in a clinically relevant timeframe, are available.

Drug Disposition and Pharmacokinetics,   By Stephen H. Curry and Robin Whelpton
© 2011 John Wiley & Sons, Ltd

Drug concentrations may also be measured to assess compliance (whether the patient is taking the drug as instructed) or to investigate adverse effects, drug–drug interactions, or acute poisoning. An additional objective of monitoring may be to minimize overall long-term exposure consistent with a useful response. The value of TDM is demonstrated most obviously by data on theophylline, phenytoin, gentamicin, digoxin, ciclosporin and lithium.

19.2   General considerations

19.2.1   *Samples and sampling*

Whole blood, plasma and serum are not the same thing. Plasma and serum are usually suitable for TDM, and, provided that the samples have been collected and stored correctly, there are generally no significant differences in the concentrations of drugs in these two fluids. However, standardization for any particular example is desirable. Traditionally, serum is used for lithium assays to avoid potential contamination by lithium heparin that may be used as the anticoagulant. Blood and plasma concentrations may be markedly different depending on the degree of partitioning into or binding to red cells. When drugs are extensively distributed into erythrocytes, then lysed whole blood (easily achieved by freezing the sample) may be better as even a small degree of haemolysis may significantly elevate plasma concentrations. Whole blood should be used for the immunosuppressants ciclosporin, sirolimus, and tacrolimus because they redistribute between plasma and erythrocytes once the sample has been collected.

Saliva is an ultrafiltrate of plasma with the addition of certain digestive enzymes and other components. There has been interest in the collection of saliva for TDM purposes because collection is non-invasive and salivary analyte concentrations are said to reflect non-protein bound plasma concentrations (Section 3.3.4). However, reliable saliva or oral fluid collection requires a co-operative individual and even then is not without problems. Saliva is a viscous fluid and thus is difficult to pipette. Some drugs, medical conditions, or anxiety, for example, can inhibit saliva secretion and so the specimen may not be available from all individuals at all times. Use of acidic solutions such as dilute citric acid to stimulate salivary flow alters saliva pH and thus may alter the secretion rate of ionizable compounds. Additionally, stimulated saliva flow can result in diluted saliva and hence reduced drug concentrations. Saliva is a useful medium to sample from children, for example those attending epilepsy clinics. Collection is non-invasive and children are less inhibited about spitting into a pot than the majority of adults.

19.2.2   *What should be measured?*

Generally, the parent drug is measured, or in the case of a prodrug, the active metabolite. Aspirin is rapidly hydrolysed to salicylate *in vivo*, and it is usually salicylate that is measured. For some drugs, an active metabolite may make a significant contribution to the overall clinical and/or toxicological effects. Some TDM protocols advocate using a total value (i.e. parent drug plus active metabolite(s)), others may consider the concentration of metabolite separately. An immunoassay may measure parent drug and metabolite and the result will depend on the relative concentrations of drug and metabolite, and the cross-reactivity of the antibody for the metabolite. This may or may not be an advantage of using immunoassay. Different immunoassays may give different results and this is one of the reasons why clinics, based upon their experience, may define their own reference ranges.

To measure the concentration of an active metabolite as well as that of the parent drug will probably require a chromatographic method. A further advantage of knowing the relative concentrations of drug and metabolite is that they may give an indication of when the sample was taken relative to the time of the last dose of drug. A higher than expected drug/metabolite ratio may indicate that the time between the dose having been taken and the sampling is shorter than the prescribed interval.

19.2.3 Timing of sample collection

Timing of the sample collection is crucial as consideration of typical multiple-dose plasma concentration–time curves (e.g. Figure 4.11) will indicate. There are likely to be major differences between peak and trough concentrations, and published concentrations will usually be based on samples collected at a defined time after the last administered dose. Furthermore, the time at which peak concentrations occur is more difficult to define and may be variable within and between patients. A very common, almost standard TDM guideline is to sample immediately before the next dose, or the following morning after an evening dose ('pre-dose' or 'trough' sample) to allow for absorption and distribution to tissues to be completed before sampling (but see individual examples, later). Trough concentrations are expected to be more repeatable than peak concentrations. Generally, samples are not taken until steady-state conditions have been achieved, approximately five times the elimination half-life. An exception to this 'rule' is carbamazepine which undergoes extensive auto-induction (Section 17.4.1). The recommended delay between initiation of therapy or dosage adjustment is 20 days, to allow for the reduction in half-life and stabilization at a new (lower) steady-state concentration. Also, because phenytoin exhibits dose-dependent kinetics, exceptionally long time intervals are needed after changing the dosing regimen before the next plasma sample is taken. For some drugs, such as immunosuppressants, measurement of the *AUC* is considered a better indicator of overall exposure but this raises practical difficulties with regard to the large numbers of samples that would have to be collected and analysed. Noticing an empirical relationship between the peak concentration and *AUC* of ciclosporin, Jorga *et al.* (2004) suggested that peak (i.e. 2 hour post-dose sampling) is possibly a better indicator of optimal dosage than pre-dose or 4 hour post-dose sampling.

19.2.4 Analyses

A requirement of TDM is that drug concentrations can be measured and the results returned quickly to the clinician. In some instances this means while the patient is visiting the clinic. To achieve this rapid turnaround, immunoassays, point-of-care testing (POCT) kits, which are often based on immunoassays, and chromatographic techniques such as high performance liquid chromatography (HPLC), sometimes coupled to mass-spectrometry, with short and/or narrow bore columns are employed.

Manufacture of a number of non-isotopic immunoassays compatible with high-throughput clinical chemistry analysers has meant that certain TDM assays are widely available. It is important that the limitations and cross-reactivities of such assays are understood. Particular difficulties have arisen with digoxin immunoassays because of cross-reactivity with cortisol, spironolactone and substance(s) referred to as 'digoxin-like immunoreactive substances' (DLIS) which were first observed in patients with a variety of volume-expanded conditions: namely, diabetes, uraemia, essential hypertension, liver disease, and pre-eclampsia. DLIS cross-react with many antidigoxin antibodies and may falsely elevate plasma digoxin concentrations.

In fact, POCT kits are not often available for therapeutic drugs although there is one based on colorimetric analysis of lithium available in the United States. A kit for serum theophylline was introduced but does not appear to have been adopted widely. Most POCT kits have been developed and marketed for detecting drugs of abuse.

Chromatographic methods require extensive resources in terms of hardware and operator expertise. However, chromatographic assays are important in the case of amiodarone (where it has proved impossible to produce an antibody that does not cross-react significantly with thyroxine and tri-iodothyronine), antiviral drugs, immunosuppressants, many psychoactive compounds, and generally where active metabolites should be measured as well as the parent compound. Examples include carbamazepine/carbamazepine-10,11-epoxide, procainamide/*N*-acetylprocainamide, and amitriptyline/nortriptyline.

Whatever technique is used when providing a TDM service, adherence to the principles of quality management (proper method implementation and validation, and adherence to internal quality control and external quality assessment procedures) is essential as treatment decisions may be based on the results.

19.2.5   *Reference ranges*

A key requirement of TDM is that the clinical effects of a drug are in some way related to its plasma concentration even if the relationship between the plasma concentration and the concentration of drug at its site of action is complex. Drugs with a narrow therapeutic window will have a small range of concentrations below which the desired effect is suboptimal, and above which adverse effects may become so serious as to indicate dose reduction. Extreme examples are the immunosuppressants used to prevent organ rejection after transplantation. Too little drug may result in loss of the new organ, but too much drug is likely to result in infection, which is often life-threatening. In the longer term use of these drugs may result in development of malignancy.

Because patients vary so much in how they respond to drugs, the term *therapeutic range* is usually applied to an individual, that is, it is the range of concentrations applicable to *that* patient. To reflect this, the term *reference range* is recommended when helping interpret TDM data, although the expressions 'therapeutic range' or 'target range' may also be encountered. It is often useful to take a sample for analysis when treatment is deemed to be satisfactory, thereby establishing a target concentration for that patient – leading to individualized therapy.

Furthermore, it is important patient treatment is based on sound clinical judgement and not solely on the basis of a laboratory TDM result; to quote Reynolds and Aronson (1993): 'Treat the patient, not the plasma drug concentration'. However, for some drugs such as lithium, it is usual to adjust the dosage to achieve concentrations in the the target range.

19.3   Specific examples

19.3.1   *Antiasthmatic drugs*

There is usually no reason to monitor those bronchodilators, such as salbutamol and ipratropium, that are commonly taken by inhalation, as the clinical effect is easily assessed and the drugs are relatively non-toxic, even if taken in overdose. Also, they may not reach plasma in detectable quantities. In contrast, theophylline is administered orally, and this drug is monitored to minimize the risk of adverse effects that occur at relatively high plasma concentrations. Theophylline, a metabolite of caffeine, is metabolized to 3-methyl-xanthine, which is itself further metabolized by xanthine oxidase (Figure 19.1). Theophylline decays according to a two-compartment model with first-order kinetics although the two phases are rarely seen with oral dosing. The half-life of the terminal phase varies from 3–13 hours, providing for much variation in plasma

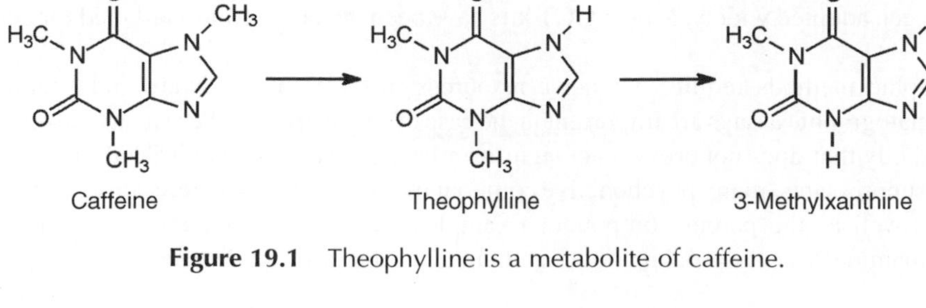

Figure 19.1   Theophylline is a metabolite of caffeine.

concentrations. When the half-life is short, such as in patients in whom the half-life is 5 hours, the fall from peak to trough within each dosage interval can be more than 50% of the peak. Because the therapeutic range is limited to peaks of approximately twice the troughs, this can result in fluctuation involving peaks and troughs outside the therapeutic range, unless dosing is inconveniently frequent. This problem is solved by the judicious use of controlled-release formulations, which reduces the extent of the fluctuation. In children, the C^{ss} concentrations are relatively close to the Km, and thus in children theophylline commonly exhibits dose-dependent kinetics, so that relatively small increases in dose result in relatively large increases in theophylline concentrations in plasma.

A variety of effects correlate with theophylline plasma concentrations, including improvement in forced expiratory volume, reduced frequency of asthma attacks, and increases in forearm blood flow. The adult reference range is $8-20\,\mathrm{mg\,L^{-1}}$. Adverse effects (anorexia, nausea, vomiting, nervousness and anxiety) become increasingly evident towards $20\,\mathrm{mg\,L^{-1}}$. Above $20\,\mathrm{mg\,L^{-1}}$ sinus tachycardia and arrhythmias become a problem, and above $40\,\mathrm{mg\,L^{-1}}$, local or generalized seizures and cardiorespiratory arrest are possible. Because prevention of unwanted effects is the most important reason for TDM with theophylline, it is usual to monitor peak concentrations at approximately 2–3 hours post-dosage. There is usually no need for monitoring of theophylline metabolites, but theophylline can be methylated to caffeine by neonates, although not by young children or adults. When theophylline and caffeine are used to treat neonatal apnoea the reference ranges are $6-12$ and $10-30\,\mathrm{mg\,L^{-1}}$, respectively.

19.3.2 Anticonvulsant drugs

There are clear indications for monitoring anticonvulsant drugs. They are used prophylactically to prevent seizures, which occur at irregular intervals so it is difficult to assess the efficacy of the treatment, in that prevention of incidents, rather than a graded effect, is sought. In general anticonvulsant drugs have low therapeutic indices and the toxicity is often difficult to access. The drugs are metabolized via the cytochrome P-450 pathways and phase 2 glucuronidation, so there are large interpatient differences in the degree of metabolism. The enzymes are prone to induction or inhibition leading to potential drug-drug interactions. Several of the class (especially phenytoin) exhibit non-linear kinetics (Section 4.3.), are highly protein bound and many have pharmacologically active metabolites, including primidone $\rightarrow$ phenobarbital, carbamazepine $\rightarrow$ 10,11-carbamazepine epoxide and clobazam $\rightarrow$ N-desmethylclobazam. Non-linear kinetics result in the time to the pharmacokinetic steady-state being variable (i.e. relatively long at relatively high doses), and, clinically, a pharmacokinetic steady-state with minimal or zero fluctuation between peaks and troughs is sought. Epilepsy is usually diagnosed at an early age, but is a chronic condition that may require lifetime treatment. Thus it is almost certain that treatment will have to be changed as the patient ages and over 30% of patients are treated concomitantly with two or more anticonvulsant drugs.

The earliest studies related to monitoring of anticonvulsants concerned phenytoin and were published in the 1960s when unsuccessful control of seizures was linked to inadequate plasma concentrations in long-term therapy. At the same time, a considerable factor affecting phenytoin concentrations was shown to be concomitant use of other drugs. Furthermore, decreasing paroxysmal activity in the EEG was shown to correlate with increasing drug concentrations, thereby establishing a dose–response relationship. In the 1970s, studies showed concentration–effect relationships for several of the available anticonvulsants including phenobarbital, primidone and ethosuximide.

Since 1990, the number of antiepileptic drugs available has more than doubled and it may be prudent to add these to the list of drugs that should be monitored. Gabapentin, levetiracetam, pregabalin, topiramate, and vigabatrin are eliminated largely or completely unchanged in urine. Hence, plasma concentrations may be affected by alterations in renal function. Gabapentin shows dose-dependent bioavailability. Concomitant use of enzyme-inducing drugs can affect topiramate concentrations markedly. For the newer drugs that are largely metabolized prior to excretion (felbamate, lamotrigine, oxcarbazepine, tiagabine, and zonisamide), interpatient variability in pharmacokinetics is just as important in dose adjustment as for many older anticonvulsants.

In newly diagnosed patients there is no clear evidence to support the use of TDM with the aim of reaching predefined target ranges in dose optimization with anticonvulsant monotherapy. However, reference ranges have been proposed (Table 19.1) that might guide dosing when necessary. The dose should not be adjusted if the patient is seizure-free but the serum concentration is below the lower limit. In fact it has been suggested that phenytoin, phenobarbital, carbamazepine and valproic acid should only have upper values.

Table 19.1   Bioavailability, protein binding and reference ranges for selected anticonvulsant drugs (adapted from Patsalos *et al.*, 2008)

Drug	Oral availability (%)	Serum protein binding (%)	Time to steady-state (days)	Reference range (mg L^{-1})
Carbamazepine	≤85	75	2–4[a]	4–12
−10,11-expoxide				0.5–2.5
Clobazam	≥95	85	7–10	0.03–0.3
desmethyl–				0.3–3
Clonazepam	≥95	85	3–10	0.02–0.07
Ethosuximide	≥90	0	7–10	40–100
Felbamate	>90	25	3–4	30–60
Gabapentin	<60[b]	0	1–2	2–20
Lambotrigine	≥95	55	3–6	2.4–15
Levetiracetam	≥95	0	1–2	12–46
Oxcarbazepine[c]	90	40	2–3	3–35
Phenobarbital	≥95	55	12–24	10–40
Phenytoin	≥80	90	5–17	10–20
Primidone	≥90	10	2–4	5–10[d]
Tiagabine	≥90	96	1–2	0.02–0.2
Topiramate	≥80	15	4–5	5–20
Valproic acid	≥90	90[e]	2–4	50–100
Vigabatrin	≥60	0	1–2	0.8–36
Zonisamide	≥65	50	9–12	10–40

[a]Initially; steady-state takes up to 5 weeks because of autoinduction. [b]Bioavailability decreases with increasing doses. [c]Prodrug; all values refer to active metabolite. [d]Phenobarbital should be monitored as well. [e]Binding decreases with increasing concentrations.

Some authorities promote the idea of measuring unbound concentrations of anticonvulsants in TDM protocols, but this is technologically more complex and is relatively expensive. Different ranges then apply. The process of dose-adjustment with phenytoin is considered in a later section.

19.3.3   Antidepressants

Successful use of antidepressant drugs is complicated because:

- Some 20–40% of patients respond to placebo within 3–4 weeks of treatment.
- Some patients remain unresponsive irrespective of dose.
- The rate of metabolism can vary by 30–40-fold between subjects.
- Compliance is a problem.

- Patients may become tolerant to some of the unwanted effects.
- The mechanism of action is not understood and clinical effects are not usually seen until two or more weeks after initiation of treatment.

Also, several of the drugs are metabolites of other antidepressants. The first example of this was the prescribing of imipramine, which was quickly followed by the introduction of desipramine, but this mode of new product discovery has continued into the 21st century with venlafaxine and desvenlafaxine. In such cases TDM must often involve pairs of compounds of different potencies and with different pharmacokinetic properties. There have been attempts to relate the antidepressant response to objective clinical pharmacological tests, such as tyramine-induced hypertension and disturbance of accommodation, but these have not found widespread application in patient care. It is sometimes thought with these drugs that there is an important dose-related response in each subject that does not translate into a population-wide phenomenon. The small therapeutic windows exhibited by the tricyclic antidepressants (imipramine, nortriptyline, etc.) make these drugs candidates for therapeutic monitoring, however establishing agreed reference ranges has proved difficult. Several studies in the 1970s produced conflicting results but did usefully show that the concentration–response relationships could be an inverted U-shape (Curry, 1980). A typical example is shown in Table 19.2, albeit with relatively small numbers of patients.

Table 19.2 Relationship between nortriptyline plasma concentrations and response – global rating as 'amelioration score (Asberg *et al.*, 1971)

Plasma concentration ($\mu g\,L^{-1}$)	Number of patients	Amelioration score	
		Mean	SEM
≤ 49	5	0.4	1.2
50–79	10	6.2	0.8
80–109	4	6.1	1.4
110–139	5	5.0	2.0
≥ 140	5	1.2	1.9

Even today there is no consensus over the role of TDM in psychiatric medicine. In practice, TDM of TCAs and also of selective serotonin reuptake inhibitors (SSRIs) and other newer antidepressants such as venlafaxine is mainly concerned with assessing whether treatment failure is due to poor adherence, ultra-rapid metabolism, or drug interactions leading to induction of metabolizing enzymes. Compliance is a problem, a study showing that during a 3-month treatment with SSRIs, 72.5% of the patients missed at least 1 dosing day, and 29% of the patients had missed 2 or more days, consecutively. Many of the class, particularly the SSRIs are metabolized by CYP2D6 and so the levels are affected by genetic differences (Section 10.5.1) and prone to drug–drug interactions. It as been suggested that phenotyping of patients should also be conducted.

Although reference ranges have been suggested (Table 19.3) a survey of 33 UK laboratories showed a high degree of variability (Wilson 2003). The range of values was greater than 100-fold for some compounds, particularly for the minimum effective concentrations. For example the lower limit for paroxetine ranged from 1 to 170 nmol L^{-1} (mean 70 nmol L^{-1}).

Table 19.3   Reference ranges for plasma concentrations of selected antidepressants and active metabolites in adults

Drug	Reference range (mg L^{-1})
Amitriptyline + nortriptyline	0.08–0.25
Citalopram	0.05–0.5
Clomipramine + norclomipramine	<1.0
Desipramine	0.08–0.16
Escitalopram	0.025–0.25
Fluoxetine	0.04–0.45
norfluoxetine	0.04–0.45
Fluvoxamine	0.16–0.22
Imipramine + desipramine	0.15–0.25
Nortriptyline	0.05–0.15
Paroxetine	0.01–0.05
Sertraline	0.03–0.19
Trimipramine + nortrimipramine	<0.50
Venlafaxine	0.05–0.5
O-desmethylvenlafaxine	0.05–0.5

19.3.4   *Antimicrobial agents*

The reasons for monitoring antimicrobial agents are much the same as for any other class of drugs. They are to check (i) that the concentrations are sufficient for efficacy but below those associated with toxicity, (ii) the bioavailability following a switch from i.v. to oral dosing, (iii) for compliance, particularly in patients receiving drugs for tuberculosis/multidrug resistant TB.

   Antimicrobial drugs can be divided into those whose effect is primarily related to the concentration of drug (i.e. the higher the concentration the greater the kill), and those for which the time of exposure is important (i.e. increased time of exposure increases the kill). Consequently, three parameters may be assessed: (i) the time that the concentration is a above the minimum inhibitory concentration (MIC, which is measured *in vitro* for the pathogen under consideration), (ii) the ratio of the peak concentration to the MIC, and (iii) the area under the 0–24 hour plasma concentration to MIC ratio (Figure 19.2).

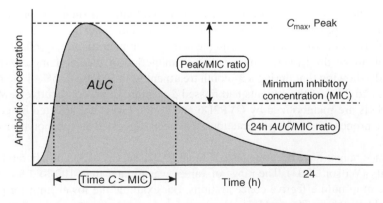

Figure 19.2   Schematic representation of parameters that might be assessed for therapeutic monitoring of antimicrobial drugs.

The process of setting up a dosing regimen for gentamicin is considered later. Gentamicin can be described as having the remarkable property of an inverse safety margin – peaks need to be above $5 \, \text{mg} \, \text{L}^{-1}$ in order to achieve a successful bacterial kill, but troughs need to be below $2 \, \text{mg} \, \text{L}^{-1}$ in order to prevent toxicity. Gentamicin and other aminoglycosides fall into the concentration-dependent category and the aim is to achieve the maximum tolerable concentrations – the appropriate parameters are C_{max}/MIC or AUC/MIC. For the β-lactam antibiotics the duration of exposure is important and the length of time above the MIC is the appropriate parameter. In the case of vancomycin, the AUC/MIC ratio is used as it is the amount of drug to which the pathogens are exposed that is important.

Monitoring of some antimicrobials is carried out in most, if not all, in-patients. These include the antibacterials, gentamicin, vancomycin, streptomycin and antifungals, itraconazole and flucytosine. The aim with gentamicin is a C_{max}/MIC ratio of 8–10 with trough concentrations less than $1–2 \, \text{mg} \, \text{L}^{-1}$. Vancomycin trough concentrations should be below $10–12 \, \text{mg} \, \text{L}^{-1}$ before the next dose is given. Ototoxicity may occur at concentrations $>80 \, \text{mg} \, \text{L}^{-1}$. Optimum concentrations of flucytosine have not been established but myelosuppression may occur above $100 \, \text{mg} \, \text{L}^{-1}$.

19.3.5 Cardioactive drugs

19.3.5.1 Digoxin

Digoxin plasma concentrations in plasma are very low $(0.1–5 \, \mu\text{g} \, \text{L}^{-1})$ and are usually measured by immunoassay (IA). The drug is excreted to a considerable extent as unchanged drug in urine, and its plasma concentrations are therefore influenced by renal function. Also, there are formal protocols for dosage adjustment based on objective measurements of renal function. Other factors affecting digoxin concentrations are bioavailability, age, weight and plasma albumin. Pharmacodynamic influences include ion balance (potassium, magnesium, calcium), acidosis and alkalosis, and oxygen tension. Hypokalaemia and hypomagnesaemia may increase the myocardial sensitivity to digoxin.

At one time, clinical practice involved increasing the digoxin dose until unwanted effects were seen, and then cutting back to a dose just below the threshold for these unwanted effects. This process was termed 'digitalization'. It is now realized that the condition sometimes being treated in many patients (arrhythmias) and the most prominent unwanted effects have so much in common that it can be difficult to determine whether a patient receiving digoxin and experiencing arrhythmia is underdosed or overdosed. Redfors (1972) demonstrated a correlation between slowing of ventricular rate and digoxin concentrations in patients with atrial fibrillation. This was a within patient study. In patients with congestive heart failure who have been prescribed cardiac glycosides for their positive inotropic effect, there is no easy way of accessing efficacy and the manifestation of toxicity (anorexia, nausea, vomiting and cardiac arrhythmia) are also signs and symptoms of the disease. Thus, plasma level monitoring may aid dosage adjustment in such circumstances. Having said that, the therapeutic index is very low $(0.8–2 \, \mu\text{g} \, \text{L}^{-1})$ and there is considerable overlap between toxic and non-toxic concentrations as can be seen from Figure 19.3, although children can tolerate relatively high levels.

Digoxin has two-compartment distribution properties, and the drug penetrates slowly to its receptors, so monitoring is based on trough concentrations, with plasma samples collected some 8–24 hours after the time of the dose that precedes them. Digitoxin concentrations are also sometimes monitored when that drug is prescribed.

19.3.5.2 Antidysrythmic drugs

Drugs under this broad heading include the membrane stabilizing drugs (sodium channel blockers: lidocaine, phenytoin, quinidine and procainamide), β-adrenoceptor blockers and calcium channel antagonists, such as

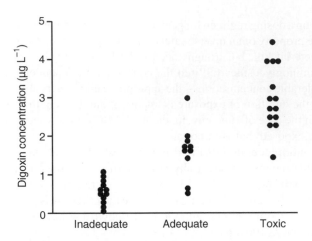

Figure 19.3   Plasma digoxin concentrations in patients judged to be inadequately digitalized, adequately digitalized and showing symptoms of toxicity. (Redrawn after Oliver *et al.*, 1971.)

verapamil. Reference ranges for some of this class of drugs are presented in Table 19.4. A relationship between the degree of antagonism of isoprenaline-induced tachycardia by propranolol and concentrations of the drug in plasma has been shown.

Table 19.4   Reference ranges for selected antidisrhythmic drugs (Flanagan *et al.*, 2008)

Drug [metabolite]	Reference range in an adult (mg L^{-1} plasma)
Amiodarone	0.5–2
[Noramiodarone]	[0.5–2]
Atenolol	0.2–0.6
Disopyramide	2.0–5.0
[Nordisopyramide]	[<5.0]
Flecainide	0.2–0.7
Lidocaine	1.5–5.0
Metoprolol	0.2-0.8
Procainamide [+ acecainide]	10–30 (procainamide only, 4–8)
Propranolol	0.01–0.1
Quinidine	2.5–5.0
Sotalol	0.8–2.0 (β-blockade), 2.5–4.0 (antiarrhythmic)
Verapamil	0.1–0.2
[Norverapamil]	[0.1–0.2]

Metabolites that are usually measured are shown thus [ ].

TDM can sometimes be useful in the case of amiodarone to monitor adherence and toxicity, and to monitor adherence to sotalol and to other β-adrenoceptor blockers such as atenolol and propranolol. Use of the calcium channel blockers verapamil and diltiazem is normally assessed by monitoring haemodynamic effects. Diltiazem and *N*-desmethyldiltiazem, desacetyldiltiazem and *N*-desmethyldesacetyldiltiazem are unstable in plasma and all may be pharmacologically active. The rate of metabolism of procainamide to *N*-acetylprocainamide (acecainide) is influenced by acetylator status (Section 10.3)

19.3.6 *Immunosuppressives*

Monitoring of immunosuppressants, used to prevent organ rejection after transplantation, is extremely important to ensure their safe and effective use. Too little drug may result in loss of the new organ, but too much drug is likely to result in infection, which is often life-threatening. Indeed, monitoring of sirolimus is a requirement of drug licensing within the European Union.

Ciclosporin, mycophenolic acid (the active metabolite of mycophenolate mofetil), sirolimus, and tacrolimus are monitored (Table 19.5). However, some patients experience acute rejection episodes or postoperative complications even when the blood concentrations are within the reference range. *AUC* values may give more useful information than measurements at a single time point particularly for mycophenolic acid. Clearly, serial blood sampling is likely to be impractical in an out-patient setting, hence the suggestion that samples collected 2 hours post-dose (so called 'peak' concentrations) are a suitable surrogate for *AUC* (Section 19.1.3). This has not been adopted universally and, in any event, peak sampling is difficult to achieve in practice. Interpretation of either 'trough' or 'peak' results is complicated because:

- There may be considerable differences between the results obtained with immunoassays when compared with chromatographic methods.
- Immunosuppressants are often used in combination to reduce the risk of toxicity from individual compounds hence the concentrations attained during optimal treatment are lower then when the drugs are used alone.
- The amount of immunosuppression required for maintenance treatment varies widely depending on the engrafted organ.

Table 19.5 gives reference ranges that are applicable to therapy with single immunosuppressants used after renal transplantation.

Table 19.5 Some immunosuppressive drug TDM assays (Flanagan *et al.*, 2008)

Drug	Reference range in adults $(mg\,L^{-1})^a$
Ciclosporin	0.04–0.25 (trough, whole blood)[b]
Mycophenolic acid	2.5–4.0 (trough, plasma)
Sirolimus	0.003–0.015 (trough, whole blood)
Tacrolimus	0.001–0.012 (trough, whole blood)

[a]Single immunosuppressant, renal transplant patients.
[b]IA may be unreliable in some patient groups (e.g. neonates, renal failure, hepatic failure).

19.3.7 *Lithium*

Lithium was the first drug to be monitored with a view to better clinical control of symptoms, and it remains the one drug for which knowledge of pharmacokinetics is almost essential if optimum therapy is to be achieved, in regard to both efficacy and safety. Lithium is interesting in that:

- It has a narrow margin of safety.
- It is an inorganic ion, and so has no complicating metabolites.
- It has predictable pharmacokinetic properties which allow definition of an individual dosage regimen: each patient may be different but the changes within a subject are predictable.
- It is nephrotoxic, which increases the risk of accumulation.

Serum lithium concentrations are very variable between individuals because of variations in the rate and extent of absorption, renal clearance and apparent volume of distribution. The elimination half-life ranges from 7–41 hours. Because of the fluctuations in concentrations within a dosage period, and diurnal variation, it has been recommended that samples be taken 12 hours after the last dose and at the same time of day, preferably in the morning. The concentrations are referred to as 'standardized 12 hour' concentrations and are thus the trough concentrations in a twice daily dosing regimen.

With lithium the reference range is a true target range. Recommended concentrations are 0.8–1.2 mmol L^{-1} for the treatment of acute mania, whilst 0.4–0.8 mmol L^{-1} is usually suitable for the prophylaxis of unipolar or bipolar illness. Toxicity is usually associated with serum concentrations >1.4 mmol L^{-1} and a vicious circle of toxicity involving renal damage and subsequent drug accumulation can be set in motion in the range 1.5–2 mmol L^{-1}. Concentrations over 2 mmol L^{-1} are associated with a high incidence of toxicity and death can occur above 3 mmol L^{-1}. The problem of fluctuating serum concentrations, which can lead to inattention, lethargy, ataxia and tremor (thought to be associated with peak concentrations) can be reduced by the use of sustained-release preparations.

Situations which can lead to increased lithium concentrations include reduced glomerular filtration, sodium depletion, diuretics (principally thiazides) and non-steroidal anti-inflammatory drugs. Thus, fever and diarrhoea lead to toxicity. As with theophylline, the half-life in an individual can be so short that the fluctuation at the pharmacokinetic steady-state is at risk of inducing concentrations above and below the desired range, and controlled-release medication has a role to play in reducing this level of fluctuation.

19.3.8   Neuroleptics

There is little need for routine monitoring of the established ('typical') antipsychotics such as chlorpromazine and haloperidol, as these are not considered as toxic as the tricyclic antidepressants. However, it can be difficult to distinguish the signs of overdose from some of the signs of the disease being treated with some of these drugs, and studies have shown inverted U-shape plasma concentration–response relationships with high concentrations being associated with poor or no response. Such an observation can also indicate inappropriate choice of drug or misdiagnosis. Indications for monitoring include patients who do not respond to high doses to confirm correspondingly high concentrations and those suspected of non-compliance. Testing for compliance requires establishment of pharmacokinetic steady-state plasma concentrations during successful therapy for comparison. Large fluctuations in concentrations on repeat testing may indicate that there are periods of non-compliance. Determining 'effective' plasma concentrations in patients undergoing successful therapy may be helpful in providing a 'target value' for that patient that can be used for subsequent treatment on readmission, when there may be a drug interaction or when the formulation is changed.

The ability to monitor anti-schizophrenia medication using plasma concentration measurements became possible at a time when large numbers of psychiatric patients were kept in hospitals for as much as two years for the treatment of acute episodes and sometimes for the rest of their lives if treatment was unsuccessful. Also, many patients on phenothiazine drugs at this time suffered from a debilitating condition termed 'tardive dyskinesia', which resembled some aspects of Parkinson's disease, with both inappropriate spontaneous movement disorders and also extreme difficulty in co-ordinating normal movement. This dyskinesia was termed 'tardive' because it developed slowly with long-term treatment. It was sometimes irreversible. It occurred because there had been a treatment concept not unlike that of digitalization with digoxin, leading to dose increases until unwanted effects were observed, then reduction of dosage to just below the threshold for what was presumed to be the desired effect. Unfortunately, the dose reduction stage of this process did not always work, and the patients became victims of what was effectively planned overdosing. The insight into such problems that was possible as the result of pharmacokinetic studies contributed, in part, to a complete change in prescribing philosophy, away from 'as much as could be tolerated', to 'treatment with the lowest dose consistent with a useful effect', and the acute phase of schizophrenia was dramatically shortened, and the risk of a need for long-term treatment was greatly reduced. At the same time, social pressures led to the closing of long-term psychiatric facilities, and this was made possible, by the new emphasis on drug exposure governed by pharmacokinetic principles, including the use of sustained-release preparations such as fluphenazine decanoate.

Monitoring newer, second generation or 'atypical' neuroleptics, notably clozapine and to an extent olanzapine, can help by assessing compliance, guiding dose adjustment, and guarding against toxicity. With clozapine dose assessment is complicated because: (i) there is a 50-fold inter-patient variation in the rate of clozapine metabolism, (ii) alteration in smoking habit can have a dramatic effect on clozapine dose requirement (on average ±50%), and (iii) the clinical features of clozapine overdosage can mimic those of the underlying disease. A range of 350–500 $\mu g \, L^{-1}$ is considered suitable for clozapine, although the upper limit is ill-defined. Plasma concentrations of the desmethyl metabolite are approximately 70% of those of the parent drug during normal therapy so monitoring norclozapine can help in accessing compliance or correct timing of sample collection. With olanzapine, a 12-hour post-dose plasma concentration of 20 $\mu g \, L^{-1}$ has been suggested.

19.3.9 Thyroxine

The normal thyroid gland produces both tri-iodothyronine (T3) and tetra-iodothyronine (T4), which control a wide variety of body functions connected with maintenance of normal metabolic activity. Increases or decreases in these hormones cause changes in blood concentrations of thyroid-stimulating hormone (TSH), which is secreted by the pituitary gland to, in turn, control the thyroid gland. Thus increases in T3 and/or T4 in the blood lead to decreases in TSH. Therapy of thyroid disorders often involves administration of thyroxine, which is T4. Thyroxine dosing is mostly monitored by means of TSH measurement, rather than by direct T3 and/or T4 measurement, so TSH achieves the status of monitoring of therapy using an endogenous biochemical (biomarker). Quite small (10%) changes in thyroxine dose can lead to dramatic (doubling or halving) changes in TSH.

19.4 Dose adjustment

For the most part, dosing adjustment in response to concentration measurements is straightforward, linear pharmacokinetic properties and the fact that most therapy occurs within a linear range of effect in relation to concentration, dictate that linear dose adjustments can be made. For example, if the desired concentration is 15 mg L^{-1}, and the measured concentration is 10 mg L^{-1}, then the dose can be increased by 50%. However, two examples, gentamicin and phenytoin deserve special attention in this context.

19.4.1 Gentamicin

In devising a gentamicin dosing regimen optimized for a particular patient, an early dose is first used for calibration purposes. This dose is infused for 30–60 minutes. This is followed by a 30–60 minute interval during which equilibration among the compartments takes place, after which a first blood sample for drug analysis is collected ('peak'). A second sample is collected just before the next dose is administered ('trough'), and a half-life is calculated from these two-points. Simple pharmacokinetic calculations are used to adjust the peak to be between 5 and 12 mg L^{-1} (depending on the characteristics of the infection involved), and the trough to <2 mg L^{-1}. The peak is adjusted by changing the dose, and the trough is adjusted by changing the dosing interval. Finally, the dose and interval are fine-tuned to meet both the medical need and the need for a convenient dosing interval, which needs to be a simple fraction of 24 hours. Figure 19.4 shows a typical pattern of development of an optimal gentamicin dosing interval.

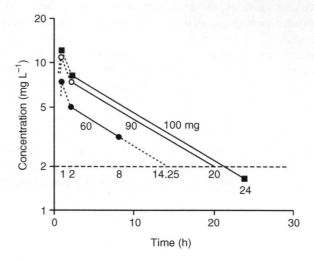

Figure 19.4   Development of a gentamicin dosage regimen. The graph shows the concentration in plasma *versus* time data following a test dose of 60 mg, and assays at 2 and 8 h after starting the 1 hour infusion. A 90 mg dose was proposed, predicting a peak concentration of 7.5 and a trough of 2.0 mg L^{-1} at approximately 20 h. A 100 mg dose gives an acceptable peak and a trough <2.0 mg L^{-1} at a convenient dosing interval of 24 h.

19.4.2   *Phenytoin*

Figure 4.13(a) shows how phenytoin concentrations vary with dose. Equation 4.44 shows how respect for the Michaelis–Menten kinetics that apply to this drug can be used to plot a straight line relationship between daily dose, and daily dose divided by C^{ss}, and this is illustrated by Figure 19.5(a). This

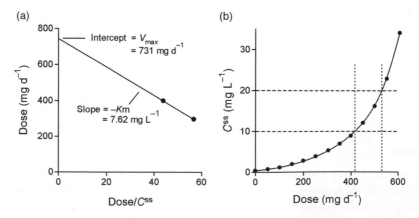

Figure 19.5   (a) Simulated experiment for determination of V_{max} and Km in a patient treated with two doses of phenytoin (300 and 400 mg per day), giving C^{ss} values of 5.3 and 9.2 mg L^{-1}, respectively. (b) Calculated steady-state concentrations as function of daily dose using the V_{max} and Km values derived in (a).

relationship gives a slope of $-K$m and an intercept on the y-axis (daily dose) of V_{max}. Note that this is V_{max} with units very different from those conventionally used in enzymology or for maximum rate of change of concentrations in blood. Data are first collected by dosing to pharmacokinetic steady-state with one dosing regimen, measuring C^{ss}, then dosing to a new steady-state with another regimen, and measuring C^{ss} again. Up to 6 weeks may be needed to reach steady-state in each case. Knowing V_{max} and Km for the patient, the C^{ss} predicted to occur with any regimen from 0 to V_{max} mg phenytoin per day can be calculated. A regimen giving C^{ss} between 10 and 20 mg L^{-1} can then be devised. Figure 19.5(b) shows a graph of C^{ss} against daily dose for the patient of Figure 19.5(a) in whom V_{max} was 731 mg day^{-1} and Km was 7.6 mg L^{-1}.

This approach does not meet all clinical needs, mainly because of the long times needed to reach steady-state. In fact, the time to reach steady-state can only be calculated, and then only as a time to reach a specific fraction of C^{ss}, if V_{max} and Km are known in advance – clearly a practical impossibility. Additionally, this approach fails if the values of Km and V_{max} change during the time over which the investigation takes place, for example because of enzyme induction. So, clinicians will tend to apply a rule of thumb in choosing the second dosing regimen, along the following lines, in the hope of choosing the optimum dose at the earliest possible time:

- If the measured initial concentration is <7 mg L^{-1}, then increase the dose by 100 mg day^{-1} or more.
- If the measured concentration is in the range 7–12 mg L^{-1}, then increase the dose by 50–100 mg day^{-1} if there is an obvious clinical need.
- If the measured concentration is >12 mg L^{-1} then increase the dose by 30–50 mg day^{-1}, again if there is an obvious clinical need.

These adjustments obviously precede the second sample, and measured serum concentrations will validate them or lead to further adjustments.

Several other approaches to this computation have been devised, notably by Graves and Cloyd, by Mullen and by Vozeh and Sheiner (see Bauer, 2001). All can be computerized, and all give valuable and complementary outcomes.

19.5   Summary

It should be remembered that not all drugs need monitoring. There is little point in investing in TDM if the results will not answer a clinical question. Ideally, a relationship between plasma concentrations and therapeutic outcome will have been established but even if this is not clear, ensuring compliance and individualizing patient treatment may be a benefit. High degrees of analytical accuracy and precision are required if the results are to be meaningful and, of course, the results have to be available when required. Having said that, previously we described the evolution of the discipline of pharmacokinetics from an initial phase (1900–1950) dominated by the scientific curiosity of the likes of Teorell, Dost and Widmark, a phase related to optimum design of dosage forms, dominated by such people as Nelson, Levy, Riegelman, Garrett, Gibaldi and Perrier, and many others, a phase of application in clinical pharmacology inspired by the writings of Brodie and the many scientists in Europe and North America, as well as other places, who have taken an interest in the field, and a recent phase of application in new drug design in the pharmaceutical industry. The use of pharmacokinetics in optimizing therapy with gentamicin and phenytoin would seem to be a high point in the validation of the subject in both pure and applied senses, although it is to be hoped that future drugs will not have the properties that have made this elegant work necessary.

References and further reading

Aronson JK, Hardman M, Reynolds DJ. ABC of monitoring drug therapy. Phenytoin. *BMJ* 1992; 305: 1215–8.

Asberg M, Cronholm B, Sjoqvist F, Tuck D. Relationship between plasma level and therapeutic effect of nortriptyline. *Br Med J* 1971; 3: 331–4.

Bauer LA. *Applied Clinical Pharmacokinetics* New York, McGraw Hill, 2001.

Baumann P, Hiemke C, Ulrich S, Gaertner I, Rao ML, Eckermann G, *et al*. Therapeutic monitoring of psychotropic drugs: an outline of the AGNP-TDM expert group consensus guideline. *Ther Drug Monit* 2004; 26: 167–70.

Curry SH. *Drug Disposition and Pharmacokinetics*, 3rd edn. Oxford: Blackwell Scientific, 1980.

Edwards LD, Fletcher AJ, Fox AW, Stonier PD. *Principles and Practice of Pharmaceutical Medicine*, 2nd edn. Chichester: John Wiley & Sons, 2007.

Flanagan RJ, Brown NW, Whelpton R. Therapeutic drug monitoring (TDM). *CPD Clin Biochem* 2008; 9: 3–21.

Jorga A, Holt DW, Johnston A. Therapeutic drug monitoring of cyclosporine. *Transplant Proc* 2004; 36(2 Suppl): 396S–403S.

Oliver GC, Parker BM, Parker CW. Radioimmunoassay for digoxin. Technical and clinical application. *Am J Med* 1971; 51: 186–92.

Olsen CK, Brennum LT, Kreilgaard M. Using pharmacokinetic-pharmacodynamic modelling as a tool for prediction of therapeutic effective plasma levels of antipsychotics. *Eur J Pharmacol* 2008; 584: 318–27.

Patsalos PN, Berry DJ, Bourgeois BF, Cloyd JC, Glauser TA, Johannessen SI, Leppik IE, Tomson T, Perucca E. Antiepileptic drugs – best practice guidelines for therapeutic drug monitoring: a position paper by the subcommission on therapeutic drug monitoring, ILAE Commission on Therapeutic Strategies. *Epilepsia* 2008; 49: 1239–76.

Peck CC, Conner DP, Murphy MG. *Bedside Clinical Pharmacokinetics*. Washington: Applied Therapeutics Inc, 1989.

Redfors A. Plasma digoxin concentration – its relation to digoxin dosage and clinical effects in patients with atrial fibrillation. *Br Heart J* 1972; 34: 383–91.

Remington G, Shammi CM, Sethna R, Lawrence R. Antipsychotic dosing patterns for schizophrenia in three treatment settings. Psychiatr Serv 2001; 52: 96–8.

Reynolds DJ, Aronson JK. ABC of monitoring drug therapy. Making the most of plasma drug concentration measurements. *Bmj* 1993; 306: 48–51.

Van Putten T, Marder SR, Wirshing WC, Aravagiri M, Chabert N. Neuroleptic plasma levels. *Schizophr Bull* 1991; 17: 197–216.

White JR, Garrine MW. *Basic Clinical Pharmacokinetics Handbook*. Vancouver: Applied Therapeutics Inc., 1994.

Wilson JF. Survey of reference ranges and clinical measurements for psychoactive drugs in serum. *Ther Drug Monit* 2003; 25: 243–7.

Winter MF. *Basic Clinical Pharmacokinetics*, 2nd edn. Spokane: Applied Therapeutics Inc., 1987.

Yeung PK, Hubbard JW, Korchinski ED, Midha KK. Pharmacokinetics of chlorpromazine and key metabolites. *Eur J Clin Pharmacol* 1993; 45: 563–9.

Appendix

Mathematical Concepts and the Trapezoidal Method

1 Algebra, variables and equations

Algebra is a way of describing relationships in general terms, usually as equations. For example, Wilhelm Beer observed that the optical absorbance, A, of a dilute solution was directly proportional to the concentration, C, of the solute:

$$A \propto C \tag{A1}$$

To write the relationship as an equation we need a constant of proportionality, k:

$$A = kC \tag{A2}$$

If C is in mol L^{-1}, and the path length is 1 cm, then $k = \varepsilon$, the molar absorptivity of the solute. If the solvent also absorbs light, the background absorption, b can be added to the equation:

$$A = kC + b \tag{A3}$$

A is known as the *dependent* variable because it changes as a result of changes in C, the *independent* variable. Equation A3 can be represented graphically by plotting A against C. The independent variable is plotted along the bottom (x-axis) and the dependent variable along the y-axis (Figure A1).

When $C = 0$, $y = b$, so the value of b can be obtained from the intercept of the line with the y-axis. The value of k is obtained from the slope of the line. Estimates of the intercept and slope are best derived from least squares regression analysis which can be done on many hand-held calculators, an Excel spreadsheet, as well as specifically designed regression software programs. If the line is a calibration line then it usual to rearrange Equation A3, so that concentrations can be calculated from measured absorbance values. Subtracting b from both sides and dividing both sides by k gives:

$$C = \frac{A - b}{k} \tag{A4}$$

2 Indices and powers

When a number, a, is multiplied by itself a number of times, n, the product can be written a^n. For example:

$$a \times a = a^2 \tag{A5}$$

Drug Disposition and Pharmacokinetics, By Stephen H. Curry and Robin Whelpton
© 2011 John Wiley & Sons, Ltd

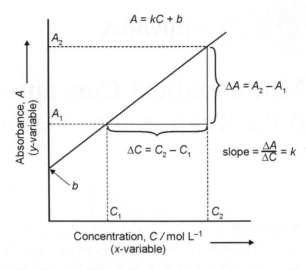

Figure A1 Straight line representation of $A = kC + b$.

and, in this case, a is said to be 'squared'. Similarly,

$$a \times a \times a = a^3 \tag{A6}$$

the result is 'a cubed' or 'a to the 3'. Multiplying a^2 and a^3:

$$a \times a \times a \times a \times a = a^5 \tag{A7}$$

Note that the result is $a^{(2+3)}$ and a general rule can be written:

$$a^n \times a^m = a^{(n+m)} \tag{A8}$$

Using similar logic it can be shown that:

$$\frac{a^n}{a^m} = a^{(n-m)} \tag{A9}$$

Thus:

$$\frac{1}{a^n} = a^{-n} \tag{A10}$$

and:

$$(a^n)^m = a^{nm} \tag{A11}$$

A fractional index indicates that a root should be taken:

$$a^{\frac{1}{n}} = \sqrt[n]{a} \tag{A12}$$

3   Logarithms

Tables of logarithms and antilogarithms are used to simplify multiplication and division of complex numbers using only addition and subtraction. In common logarithms, (base 10) each number is expressed as 10 raised to the appropriate power. Logarithms have additional importance in pharmacology for contracting the range of numbers and in the linear transformation of data. The base of the logarithm can be indicated e.g. '$\log_{10}$' but generally common logarithms are written 'log'. Logarithms to the base e (e $\approx 2.718\ldots$) are referred to as natural logarithms, and can be written $\log_e$ or more commonly, simply, ln. A word of warning, in many computer languages, and in engineering parlance, 'log' means '$\log_e$'. To convert a logarithm to a number the base of the logarithm is raised to the logarithm. For example,

$$\log 2 = 10^2 = 100$$

and if:

$$x = \ln y, \quad \text{then } y = e^x \tag{A13}$$

Negative numbers cannot have logarithms, but note that

$$\log \frac{1}{x} = -\log x \tag{A14}$$

Furthermore, it should be noted that although logarithms are dimensionless, it is necessary, to indicate the units of the original number. So for a concentration of $C\,\mathrm{mg\,L^{-1}}$, the logarithm should be written: $\log(C/\mathrm{mg\,L^{-1}})$.

4   Calculus

4.1   Differentiation

Calculus was invented to deal with slopes of curves and areas under them. When the rate of change of a measurement (e.g. y as a function of x) is changing, we may wish to determine the average rate of change of y with respect to x. This is done by differentiation. Suppose δy represents a small increase in y which occurs while x increases by δx, then $\delta y/\delta x$ is the mean gradient of the graph over the small range examined. If we reduce the values of δy and δx towards zero, then the line showing the gradient tends towards a tangent. The limiting value of $\delta y/\delta x$ is called dx/dy or the differential coefficient of y with respect to x. The process of finding the limit of dy/dx is *differentiation*, and dy/dx is a measure of the slope of the tangent at a given value of x [Figure A2(a)]. If the y-variable is time, t, and the x-variable is concentration, C, then dC/dt is the instantaneous rate of change in concentration at that time.

For equations of the type: $y = ax^n + b$

$$\frac{dy}{dx} = anx^{(n-1)} \tag{A15}$$

Thus for a straight line, $y = ax + b$, $n = 1$, so

$$y = ax^1 + b$$
$$\frac{dy}{dx} = ax^{(1-1)} = ax^0 = a \tag{A16}$$

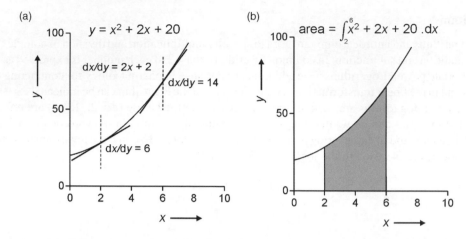

Figure A2 (a) Differentiation allows the slope at any value of x to be calculated. (b) Integration between $x = 2$ and $x = 6$ allows the area depicted by the shading to be calculated (see text for details).

i.e. the slope of the straight line is a. Reciprocals of x are treated the same way:

$$y = \frac{1}{x^n} = x^{-n}$$
$$\frac{dy}{dx} = -nx^{-n-1} = -nx^{-(n+1)}$$
(A17)

However the differential of $\ln x$ is a special case:

$$y = \ln x$$
$$\frac{dy}{dx} = \frac{1}{x}$$
(A18)

4.2 Integration

The reverse of differentiation is integration. If dy/dx is known we may wish to find y in terms of x. Thus, if dy/dx is some function of x, written as $f(x)$, then y is the integral of $f(x)$ with respect to x:

$$y = \int f(x).dx$$
(A19)

If $y = ax^n$, then:

$$y = \int ax^n.dx = a \int x^n.dx = \frac{ax^{(n+1)}}{n+1} + c$$
(A20)

Note the appearance of c, sometimes referred to as a constant of integration, which is necessary because constants are 'lost' on differentiation, see Equation A16. The value is found by substituting $x = 0$, when $y = c$.

The integral of $1/x$ is of particular importance because of the form of the rate equation of first-order reactions:

$$y = \int \frac{1}{x} . dx = \ln x + c = ce^x \qquad (A21)$$

A quantity written as a power of e is an exponential function, and the above may be written $c \exp(x)$. The quantity increases more and more rapidly as the power increases – *exponential growth*. If the power is negative, e.g. $y = e^{-x}$, then this represents *exponential decay* and y becomes ever nearer to 0 as x increases, but only reaches 0 when x is infinite; y is said to *asymptote* to 0.

4.2.1   Areas under curves

Integration is important for calculating areas under curves. Using the equation of Figure A2 as an example, the area under the curve (AUC) from 2 and 6 is obtained by integrating between the limits:

$$AUC_{(2-6)} = \int_{2}^{6} x^2 + 2x + 20 . dx = \frac{x^3}{3} + x^2 + 20x \Big|_{2}^{6} \qquad (A22)$$

$AUC_{(2-6)}$ is the difference between the value obtained by substituting $x = 6$ and $x = 2$.

$$AUC_{(2-6)} = \left[\frac{6^3}{3} + 6^2 + 20(6) \right] - \left[\frac{2^3}{3} + 2^2 + 20(2) \right]$$

$$AUC_{(2-6)} = 228 - 46.7$$

$$AUC_{(2-6)} = 181.3$$

Note that because we are using the difference, the constant of integration cancels and need not be included. In this example, the numbers were dimensionless and so *AUC* has no units. However, in most practical instances the x- and y-values will represent variables with units and so anything derived from them should have the appropriate units. The slope of a ln(concentration) against time plot has units of time to the minus 1 (T^{-1}). The area under the curve of a plasma concentration (mg L^{-1}) against time (h) plot has units of mg h L^{-1}, for example.

4.2.2   Calculating AUC values: the trapezoidal method

Probably the simplest way to obtain the area under a curve is the *trapezoidal method*. The plasma concentration–time curve is plotted and each segment between adjacent collection time points treated as a trapezium [Figure A3(a)]. The area of the trapezium is the average length of two sides multiplied by the width between them, so for the nth trapezium the area, A_n, is:

$$A_n = \frac{C_n + C_{(n+1)}}{2} (t_{(n+1)} - t_n) \qquad (A23)$$

The area from $t = 0$ to the time of last plasma sample, $AUC_{(0-t_z)}$ is obtained by adding the areas of all the trapeziums. The remaining area, from the time of the last collected sample to infinity is extrapolated using

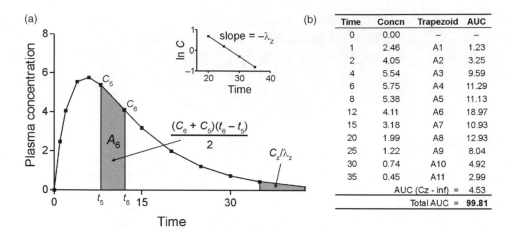

Figure A3 (a) Example of the trapezoidal method of calculating the area under a concentration versus time curve. The rate constant is calculated from the slope of the terminal points of ln C versus time plot (inset). (b) Printout from an Excel spreadsheet showing the areas of individual trapeziums and the total which is in good agreement with the theoretical value of 100 for this example.

C_z/λ_z, where λ_z is the rate constant of the terminal decay phase. This can be estimated from a ln(concentration) versus time plot of the terminal data [Figure A3(a) inset]. The areas are conveniently calculated from the concentration–time data using a spreadsheet [Figure A3(b)].

Points to note are:

- It is the area under the concentration–time plot, NOT the ln (concentration)–time plot that must be used.
- Although the first segment on the oral plot is a triangle the formula for a trapezium gives the correct area because length of one side is 0.
- The greater the number of trapeziums the greater will be the accuracy of the calculation.
- Ideally the extrapolated area should be <5% of the total.
- The units of AUC are concentration $\times$ time e.g. $mg\,h\,L^{-1}$.

Acknowledgements

The Authors would like to acknowledge the following for permission to reprint some of the figures contained in this book:

Adis (Wolters Kluwer)
figure 7.6

The American Medical Association
figure 8.6

The American Society for Pharmacology and Experimental Therapeutics
figures 2.10; 2.11a; 10.3

The British Medical Journal
figure 10.1 and table 19.2

Elsevier
figures 2.23; 7.1b; 8.4

Lippincott
figures 13.5; 13.7

Nature Publishing Group
figures 6.10; 10.4; 12.4; 13.9

The Pharmaceutical Press
figures 2.18; 2.19;

Rila Publications
tables 19.4 and 19.5

John Wiley and Sons Inc.
figures 2.14; 2.21; 2.22; 4.5; 6.13; 11,1; 11.2; 12.1; 12.3; 14.1; 14.2; 14.4; 15.1; 15.4; 17.5; 18.7
    and tables 7.3 and 15.2

Index

Uptake 1   39
Urokinase   2

Vaginal secretions   75
Valproate/valproic acid   49, 56, 75, 93, 174, 193–4,
    295, 334
Vancomycin   178, 337
Vasopressin   280, 284, 289–90
Vecuronium   179–80
Vena cava   31, 40, 146
Venlafaxine   73–4, 194, 335–6
  O-Desmethyl-   74, 336
Verapamil   178–80, 183–4, 224, 260, 305, 338
  Nor-   338
Victim drug   294, 297, 308–10
Vigabatrin   333–4
Viral hepatitis   215, 223
Vitamin K   128, 172, 198, 210, 232
Volume   34
  blood   33
  intracellular   33
  of distribution   1, 16, 33–4, 78, 80–1, 86, 89, 93,
      96, 154, 177, 218, 225, 262, 270–1, 286, 288, 297,
      309, 340
    allometric exponent   272, 274–5
    area (V_{area})   103–4, 113
    at steady-state (V_{ss})   103, 115–116, 275–7
    by extrapolation (V_{extrap})   104
    effect of plasma protein binding   49

  in congestive heart failure   22, 220
  in the elderly   210
  of the central compartment   103
  of the effect compartment   259
  of tissues   152–3, 260

Warfarin   5, 146, 172, 193, 198, 210, 298–9, 304, 329
  concentration in plamsa   297
  dose-response relationship   242, 243
  effect of age on   210
  effects   254
  hepatic extraction   142
  7-Hydroxy-   298
  in disease   220, 224, 231–2
  interaction with choral   310
  interaction with choral hydrate   306
  interaction with orlistat   171
  interaction with phenylbutazone   22
  interaction with pheobarbital   298
  intrinsic clearance   144
  metabolism   58, 172, 298
  PK/PD models   261–3
  protein binding   47–8, 146, 298
  time of maximum effect   249
Weighted-regression   111

Xanthine oxidase   57, 195, 332

Zonisamide   334